T0281522

Lecture Notes in Computer Science 14124

The series Lecture Notes in Computer Science (LNCS), including its subseries Lecture Notes in Artificial Intelligence (LNAI) and Lecture Notes in Bioinformatics (LNBI), has established itself as a medium for the publication of new developments in computer science and information technology research, teaching, and education.

LNCS enjoys close cooperation with the computer science R & D community, the series counts many renowned academics among its volume editors and paper authors, and collaborates with prestigious societies. Its mission is to serve this international community by providing an invaluable service, mainly focused on the publication of conference and workshop proceedings and postproceedings. LNCS commenced publication in 1973.

Jaques Blanc-Talon · Patrice Delmas ·
Wilfried Philips · Paul Scheunders
Editors

Advanced Concepts for Intelligent Vision Systems

21st International Conference, ACIVS 2023 Kumamoto, Japan, August 21–23, 2023 Proceedings

Editors
Jaques Blanc-Talon
DGA TA
Toulouse, France

Patrice Delmas
University of Auckland
Auckland, New Zealand

Wilfried Philips
Ghent University
Ghent, Belgium

Paul Scheunders
University of Antwerp
Wilrijk, Belgium

ISSN 0302-9743 ISSN 1611-3349 (electronic)
Lecture Notes in Computer Science
ISBN 978-3-031-45381-6 ISBN 978-3-031-45382-3 (eBook)
https://doi.org/10.1007/978-3-031-45382-3

This Springer imprint is published by the registered company Springer Nature Switzerland AG
The registered company address is: Gewerbestrasse 11, 6330 Cham, Switzerland

Paper in this product is recyclable.

Preface

These proceedings gather the papers of the Advanced Concepts for Intelligent Vision Systems (ACIVS) conference which was held in Kumamoto, Japan from August, 21 to August 23, 2023.

This event was the 21st ACIVS. After the very first event, in Germany in 1999, ACIVS has become a larger and independent scientific conference. However, the seminal distinctive governance rules have been maintained:

- To update the conference scope on a yearly basis. While keeping a technical backbone (the classical low-level image processing techniques), we have introduced topics of interest like - chronologically - image and video compression, 3D, security and forensics, evaluation methodologies in order to fit the conference scope to our scientific community's needs. In addition, speakers usually give invited talks on hot issues;
- To remain a single-track conference in order to promote scientific exchanges within the audience;
- To grant oral presentations a duration of 25 minutes and published papers a length of 12 pages, which is significantly different from most other conferences.

The second and third items imply a complex management of the conference; in particular, the number of time slots is smaller than in larger conferences. Although the selection between the two presentation formats is primarily determined by the need to compose a well-balanced program, papers presented during plenary and poster sessions enjoy the same importance and publication format.

The first item is strengthened by the fame of ACIVS, which has been growing over the years: official Springer records show a cumulative number of downloads at the first of January, 2019, of about 1,200,000 (for ACIVS 2005–2018 only). Due to the COVID-19 pandemic, the conference activity was shut down in 2021 and 2022.

This year's event also included invited talks of leading scientists from IROAST (International Research Organization for Advanced Science and Technology), based in Kumamoto, Japan. We would like to sincerely thank all of them for enhancing the technical program with their presentations.

ACIVS attracted submissions from many different countries, mostly from Asia, but also from the rest of the world: Belgium, Canada, China, France, Germany, Greece, India, Israel, Japan, Mexico, New Zealand, Poland, Singapore, Slovenia, Spain, South Africa, Sweden, Taiwan, Tunisia, Vietnam and the USA. From 49 submissions, 29 were selected for oral presentation and 8 as posters. The paper submission and review procedure was carried out electronically and a minimum of three reviewers were assigned to each paper. A large and energetic Program Committee (61 people), helped by additional referees, completed the long and demanding reviewing process. We would like to thank all of them for their timely and high-quality reviews, achieved in quite a short time and during the summer holidays.

Also, we would like to thank our sponsors (in alphabetical order), IROAST and the city of Kumamoto from Japan and Springer for their valuable support.

Finally, we would like to thank all the participants who trusted in our ability to organize this conference for the 21st time. We hope they attended a different and stimulating scientific event and that they enjoyed the atmosphere of the ACIVS social events in the city of Kumamoto.

A conference like ACIVS would not be feasible without the concerted effort of many people and the support of various institutions. We are indebted to the local organizers for having smoothed all the harsh practical details of an event venue, and we hope to return in the near future.

Acivs 2023 was organized by the University of Auckland, New Zealand.

August 2023

Jacques Blanc-Talon
Patrice Delmas
Wilfried Philips
Paul Scheunders

Organization

Steering Committee

Jacques Blanc-Talon	DGA, France
Patrice Delmas	University of Auckland, New Zealand
Wilfried Philips	Ghent University - imec, Belgium
Paul Scheunders	University of Antwerp, Belgium

Organizing Committee

Patrice Delmas	University of Auckland, New Zealand
Toshifumi Mukunoki	Kumamoto University, Japan

Program Committee

Hamid Aghajan	Ghent University - imec, Belgium
João Ascenso	Instituto de Telecomunicações, Portugal
Sebastiano Battiato	University of Catania, Italy
George Bebis	University of Nevada, USA
Fabio Bellavia	Università degli Studi di Palermo, Italy
Jenny Benois-Pineau	University of Bordeaux, France
Janusz Bobulski	Czestochowa University of Technology, Poland
Egor Bondarev	Technische Universiteit Eindhoven, The Netherlands
Salah Bourennane	École Centrale de Marseille, France
Catarina Brites	Instituto de Telecomunicações, Portugal
Vittoria Bruni	University of Rome La Sapienza, Italy
Odemir Martinez Bruno	University of São Paulo, Brazil
Giuseppe Cattaneo	University of Salerno, Italy
Jocelyn Chanussot	Université de Grenoble Alpes, France
Kacem Chehdi	ENSSAT, France
Pamela Cosman	University of California, USA
Mihai Datcu	DLR (German Aerospace Center), Germany
Patrice Delmas	University of Auckland, New Zealand
Stéphane Derrode	École Centrale de Lyon, France
Israel Mendonça dos Santos	Kumamoto University, Japan

Christine Fernandez-Maloigne	Poitiers University, France
Jérôme Gilles	San Diego State University, USA
Daniele Giusto	Università degli Studi di Cagliari, Italy
Christine Guillemot	Inria, France
Monson Hayes	George Mason University, USA
Dimitris Iakovidis	University of Thessaly, Greece
Yuji Iwahori	Chubu University, Japan
Hamid Krim	North Carolina State University, USA
Bogdan Kwolek	AGH University of Science and Technology, Poland
Ting-Lan Lin	National Taipei University of Technology, Taiwan
Ludovic Macaire	Université de Lille, France
Jean-Michel Morel	ENS, France
Toshifumi Mukunoki	Kumamoto University, Japan
Adrian Munteanu	Vrije Universiteit Brussel, Belgium
Shigeki Nakauchi	Toyohashi University of Technology, Japan
María Navascués	Universidad de Zaragoza, Spain
António J. R. Neves	University of Aveiro, Portugal
Jennifer Newman	Iowa State University, USA
Jun Otani	Kumamoto University, Japan
Jussi Parkkinen	University of Eastern Finland, Finland
Rudi Penne	University of Antwerp, Belgium
Giovanni Ramponi	University of Trieste, Italy
Florent Retraint	Université de Technologie de Troyes, France
Luis Salgado	Universidad Politécnica de Madrid, Spain
Natalia Schmid	West Virginia University, USA
Guna Seetharaman	U.S. Naval Research Laboratory, USA
Changming Sun	CSIRO, Australia
Tamas Sziranyi	University of Technology and Economics, Hungary
Jean-Philippe Thiran	EPFL, Switzerland
Nadège Thirion-Moreau	Université de Toulon, France
Sylvie Treuillet	Université d'Orléans, France
Alex Vasilescu	UCLA, USA
Luisa Verdoliva	Università degli Studi di Napoli Federico II, Italy
Serestina Viriri	University of KwaZulu-Natal, South Africa
Bing Xue	Victoria University, New Zealand
Gerald Zauner	University of Applied Sciences Upper Austria, Austria
Josiane Zerubia	Inria, France
Djemel Ziou	Sherbrooke University, Canada

Reviewers

Hamid Aghajan	Ghent University - imec, Belgium
João Ascenso	Instituto de Telecomunicações, Portugal
George Bebis	University of Nevada, USA
Fabio Bellavia	Università degli Studi di Palermo, Italy
Jacques Blanc-Talon	DGA, France
Janusz Bobulski	Czestochowa University of Technology, Poland
Philippe Bolon	University of Savoie, France
Salah Bourennane	École Centrale de Marseille, France
Catarina Brites	Instituto de Telecomunicações, Portugal
Giuseppe Cattaneo	University of Salerno, Italy
Jocelyn Chanussot	Université de Grenoble Alpes, France
Patrice Delmas	University of Auckland, New Zealand
Stéphane Derrode	École Centrale de Lyon, France
Israel Mendonça dos Santos	Kumamoto University, Japan
Christine Fernandez-Maloigne	Poitiers University, France
Alfonso Gastelum-Strozzi	Universidad Nacional Autónoma de México, Mexico
Jérôme Gilles	San Diego State University, USA
Daniele Giusto	Università degli Studi di Cagliari, Italy
Monson Hayes	George Mason University, USA
Dimitris Iakovidis	University of Thessaly, Greece
Yuji Iwahori	Chubu University, Japan
Bogdan Kwolek	AGH University of Science and Technology, Poland
Ting-Lan Lin	National Taipei University of Technology, Taiwan
Ludovic Macaire	Université de Lille, France
Adrian Munteanu	Vrije Universiteit Brussel, Belgium
Shigeki Nakauchi	Toyohashi University of Technology, Japan
Jennifer Newman	Iowa State University, USA
Jussi Parkkinen	University of Eastern Finland, Finland
Rudi Penne	University of Antwerp, Belgium
Florent Retraint	Université de Technologie de Troyes, France
Paul Scheunders	University of Antwerp, Belgium
David Soriano Valdez	UNAM, Mexico
Changming Sun	CSIRO, Australia
Tamas Sziranyi	University of Technology and Economics, Hungary
Jean-Philippe Thiran	EPFL, Switzerland
Nadège Thirion-Moreau	Université de Toulon, France
Luisa Verdoliva	Università degli Studi di Napoli Federico II, Italy

Serestina Viriri	University of KwaZulu-Natal, South Africa
Damien Vivet	ISAE-SUPAERO, France
Alexander Woodward	RIKEN Center for Brain Science, Japan
Gerald Zauner	University of Applied Sciences Upper Austria, Austria
Pavel Zemcik	Brno University of Technology, Czech Republic
Josiane Zerubia	Inria, France
Djemel Ziou	Sherbrooke University, Canada

Contents

A Hybrid Quantum-Classical Segment-Based Stereo Matching Algorithm

Shahrokh Heidari[1,2(✉)] and Patrice Delmas[1,2]

[1] School of Computer Science, The University of Auckland, Auckland, New Zealand
Shei972@aucklanduni.ac.nz

[2] Intelligent Vision Systems Lab, The University of Auckland, Auckland, New Zealand

Abstract. Our contribution introduces a hybrid quantum-classical Stereo Matching algorithm that demonstrates the potential of using Quantum Annealing in Computer Vision in the future once quantum processors have enough qubits to support practical, real-world Computer Vision applications. The classical component of our approach involves dividing the input image into homogeneous color segments and using a local Stereo Matching technique to estimate their respective initial disparity planes. In the quantum component, we assign a label to each segment by the estimated disparity planes. Such a labeling problem is classically intractable. The outcomes of our experiments on the D-Wave quantum computer indicate that our method produces results that compare well with the ground truth. Nonetheless, the precision of our approach is greatly influenced by the quality of the initial image segmentation, which is a common challenge for all classical Stereo Matching methods that rely on segmentation-based techniques.

Keywords: Quantum Annealing · Stereo Matching · QUBO · D-Wave

1 Introduction

Many early Computer Vision (CV) problems can be modeled by a labeling problem where a set of image features (such as pixels, edges, or segments) is labeled by some quantities (such as disparities in motion and stereo, or intensities in image restoration) [28]. Such labeling problems are generally represented in terms of naturally intractable optimizations. Given the intrinsically tricky nature of CV problems, researchers have always been looking for efficient algorithms to approximate the optimal solution as fast and accurately as possible [16], from the classical algorithms in the 90 s, such as Simulated Annealing [7], Mean-field Annealing [12], and Iterated Conditional Modes [3] to the recent state-of-the-art minimization algorithms, such as Swap move [6], Expansion move [6], Max-product Belief Propagation [11] and Tree-Re-Weighted message passing

J. Blanc-Talon et al. (Eds.): ACIVS 2023, LNCS 14124, pp. 1–13, 2023.
https://doi.org/10.1007/978-3-031-45382-3_1

[29]. Despite being extensively researched and even considering the most recent advances using deep learning-based strategies [25], which are computationally expensive, CV labeling is still considered an open problem with no optimal solution [16], and researchers have always been looking for alternatives to tackle the problem regarding accuracy and speed.

With the advent of quantum computations, which promise potentially lower-time complexity on certain problems than the best-classical counterparts [32], recent studies have focused on leveraging quantum properties to overcome intractable classical problems using Quantum Annealing computation [20]. D-Wave system was the first company to build a Quantum Processing Unit (QPU) that naturally approximates the ground state of a particular problem representation, namely Ising model [22]. In 2016, a Google study [9] compared D-Wave-2X QPU with two classical algorithms (Simulating Annealing and Quantum Monte Carlo algorithms) run on a single-core classical processor, showing the D-Wave QPU to be up to 100 million times faster. A later study used the state-of-the-art GPU implementations of these algorithms (on NVIDIA GeForce GTX 1080) for a more complex problem and showed that the D-Wave-2000Q QPU was 2600 times faster in finding the ground state [17]. However, the scarcity of available quantum bits (qubits) on a D-Wave QPU has always been challenging, from 128-qubits D-Wave One built in 2011 to the newly released 5000-qubit D-Wave Advantage. Therefore, large CV problems involving highly non-convex functions in a search space of many thousands of dimensions have yet to be widely studied based on D Wave quantum computers. To the best of our knowledge, the first CV minimization problem implemented on a D-Wave QPU was to solve a specific Stereo Matching problem on simplistic synthetic images [8]. An efficiency improvement in terms of the number of variables in the quantum model and applied to natural gray-scale images was proposed in [14]. Both quantum models solve the minimum cut problem on a specific graph with two terminal vertices, which can be solved efficiently in a polynomial time on classical processors. To show the real advantage of using D-Wave for CV problems, we need to model a computationally intractable problem. Motivated by a CV application (Image Restoration), we recently proposed an efficient quantum model for the minimum multi-way cut problem [13], which is an NP-hard problem.

Here, we show how a specific Stereo Matching problem (as a significant CV labeling problem) can be solved based on Quantum Annealing. Due to the limitations of the current D-Wave quantum processors, it is impossible to directly solve this problem on real-world full-sized images. Therefore, we present a hybrid quantum-classical segment-based Stereo Matching method. We use the disparity plane concept to represent the disparities of the input pixels. In the classical part, we first partition the input image into small segments and estimate the best disparity planes. Next, we label each segment by a disparity plane using an optimization approach, which is an NP-Hard problem. Quantum Annealing carries out the minimization.

The rest of the paper is organized as follows. Section 2 explains Quantum Annealing and how an optimization problem can be solved using a D-Wave QPU.

Section 3 describes the Stereo Matching problem as an essential CV labeling problem. Our hybrid quantum-classical approach for Stereo Matching is given in Sect. 4. The experimental set-up and results are discussed in Sect. 5. Finally, Sect. 6 provides the conclusion.

2 Quantum Annealing

Quantum Annealing model [10] is an alternative equivalent [2] to the Quantum Gate model. In this model, quantum bits (qubits) are particles in a quantum dynamical system that evolve based on special forces acting on them. These forces are either internal (from interactions among qubits) or external (from other sources). Each state of a register of qubits has energy based on the applied forces. A time-dependent Hamiltonian is a mathematical description of the system that gives the system's energy and characterizes the forces at any time [22]. Quantum Annealing is the process of finding a state of the system with the lowest energy based on the time-dependent Hamiltonian. Therefore, Quantum Annealing is a computing paradigm that efficiently solves optimization problems and approximates the optimum solution. A Quantum Annealing algorithm is generally described by a time-dependent Hamiltonian $H(t)$ that has three components as follows [22]: Initial Hamiltonian H_I, where all qubits are in a superposition state. *Problem Hamiltonian* H_p, where the internal and external forces are defined to encode the objective function. The lowest-energy state of H_p is the solution that minimizes the objective function. *Adiabatic path* $s(t)$, which is a smooth function that decreases from 1 to 0 (such as $s(t) = 1 - \frac{t}{t_f}$, where $s(t)$ decreases from 1 to 0 as t increases from 0 to some elapsed time t_f). During quantum annealing, the *Initial Hamiltonian* is slowly evolved along the *Adiabatic path* to the *Problem Hamiltonian* as $H(t) = s(t)H_I + (1 - s(t))H_p$ [22], decreasing the influence of H_I over time to reach H_P as $s(t)$ goes from 1 to 0. Practically, a Quantum Annealer, such as a D-Wave QPU is needed to accomplish this process. One way to prepare an objective function as a *Problem Hamiltonian* to be minimized by the D-Wave QPU is the Quadratic Unconstrained Binary Optimization (QUBO) model [22]. Given a vector of n binary variables as $\mathbf{x} = (x_1, x_2, \ldots, x_n) \in \{0,1\}^n$, a QUBO model is represented as $H_{qubo}(\mathbf{x}) = \mathbf{x}^T\mathbf{Q}\mathbf{x}$, where $\{0,1\}^n$ is a set of n binary values, and $\mathbf{Q}$ is an $n \times n$ matrix that can be chosen to be upper-diagonal. Therefore, $H_{qubo}(\mathbf{x})$ can be reformulated as (1).

$$H_{qubo}(\mathbf{x}) = \sum_{i} \mathbf{Q}_{i,i}x_i + \sum_{i<j} \mathbf{Q}_{i,j}x_ix_j. \quad (1)$$

The diagonal terms $\mathbf{Q}_{i,i}$ are the linear coefficients acting as the external forces, and the off-diagonal terms $\mathbf{Q}_{i,j}$ are the quadratic coefficients for the internal forces [22].

3 Stereo Matching

One of CV's oldest yet unsolved problems is Stereo Matching, the most computationally extensive part of 3D reconstruction from digital images. In analogy to human depth perception using two eyes, a stereo vision system typically has two cameras placed horizontally, one on the left and the other on the right side, to make a binocular vision. Each camera similarly captures the image with some displacement. This displacement is called disparity which shows the differences between the actual position, e.g., coordinates of the projection in the left, respectively right, image of a 3D point in the real world. A rectification process is also used to make sure that the corresponding pixels in the left and right images are in the same line of pixels and there are only horizontal disparities. The disparity is inversely proportional to the distance between the cameras and the object itself in the real world. If a 3D point in the real world is closer, respectively further away, to our eyes, the disparity value of the corresponding projections in the images is larger, respectively smaller. When we visualize the disparities of all pixels (known as a disparity map), closer objects with larger disparity values are lighter than further-away objects with lower disparity values.

4 A Hybrid Quantum-Classical Segment-Based Stereo Matching Method

Given the left and right images, our contribution needs five steps which can be briefly described as follows:

1. **Color image segmentation:** The left image is first partitioned into homogeneous color segments.
2. **Initial disparity estimation:** A local Stereo Matching algorithm is used to initially estimate the disparities for both left and right stereo images.
3. **Disparity plane fitting:** An iterative plane fitting algorithm fits disparity planes into color segments based on step 2's estimated initial disparity values.
4. **Segment and disparity plane refinement:** Color segments are combined according to a similarity measurement from the disparity map obtained from step 3. Next, the disparity planes are updated by a plane fitting algorithm on the new combined color segments. The final outputs are a set of color segments and a set of disparity planes.
5. **Optimization by Quantum Annealing and D-Wave QPU:** An objective function labels each segment with a disparity plane. Then, we model an equivalent QUBO to this objective function that a D-Wave QPU can minimize.

The first four steps are performed on a classical computer, while a D-Wave QPU performs the last step.

Color Image Segmentation: Segment-based Stereo Matching methods are generally built on the assumption that disparities vary smoothly within a homogeneous color image segment while disparity discontinuities only coincide with segment borders [4]. It is also important that a segment does not overlap across a disparity discontinuity. Therefore, segment-based Stereo Matching methods use over-segmentation to group image pixels into homogeneous color segments. In this study, we use the Quickshift image segmentation method [27] to segment the left image into homogeneous color segments.

Initial Disparity Estimation: We use a cross-based local Stereo Matching method [33] to estimate the initial disparity values for the left and right stereo images. Let the integers d_{min} and d_{max} be the lowest and highest possible disparity values, respectively. Given a pixel (x, y) in the left image, the allocated disparity is computed as (2) based on the winner-take-all approach:

$$D(\mathrm{x}, \mathrm{y}) = \min_{d_{min} \leq d \leq d_{max}} Cost(\mathrm{x}, \mathrm{y}, d), \tag{2}$$

where $Cost(\mathrm{x}, \mathrm{y}, d)$ is the cost of allocating the disparity value d to the pixel (x, y). We define the matching cost function as two similarity functions found in the literature [5,21]: the truncated absolute difference on the color intensities (TAD_c) and the truncated absolute difference on the image gradient values (TAD_g). Let I_l and I_r be the left and right images, respectively.

$$Cost(\mathrm{x}, \mathrm{y}, d) = \frac{1}{|R_{\mathrm{xy}}|} \sum_{(\mathrm{x}', \mathrm{y}') \in R_{\mathrm{xy}}} (1-\alpha) TAD_c(\mathrm{x}', \mathrm{y}', d) + \alpha TAD_g(\mathrm{x}', \mathrm{y}', d), \tag{3}$$

$$TAD_c(\mathrm{x}', \mathrm{y}', d) = \min\left(|I_l(\mathrm{x}', \mathrm{y}') - I_r(\mathrm{x}' - d, \mathrm{y}')|, \tau_c\right), \tag{4}$$

$$TAD_g(\mathrm{x}', \mathrm{y}', d) = \min(|\nabla I_l(\mathrm{x}', \mathrm{y}') - \nabla I_r(\mathrm{x}' - d, \mathrm{y}')|, \tau_g), \tag{5}$$

where the user-defined parameter α balances the influence of the color and gradient terms, and τ_c and τ_g are the color and gradient truncation thresholds, respectively. Such a matching cost computation can handle radiometric differences in the input images because of using image gradients [15]. R_{xy} is a set of pixels forming an adaptive cross-based support region [33] for a given pixel, which relies on a linearly expanded cross skeleton for the cost aggregation step (we refer interested readers to the corresponding paper for further information [33]), and $|R_{\mathrm{xy}}|$ is the number of pixels in this region.

Disparity Plane Fitting: The next step is to fit a plane to each color segment based on the corresponding initial disparity values to compute the disparities of a segment of pixels by a plane. Take the segmented left stereo image of *Bull* dataset (taken from 2001-Middlebury stereo datasets [23]) as an example. To show the idea of disparity plane fitting, we select one of the color segments as an example (see Fig. 1a) and fit a plane to its pixels based on the corresponding initial disparity values computed from the previous step. Figure 1b shows a top view of the selected pixels in a 3D plot where the third axis gives the corresponding initial disparity values. The red points, called outliers, are the inaccurately estimated disparities caused by texture-less and occluded regions in the left stereo image, and the blue points, called inliers, are the points with accurate disparities to which we want to fit a plane. Once a plane is fitted to the blue points (Fig. 1e), the plane parameters could be used to obtain the disparity of each pixel inside the segment. For the disparity plane fitting step, we use an iterative algorithm proposed in [26] and widely used in later studies [21,31]. Let $S = \{S_1, S_2, \ldots, S_{n_s}\}$ be the set of color segments computed from the color image segmentation step, where n_s is the number of segments. The disparities of each segment can be modeled by the function $D'(\mathrm{x}, \mathrm{y}) = a_i\mathrm{x} + b_i\mathrm{y} + c_i$, where (a_i, b_i, c_i) are the fitted plane parameters, $(\mathrm{x}, \mathrm{y}) \in S_i$ for $1 \leq i \leq n_s$, and $D'(\mathrm{x}, \mathrm{y})$ is the computed disparity for the pixel (x, y) inside segment S_i. This step aims to capture a set of disparity planes (each having three plane parameters). This set of disparity planes will then be used to label each color segment in the left image with a disparity plane based on a cost function.

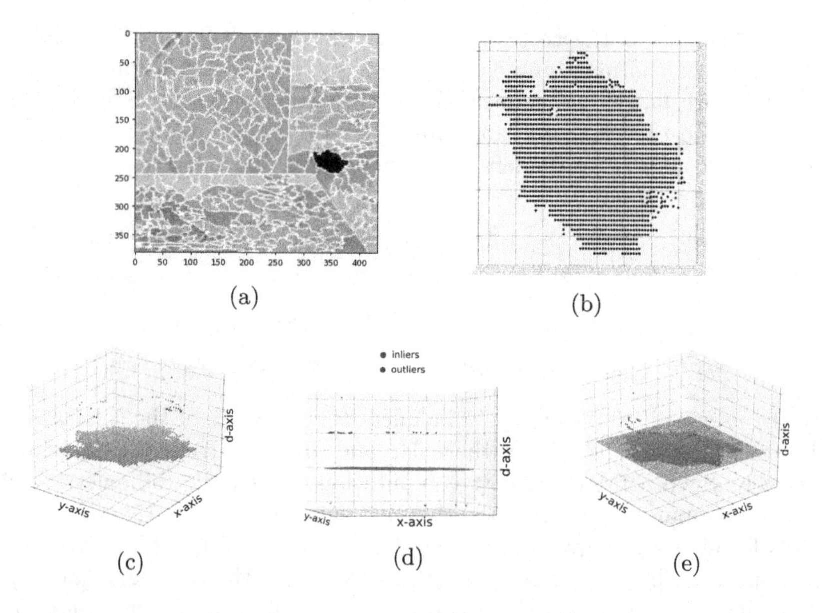

Fig. 1. A disparity-plane-fitting example. (a) The selected segment of the left stereo image of the *Bull* dataset shown in a 2D plot, (b) the selected pixels in a 3D plot (top view) where the third axis shows the corresponding initial disparity values, (c) the selected pixels in the 3D plot (corner view), (d) the selected pixels in the 3D plot (front view), (e) and the fitted plane to the selected pixels.

Segment and Disparity Plane Refinement. We have computed a set of color segments and a set of disparity planes to estimate the disparities of the left image pixels. However, the main purpose is not to have the best disparity planes for the segments but to extract all possible disparity planes to represent the scene structure accurately. An additional refinement step combines the color segments and fits new disparity planes to the updated segments. Let $G(V, E, C)$ be an undirected weighted graph where $V = \{1, 2, \ldots, n_s\}$ is the set of vertices representing the segment numbers, E is the set of the edges connecting the corresponding adjacent segments, and C is a function that allocates weights to the edges. A weight between two adjacent vertices (segments) represents how similar or dissimilar the two segments are. To weigh the edges, we first define the mean color of each segment concerning its corresponding computed disparity values using the estimated planes (i.e., we use disparity values for computing the mean value of each segment). Let Γ be a function that computes the mean value; we weigh a given edge $(u, v) \in E$ as $C(u, v) = |\Gamma(u) - \Gamma(v)|$. Next, we combine the corresponding segments for u and v if $C(u, v)$ is less than a threshold. Once the segments are combined, we rerun the disparity plane fitting algorithm on the combined segments to estimate new disparity planes.

Optimization by Quantum Annealing and D-Wave. The main purpose of this section is to label each color segment by a disparity plane to create a disparity map for the whole scene. Labeling problems can be resolved through an energy minimization scheme derived from probabilistic graphical models such as Markov Random Field. For a detailed explanation of how this energy function is defined, we refer the interested reader to the related studies [18,19,30]. This energy function is usually composed of two terms as: (i) a cost/error function between the solution and observed data, and (ii) a regularization term that enforces local or global consistency of the solution [24]. Next, we define an energy function to solve our labeling problem and then model an equivalent QUBO to be minimized by a D-Wave QPU. Before going through the explanations, we first define some notations.

- $S = \{S_1, S_2, \ldots, S_{n_s}\}$ is the set of updated segments, where n_s is the number of updated segments.
- $\rho = \{\rho_1, \rho_2, \ldots, \rho_{n_p}\}$ denotes the set of refined disparity planes, where n_p is the number planes. Each plane is a vector of three real numbers representing the plane parameters.
- $L = \{1, 2, \ldots, n_p\}$ is the set of labels.
- $G(V, E)$ is an undirected graph with the set of vertices $V = \{1, 2, \ldots, n_s\}$ representing the color segment numbers, and the set of edges E connecting the adjacent color segments.
- $\mathbf{w} = (w_1, w_2, \ldots, w_{n_s})$ is a vector of natural variables, where $w_j \in L$ for $1 \leq j \leq n_s$.

To specify each vertex $u \in V$ by a label $l \in L$, we define the energy function as:

$$F(\mathbf{w}) = \sum_{u \in V} C^u_{seg}(w_u) + \lambda \sum_{(u,v) \in E} \xi(w_u, w_v), \tag{6}$$

$$C^u_{seg}(w_u) = \sum_{(x,y)\in S_u} Cost(x, y, f(x, y, \rho_{w_u})), \tag{7}$$

$$\xi(w_u, w_v) = \begin{cases} 1, if w_u \neq w_v \\ 0, otherwise, \end{cases} \tag{8}$$

where C^u_{seg} computes the cost of allocating the estimated planes to the segments using the *Cost* function defined in (3), and the function ξ is to penalize the solution when disparity planes of adjacent segments are not equal. Indeed, the second term ensures that the disparity planes of a neighborhood of segments present some coherence and generally do not change abruptly. Given $\rho_{w_u} = (a, b, c)$ as a vector of three real numbers for plane parameters, and $(x, y) \in S_u$, the function f computes the corresponding disparity as $f(x, y, \rho_{w_u}) = ax + by + c$. Finally, λ weighs the penalties given by ξ. We aim to minimize (6) by making a QUBO model and then giving this QUBO to the D-Wave QPU to approximate the optimal solution. Since (6) is a discrete minimization, we use a method introduced by D-Wave to make an equivalent QUBO model. We encourage the readers to check [1] for further information.

Let $\mathbf{x} \in \{0, 1\}^{n_s n_p}$ be a vector of $n_s n_p$ binary variables as $\mathbf{x} = (x_{u,l})$, where $u \in V$ and $l \in L$. Indeed, we have allocated n_p binary variables to each vertex as follows:

$$\mathbf{x} = \{x_{1,1}, x_{1,2}, \ldots, x_{1,n_p}, x_{2,1}, x_{2,2}, \ldots x_{2,n_p}, \ldots, x_{n_s,1}, x_{n_s,2}, \ldots, x_{n_s,n_p}\}$$

Let our QUBO model be defined as (9), where $\beta > \left(\sum_{u\in V} \max\{C^u_{seg}(l) | l \in L\}\right) + \lambda|E|$.

$$H_{qubo}(\mathbf{x}) = \beta \left(\sum_{u\in V} (1 - \sum_{l\in L} x_{u,l})^2 \right) \tag{9}$$
$$+ \sum_{u\in V} \sum_{l\in L} C^u_{seg}(l) x_{u,l} + \lambda \sum_{(u,v)\in E} \sum_{l_1\in L} \sum_{l_2\in L} \xi(l_1, l_2) x_{u,l_1} x_{v,l_2}.$$

We set $\mathbf{x}^* = \arg\min_{\mathbf{x}} H_{qubo}(\mathbf{x})$ and define a vector of n_s natural values as $\mathbf{w}^* = (w^*_u)_{u\in V}$ where $w^*_u = l$ if $x^*_{u,l} = 1$. Then, $\mathbf{w}^* = \arg\min_{\mathbf{w}} F(\mathbf{w})$.

Once we obtain the optimal solution of minimizing F, we can compute the final disparity map as follows. Given the set of color segments S, the set of disparity planes ρ, and $\mathbf{w}^*$ as the allocated vector of labels, we have $disp(x, y) = ax + by + c$ where, $(x, y) \in S_u$, and $\rho_{w^*_u} = (a, b, c)$ for $(x, y) \in S_u$, and $u \in V$. Figure 2 shows the result of each step in the proposed hybrid quantum-classical segment-based Stereo Matching method for the left image of the *Bull* dataset.

5 Experimental Set-Up and Results

A QUBO model is embedded in the D-Wave QPU for minimization. Embedding, which is the process of mapping QUBO variables to the physical qubits on the QPU, is challenging since the number of available qubits is relatively small. Two or more qubits on the QPU might be chained together to represent a QUBO variable. The scarcity of available qubits makes it difficult to directly embed large QUBOs into the QPUs. Hybrid solvers tackle the problem using classical and quantum problem-solving approaches. D-Wave hybrid solvers accept problems with much more variables than those solved directly by the D-Wave. They can give a reliable estimate of the future D-Wave QPUs' accuracy once enough qubits are available on the hardware. In a hybrid solver, multiple quantum and classical solvers are run in parallel, and the best solution is chosen from a pool of results. To have experimental results from real-world-sized images, we used a D-Wave hybrid solver named Constrained Quadratic Model (CQM)-hybrid solver, which accepts up to 500,000 QUBO variables. We chose the stereo datasets from 2001-Middlebury stereo datasets [23], namely *Bull*, *Venus*, *Sawtooth*, and *Barn*. The final number of segments and disparity planes (after the refinement step) for the D-Wave minimization considering each dataset are as follows: *Bull*: (24, 11), *Venus*: (23, 11), *Sawtooth*: (22, 9), and *Barn*: (23, 15).

Fig. 2. From left to right: Color image segmentation, initial disparity estimation, disparity plane estimation, segment and disparity plane refinement, and D-Wave minimization result.

We also provide a disparity-variation representation with respect to the corresponding ground truth for each computed disparity map. Let *disp* and *truth* be two functions giving the final disparity map and the exact disparity map (also called ground truth), respectively. Given a pixel (x, y), let $\varphi_{\text{xy}} = |disp(\text{x}, \text{y}) - truth(\text{x}, \text{y})|$ be the absolute difference between the computed disparity value and its corresponding ground truth value. In the disparity-variation representation, a given pixel (x, y) is represented in yellow if $\varphi_{\text{xy}} > 2$, and in red if $\varphi_{\text{xy}} > 4$. Figure 3 shows experimental results for the selected datasets. Two commonly-used metrics [23] assess the accuracy of the results compared to the ground truth: *root-mean-squared error* (*rmse*) and *percentage of bad matching pixels* (*bad-B*). The *rmse* and *bad-B* can be defined as follows [23]:

$$rmse = \sqrt{\frac{1}{N} \sum_{(\text{x},\text{y}) \in P} (disp(\text{x}, \text{y}) - truth(\text{x}, \text{y}))^2}, \tag{10}$$

$$bad\text{-}B = \left(\frac{1}{N} \sum_{(\text{x},\text{y}) \in P} (|disp(\text{x}, \text{y}) - truth(\text{x}, \text{y})| > B) \right) \times 100, \tag{11}$$

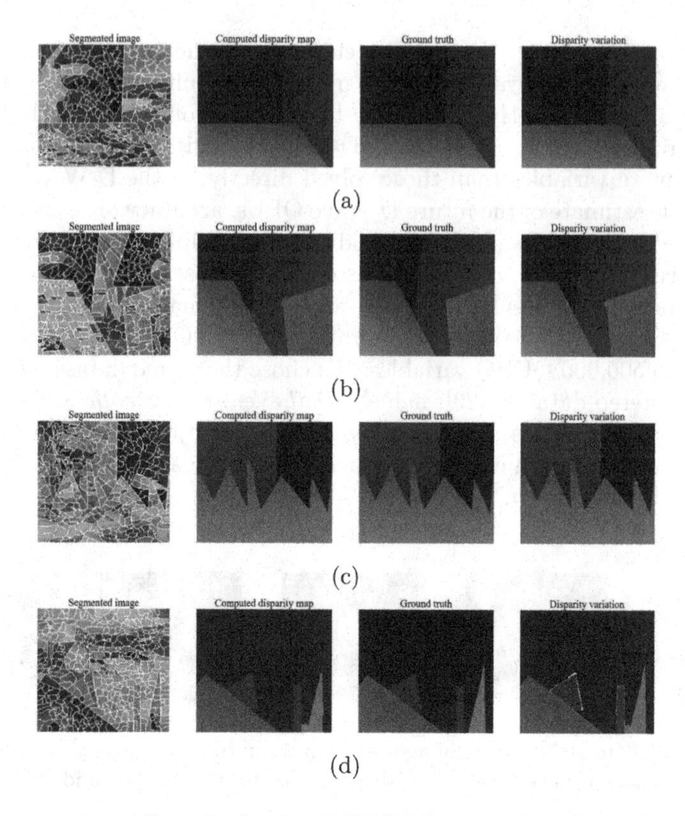

Fig. 3. The experimental results for the 2001-Middlebury stereo datasets: (a) *Bull*, (b) *Venus*, (c) *Sawtooth*, (d) *Barn*. For each dataset from left to right: the segmented left image, the computed disparity map, the corresponding ground truth, and the disparity variation. Given $(\mathrm{x}, \mathrm{y}) \in P$, it has been shown in yellow if $\varphi_{\mathrm{xy}} > 2.0$, and in red $\varphi_{\mathrm{xy}} > 4.0$.

where, B is the disparity error tolerance, and N is the number of pixels. In our evaluation, B has been set to three different values: 0.5, 1.0, and 2.0, namely *bad*-0.5, *bad*-1.0, and *bad*-2.0, respectively (Table 1).

Table 1. *rmse* and *bad-B* results for the computed disparity maps.

Stereo datasets	*rmse*	*bad-0.5*	*bad-1.0*	*bad-2.0*
Bull	0.188	0.054	0.054	0.028
Venus	0.42	8.394	0.534	0.315
Sawtooth	0.404	3.624	0.259	0.257
Barn	1.274	67.07	32.39	1.177

6 Conclusion

Our study presents a novel approach to Stereo Matching as a hybrid quantum-classical segment-based pipeline that leverages Quantum Annealing for minimization. The classical components of our method initially divide the left image into small, homogeneous color segments, followed by estimating initial disparities using a local Stereo matching method and disparity plane fitting. The quantum component then assigns labels to each segment based on the estimated disparity planes and an objective function that can be minimized using a quantum model and Quantum Annealing. It is worth noting that such a labeling problem is classically intractable. Due to the limitations of current D-Wave quantum processors, we employed a D-Wave hybrid solver for the minimization part. Despite accurate results on a Middlebury dataset, our method is susceptible to the initial segmentation since the initial disparity estimation relies on the segment boundaries. Any inaccuracies in the initial segmentation can lead to inaccurate disparity planes for the minimization part, ultimately impacting the overall accuracy of our approach. For future works, one can address this issue by ignoring the initial disparity values near the boundaries of each segment to fit more accurate disparity planes to the segment.

References

1. Discrete quadratic models (2023). https://docs.ocean.dwavesys.com/en/stable/concepts/dqm.html
2. Aharonov, D., Van Dam, W., Kempe, J., Landau, Z., Lloyd, S., Regev, O.: Adiabatic quantum computation is equivalent to standard quantum computation. SIAM Rev. **50**(4), 755–787 (2008)
3. Besag, J.: On the statistical analysis of dirty pictures. J. Roy. Stat. Soc.: Ser. B (Methodol.) **48**(3), 259–279 (1986)
4. Bleyer, M., Breiteneder, C.: Stereo matching state-of-the-art and research challenges. In: Farinella, G., Battiato, S., Cipolla, R. (eds.) Advanced Topics in Computer Vision. Advances in Computer Vision and Pattern Recognition, pp. 143–179. Springer, London (2013). https://doi.org/10.1007/978-1-4471-5520-1_6
5. Bleyer, M., Rhemann, C., Rother, C.: PatchMatch stereo-stereo matching with slanted support windows. In: BMVC, vol. 11, pp. 1–11 (2011)
6. Boykov, Y., Veksler, O., Zabih, R.: Fast approximate energy minimization via graph cuts. IEEE Trans. Pattern Anal. Mach. Intell. **23**(11), 1222–1239 (2001)
7. Černý, V.: Thermodynamical approach to the traveling salesman problem: an efficient simulation algorithm. J. Optim. Theory Appl. **45**(1), 41–51 (1985)
8. Cruz-Santos, W., Venegas-Andraca, S.E., Lanzagorta, M.: A QUBO formulation of the stereo matching problem for D-Wave quantum annealers. Entropy **20**(10), 786 (2018)
9. Denchev, V.S., et al.: What is the computational value of finite-range tunneling? Phys. Rev. X **6**(3), 031015 (2016)
10. Farhi, E., Goldstone, J., Gutmann, S., Sipser, M.: Quantum computation by adiabatic evolution. arXiv preprint quant-ph/0001106 (2000)
11. Felzenszwalb, P.F., Huttenlocher, D.P.: Efficient belief propagation for early vision. Int. J. Comput. Vision **70**(1), 41–54 (2006)

12. Geiger, D., Girosi, F.: Parallel and deterministic algorithms from MRFs: surface reconstruction. IEEE Trans. Pattern Anal. Mach. Intell. **13**(05), 401–412 (1991)
13. Heidari, S., Dinneen, M.J., Delmas, P.: An equivalent QUBO model to the minimum multi-way cut problem. Technical report, Department of Computer Science, The University of Auckland, New Zealand (2022)
14. Heidari, S., Rogers, M., Delmas, P.: An improved quantum solution for the stereo matching problem. In: 2021 36th International Conference on Image and Vision Computing New Zealand (IVCNZ), pp. 1–6. IEEE (2021)
15. Hosni, A., Rhemann, C., Bleyer, M., Rother, C., Gelautz, M.: Fast cost-volume filtering for visual correspondence and beyond. IEEE Trans. Pattern Anal. Mach. Intell. **35**(2), 504–511 (2012)
16. Kappes, J.H., et al.: A comparative study of modern inference techniques for structured discrete energy minimization problems. Int. J. Comput. Vision **115**(2), 155–184 (2015)
17. King, J., et al.: Quantum annealing amid local ruggedness and global frustration. J. Phys. Soc. Jpn. **88**(6), 061007 (2019)
18. Koller, D., Friedman, N.: Probabilistic Graphical Models: Principles and Techniques. MIT Press, Cambridge (2009)
19. Li, S.Z.: Markov Random Field Modeling in Computer Vision. Springer, Heidelberg (1995). https://doi.org/10.1007/978-4-431-66933-3
20. Lucas, A.: Ising formulations of many NP problems. Front. Phys. **2**, 5 (2014)
21. Ma, N., Men, Y., Men, C., Li, X.: Accurate dense stereo matching based on image segmentation using an adaptive multi-cost approach. Symmetry **8**(12), 159 (2016)
22. McGeoch, C.C.: Adiabatic quantum computation and quantum annealing: theory and practice. Synthesis Lect. Quantum Comput. **5**(2), 1–93 (2014)
23. Scharstein, D., Szeliski, R.: A taxonomy and evaluation of dense two-frame stereo correspondence algorithms. Int. J. Comput. Vision **47**(1), 7–42 (2002)
24. Szeliski, R., et al.: A comparative study of energy minimization methods for Markov random fields with smoothness-based priors. IEEE Trans. Pattern Anal. Mach. Intell. **30**(6), 1068–1080 (2008)
25. Tankovich, V., Hane, C., Zhang, Y., Kowdle, A., Fanello, S., Bouaziz, S.: HITNet: hierarchical iterative tile refinement network for real-time stereo matching. In: Proceedings of the IEEE/CVF Conference on Computer Vision and Pattern Recognition, pp. 14362–14372 (2021)
26. Tao, H., Sawhney, H.S., Kumar, R.: A global matching framework for stereo computation. In: Proceedings Eighth IEEE International Conference on Computer Vision, ICCV 2001, vol. 1, pp. 532–539. IEEE (2001)
27. Vedaldi, A., Soatto, S.: Quick shift and kernel methods for mode seeking. In: Forsyth, D., Torr, P., Zisserman, A. (eds.) ECCV 2008. LNCS, vol. 5305, pp. 705–718. Springer, Heidelberg (2008). https://doi.org/10.1007/978-3-540-88693-8_52
28. Veksler, O.: Efficient graph-based energy minimization methods in computer vision. Cornell University (1999)
29. Wainwright, M., Jaakkola, T., Willsky, A.: Tree consistency and bounds on the performance of the max-product algorithm and its generalizations. Stat. Comput. **14**(2), 143–166 (2004)
30. Wang, C., Komodakis, N., Paragios, N.: Markov random field modeling, inference & learning in computer vision & image understanding: a survey. Comput. Vis. Image Underst. **117**(11), 1610–1627 (2013)

31. Xiao, J., Yang, L., Zhou, J., Li, H., Li, B., Ding, L.: An improved energy segmentation based stereo matching algorithm. ISPRS Ann. Photogram. Remote Sens. Spat. Inf. Sci. **1**, 93–100 (2022)
32. Yaacoby, R., Schaar, N., Kellerhals, L., Raz, O., Hermelin, D., Pugatch, R.: A comparison between D-Wave and a classical approximation algorithm and a heuristic for computing the ground state of an Ising spin glass. arXiv preprint arXiv:2105.00537 (2021)
33. Zhang, K., Lu, J., Lafruit, G.: Cross-based local stereo matching using orthogonal integral images. IEEE Trans. Circuits Syst. Video Technol. **19**(7), 1073–1079 (2009)

Adaptive Enhancement of Extreme Low-Light Images

Evgeny Hershkovitch Neiterman(✉), Michael Klyuchka, and Gil Ben-Artzi

Department of Computer Science, Ariel University, Ari'el, Israel
neiterman@ariel.ac.il

Abstract. Existing methods for enhancing dark images captured in a very low-light environment assume that the intensity level of the optimal output image is known and already included in the training set. However, this assumption often does not hold, leading to output images that contain visual imperfections such as dark regions or low contrast. To facilitate the training and evaluation of adaptive models that can overcome this limitation, we have created a dataset of 1500 raw images taken in both indoor and outdoor low-light conditions. Based on our dataset, we introduce a deep learning model capable of enhancing input images with a wide range of intensity levels at runtime, including ones that are not seen during training. Our experimental results demonstrate that our proposed dataset combined with our model can consistently and effectively enhance images across a wide range of diverse and challenging scenarios. Code is available at https://github.com/mklyu/CEL-net.

Keywords: Computational imaging · Extreme Low light

1 Introduction

Images captured in low light are characterized by low photon counts, which results in a low signal-to-noise ratio (SNR). Setting the exposure level while capturing an image can be done by the user in manual mode, or automatically by the camera in auto exposure (AE) mode. In manual mode, the user can adjust the ISO, f-number, and exposure time. In auto exposure (AE) mode, the camera measures the incoming light based on through-the-lens (TTL) metering and adjusts the exposure values (EVs), which refers to configurations of the above parameters.

We consider the problem of enhancing a dark image captured in an extremely low-light environment, based on a single image [7]. In a dark environment, adjusting the parameters to increase the SNR has its own limitations. For example, high ISO increases the noise as well, and lengthening the exposure time might introduce blur. Various approaches have been proposed as post-processing enhancements in low-light image processing [6,14,15,19,35,37]. In extreme low light conditions, such methods often fail to produce satisfactory results.

Each of the first two authors contributed equally.

J. Blanc-Talon et al. (Eds.): ACIVS 2023, LNCS 14124, pp. 14–26, 2023.
https://doi.org/10.1007/978-3-031-45382-3_2

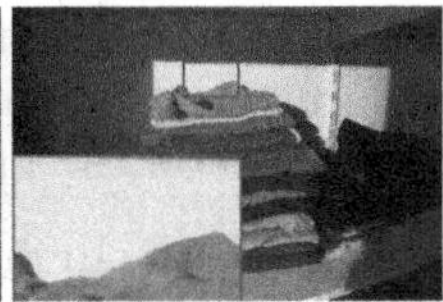

Fig. 1. At runtime, the intensity level for optimal restoration of a given dark image might be different from the trained one and can lead to dark areas or low contrast. For a wide range of output image's intensity levels, our model optimizes the enhancement of the input image. (a) left: The input, (b) center left: Ground truth, (c) center right: SID [7] enhances the image to a fixed intensity level, which is not optimal for the input image, and as a result, there are noticeable artifacts. (d) right: Our approach adapts the enhancement operation to optimally match any selected intensity levels, thereby reducing the presence of artifacts.

Recent works [7,16,23,27,32] introduce data-driven approaches to replace the traditional image signal processing pipeline and learn a direct mapping from low-exposure input images to well-lit output images. Such models are trained using a fixed intensity level for the output image. Given a dark image, they first multiply its intensity values by a constant factor to increase its brightness and then apply the enhancement model that is trained specifically for this fixed intensity level to produce high-quality image. However, during runtime, it is common for the optimal intensity level of the output image to differ from the trained one and the model outputs less visually appealing images.

We address this limitation by proposing a model that can enhance dark images across a wide range of intensity levels, including those that were not seen during training. Our model achieves this by adaptively adjusting the enhancement operation during runtime to optimally match the selected intensity level from a given range, without requiring retraining of the model. As a result, our model can significantly reduce artifacts in the output image even for previously unseen intensity levels (Fig. 1).

To enhance real-world images captured in low light, training based on only synthetic noise samples is insufficient. We have collected 1500 raw images captured with five different exposure levels in extreme low-light conditions, in both indoor and outdoor environments, and under various camera parameters. We use fixed exposure times as the exact output intensity of each enhanced image might be different from each other. Each exposure time corresponds to a distinct intensity level, and the range of exposures yields a range of intensity levels. Using the dataset, we show how to train our model such that it can successfully enhance images with different intensity levels at runtime.

Previous adaptive approaches [12,18] only considered signal-independent noise using sRGB synthetic noise samples. Here we propose an adaptive model with two input parameters to control the output intensity and address both signal-dependent and signal-independent noises. The first parameter controls the intensity level of the output image by simple multiplication. This results in an inevitable amplification of the noises and other artifacts as well. The second parameter adjusts the operation of the image signal processing (ISP) unit to enhance the degradations that are the result of the increase in the intensity, conditioned on the intensity level.

Contribution. In pursuit of advancing research in the field and facilitating the development of adaptive models, we have curated a dataset containing 1,500 raw images captured in extremely low-light conditions, comprising indoor and outdoor scenes with diverse exposure levels. We propose and train a model that can produce compelling results for restoring dark images with a wide range of optimal intensity levels, including ones that were not available during training. Our experimental results, which incorporate both qualitative and quantitative measures, demonstrate that our model along with our dataset improves the enhancement quality of dark images.

2 Related Work

Datasets. A key contribution of our work is a dataset of real-world images that enable training and evaluating multi-exposore models in exterme low light. Unlike existing datasets, we introduce a long-exposure reference image with multiple shorter exposure times for each scene, in both indoor and outdoor scenes, and directly provide the raw sensor data. Our dataset fills the gap and allows the training of an adaptive model in extreme low-light conditions by combining multiple exposures. Our dataset vs. other datasets is compared in Table 1.

Table 1. Comparison with previous datasets.

Dataset	Format	# Images	Publicly Available	Multi Exposure	Extreme Low Light
DND [22]	RAW	100	yes	no	no
SIDD [1]	RAW	30000	yes	yes	no
LLNet [21]	RGB	169	yes	no	no
MSR-Net [25]	RGB	10000	no	no	no
SID [7]	RAW	5094	yes	no	yes
SICE [5]	RGB	4413	yes	no	no
RENOIR [3]	RAW	1500	yes	no	no
LOL [8]	RGB	500	yes	no	no
DeepUPE [31]	RGB	3000	no	no	no
VE-LOL-L [20]	RGB	2500	yes	no	no
DarkVision [36]	RAW	13455	yes	no	no
Our	RAW	1500	yes	yes	yes

Adaptive Restoration Networks. Adaptive restoration networks can broadly be categorized as models that allow tuning different objectives at runtime [13] or different restoration levels of the same objective. Dynamic-Net [26] adds specialized blocks directly after convolution layers, which are optimized during the training for an additional objective. CFSNet [28] uses branches, each one targeted for a different objective. AdaFM [12] adds modulation filters after each convolution layer. Deep Network Interpolation (DNI) [29] trains the same network architecture on different objectives and interpolates all parameters. These methods are optimized for well-lit images and as we demonstrate in the experiments, struggle to enhance images captured in extreme low light conditions.

Low-light Image Enhancement. Widely used enhancement methods are histogram equalization, which globally balances the histogram of the image; and gamma correction, which increases the brightness of dark pixels. More advanced methods include illumination map estimation [11], semantic map enhancement [33], bilateral learning [10], multi-exposure [2,5,34], Retinex model [4,9,30,38] and unpaired enhancement [17]. In contrast to these methods, we consider an extreme low-light environment with very low SNR, where the scene is barely visible to the human eye. Chen [7] has introduced an approach to extreme low-light imaging by replacing the traditional image processing pipeline with a deep learning model based on raw sensor data. Wang [27] introduced a neural network for enhancing underexposed photos by incorporating an illumination map into their model, while Xu [32] presented a model for low-light image enhancement based on frequency-based decomposition. These methods are optimized to output an enhanced image with a fixed exposure. In cases where the user requires a change in the exposure (intensity level) of the output image, these methods require retraining the models, typically on additional sets of images. In contrast, we introduce an approach that enables continuous setting of the desired exposure at inference time.

Fig. 2. The multi-exposure dataset. The top two rows are images of outdoor scenes, and the bottom two rows are images of indoor scenes. From left to right are exposure times of 0.1 s, 0.5 s, 1 s, 5 s, and 10 s.

3 Our Approach

3.1 Multi-exposure Extreme Low-Light Dataset (ME2L)

We collected a total of 1500 images. In order to capture a variety of realistic low-light conditions and cover a broad range of scenes with extreme low-light conditions, the images were captured in both indoor and outdoor scenes. The images were captured over different days in multiple locations. We captured five different exposures for each of the scenes - 0.1 s, 0.5 s, 1 s, 5 s, and 10 s resulting in a range of intensity levels. Various scenes have different intensity levels. By training our model on all the images it learns how to optimize the whole range of intensity levels.

The outdoor images were captured late at night under moonlight or street lighting. The indoor images were captured in closed rooms with indirect illumination. Generally, the lowest exposure image in both indoor and outdoor scenes is completely dark and no details of the scene can be observed.

All the scenes in the dataset are static to accommodate the long exposure. For each scene, similar to [7], the settings of the camera were adjusted to optimize the longest-exposure image. We used a tripod and a mirrorless camera to capture the exact same scene without any misalignment. At each scene, after the long exposure image was optimally captured, we used a smartphone application to decrease the exposure and capture the images without touching the camera or changing the camera's parameters. After capturing the images, we manually verified that the images are aligned and the long-exposure reference images are of high perceptual quality.

The images were captured using a Sony α5100 with a Bayer sensor and a resolution of 6000 × 4000. Figure 2 shows samples from our dataset.

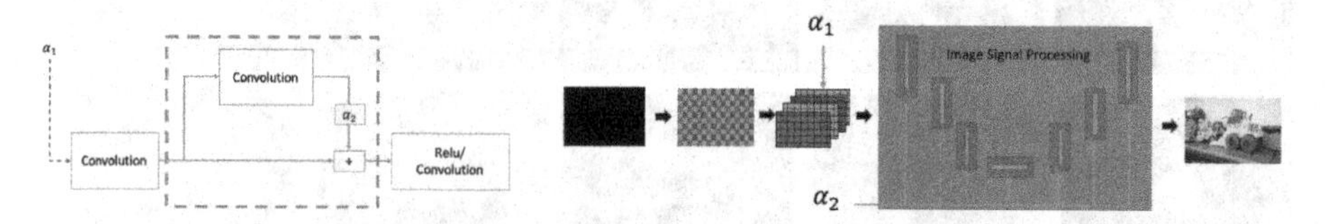

Fig. 3. The dashed red rectangle is the modulation module. The enhancement parameter α_2 represents a weighted sum between the feature map of the initial and final exposure levels. The blue dashed line is to emphasize that the operation of the modulation module is also affected by the α_1 parameters which control the brightness of the image. (Color figure online)

Fig. 4. The architecture of our network. There are two input parameters, α_1 (brightness) and α_2 (enhancement). α_1 controls the brightness of the raw input data. α_2 modulates the weights of the filters and tunes the network, which operates as an Image Signal Processing (ISP) unit. We train the model for an initial and final exposure level, where for each value of α_1 there is a single value of α_2. At inference time, each parameter can be set independently of the other.

3.2 Our Model

The goal of existing extreme low light approaches is to find a function that can map at inference time a data point from raw data space to a single data point in the sRGB space, denoted as $f : \mathcal{Y} \rightarrow \mathcal{Y}_{rgb}$, where $\mathcal{Y}_{rgb}$ is the sRGB space. This approach leads to inaccurate results in cases where the optimal intensity level of the output image is not the same as in the training and may result in noticeable artifacts. Direct change of the signal's mean by a multiplication and applying the same model does not result in the desired outcome, since the noise in raw sensor data results from two main sources: signal-dependent noise and signal-independent. The first one is referred to as shot noise, which is related to the uncertainty that is a property of the underlying signal itself, describing the photon arrival statistics. The second one is read-noise, which is the result of uncertainty generated by the electronics of the camera when the charge stored is read out. The shot noise is a Poisson random variable, whose mean is the expected number of photons per unit time interval, describing the true light intensity. The read noise is a Gaussian random variable with zero mean whose variance is fixed.

The heteroscedastic Gaussian model is a more widely acknowledged alternative to the Poisson-Gaussian model, which substitutes the Poisson component with a Gaussian distribution whose variance is signal-dependent:

$$y \sim \mathcal{N}(x, \beta_{read} + \beta_{shot}x), \tag{1}$$

where $y \in \mathcal{Y}$ is the observed (raw) intensity at a pixel in the raw data space $\mathcal{Y}$, x is the original (unknown) signal, β_{shot} is proportional to the analog gain (g_a) and digital gain (g_d) and β_{read} is proportional to the sensor readout variance (σ_r^2) and digital gain: $\beta_{read} = g_d^2\sigma_r^2$, $\beta_{shot} = g_d g_a$.

It is therefore evident from Eq. (1) that unlike previous methods, adding a single noise source (e.g. Gaussian) or using a simple multiplication to adjust the image intensity is not equivalent to acquiring an image with such original intensity. We propose an alternative approach to enhance both read and shot noises by employing two input parameters each contributing differently to Eq. (1), a modulation layer [12], and mapping of a *single* data point from raw data space to *multiple* points in sRGB, each with a different output intensity level.

Our Raw-to-sRGB pipeline is formulated as a function $f : \mathcal{Y} \times \mathcal{R} \times \mathcal{R} \rightarrow \mathcal{Y}_{rgb}$, $y_{rgb} = f(y, \alpha_1, \alpha_2; \theta)$, where α_1 is a scalar that sets the mean of the signal in Eq. (1) to the desired level by multiplication of the raw data, α_2 controls the enhancement level of the Raw-to-sRGB pipeline, θ represents the parameters of $f(\cdot)$ and $y_{rgb} \in \mathcal{Y}_{rgb}$ is the signal of the sRGB image. The function f is realized by a deep network with modulation layers. To obtain θ, we train our network in two steps. First, the base model is trained to fit the enhanced image with an initial intensity level, without any additional modifications to the existing architecture. Then we freeze the weights of the base model, and each modulation layer (g) is inserted after each existing convolutional kernel $g(w, b) \circ X$, where X is the output feature map of existing convolutional kernels in the base network and w, b are weights and bias of the modulation layer's convolutional filter kernel.

The network is then fine-tuned to fit the enhanced image with a final intensity level by learning the weights of the additional convolutional kernels. Thus, in our formulation, θ includes the parameters of both the base network and the modulation layers. During runtime, assuming w_1 is the base convolution kernel, w and b are the weights of filter and bias in each modulation layer, the output of the modulation layer is:

$$w_1 + \alpha_2 w_1 * w + \alpha_2 b, \tag{2}$$

for the given scalar $0 \leq \alpha_2 \leq 1$ representing the enhancement parameter (Fig. 3).

To control both noise sources, we set $\alpha_2 \in [0, 1]$ such that it linearly corresponds to α_1 and $\alpha_2 = 1$ corresponds to the maximum value of α_1. Our key intuition is that for $\alpha_1, \alpha_2 \to 0$, it is the trained base network (before fine tuning) that produces the most significant output, and it enhances the read noise (Eq. (1) and (2)). During training, both parameters are adjusted according to the ground-truth image. The input arrays' values are multiplied by the α_1 parameter, which represents the ratio between the input image's exposure time and the required output image's exposure time, effectively setting the intensity and noise levels of the output. The overall architecture of our network is presented in Fig. 4.

Unlike existing adaptive method, we do not operate in sRGB domain for noisy images as it limits the representation power of the architecture [1]. Instead, we operate in the raw domain and employ a U-Net [24] as our base architecture (f). It replaces the entire image signal processing (ISP) pipeline. The input is a short exposure raw image from Bayer sensor data and the output is an sRGB image. The raw Bayer sensor data is packed into four channels, the spatial resolution is reduced by a factor of two in each dimension; and the black level is subtracted. The output is a 12-channel image processed to recover the original resolution of the input image.

For testing, we set the intensity level (α_1) and the enhancement (α_2) parameters of the network to the desired exposure and ISP configuration. The input image is multiplied according to the intensity level parameter, resulting in a noisy, brighter image. The weights of the filter and bias in the modulation module after the fine-tuning phase are adjusted according to the value of the enhancement parameter.

We train the model using L1 loss and the Adam optimizer. The inputs are random 512×512 patches with standard augmentation. The learning rate is 10^{-4} for 1000 epochs and then 10^{-5} for an additional 1000 epochs, a total of 2000 epochs for the training phase. Fine-tuning the model for the final exposure level requires an additional 1000 epochs.

4 Experiments

Baselines. We compare our results with state-of-the-art adaptive methods [12,18]. Using our dataset, we train them in accordance with their authors' instructions. The inputs of the compared models were modified to operate on raw images in order to ensure fair comparisons. The **SID** [7] is the baseline model for extreme low light enhancement, and it enhances dark images to a fixed intensity level.

Evaluation Metrics. We use 70%, 10%, and 20% of the images for training, validation, and testing, respectively, with uniform sampling and equal representation for indoor and outdoor scenes in each set. The ground truth images are the corresponding long-exposure images processed by LibRaw[1] to sRGB format.

4.1 Quantitative Comparisons

Table 2 presents the PSNR and SSIM metrics for various experiments designed to evaluate the different approaches. Each section (A-D) represents a different experiment. The left column shows the different methods and their training protocols. The input for both training and testing is a dark image with an exposure time of 0.1 s. For each method, the ground truth exposure times that were used for training (1 s/5 s/10 s) are shown with each model (using the ⇒).

In Table 2A we train the SID model for every single input and output intensity (and exposure) independently. Note that SID is optimized for a single output only. By testing the model on the same exposure as trained, we obtain the

Table 2. For all methods, the input exposure for both training and testing is 0.1 s. ⇒ denotes the ground-truth images used for training. The bold are the two best results. As can be seen, our model outperforms all other methods. See text for more details.

Train/Test	1 s		5 s		10 s	
	PSNR	SSIM	PSNR	SSIM	PSNR	SSIM
A - Single Exposure Baseline						
SID [7] ⇒ 1	**38.17**	0.95	30.7	0.87	27.7	0.84
SID [7] ⇒ 5	36.82	0.94	**33.35**	**0.91**	28	0.86
SID [7] ⇒ 10	34.88	0.9	30.52	0.88	**30**	**0.88**
B - Multi Exposure Baseline						
SID [7] ⇒ 1,5,10	35.77	0.92	29.55	0.86	26.25	0.82
Retinex [30] ⇒ 1,5,10	16.29	0.08	15.15	0.12	13.67	0.16
C - Two Exposure Interpolation						
AdaFM [12] ⇒ 1,10	37.86	0.85	30.51	0.73	26.95	0.72
CResMD [18] ⇒ 1,10	36.37	0.8	21.63	0.46	26.52	0.64
Ours ⇒ 1,10	**38.17**	**0.95**	**32.35**	**0.89**	**29.67**	**0.87**
D - Two Exposure Extrapolation						
AdaFM [12] ⇒ 1,5	37.86	0.85	31.12	0.76	25.98	0.7
CResMD [18] ⇒ 1,5	34.97	0.73	23.73	0.59	16.17	0.17
Ours ⇒ 1,5	**38.17**	**0.95**	**31.78**	**0.89**	**28.65**	**0.86**

[1] www.libraw.org.

optimal achievable restoration accuracy as the model is specialized on a single intensity level. Testing on other exposures (e.g., training on 5 s and testing on ground truth image of 10 s by setting α_1 to the optimal value) shows that the resulting enhanced image quality is significantly reduced, which is the key limitation of single-output methods. The goal of our approach is to overcome this and achieve high restoration quality over the continuous range of possible exposure times with a single model.

Table 2B evaluates the ability to train single-output approach [7] to generalize to multiple output exposures. We train the model based on all possible output exposures and evaluate its ability to enhance specific exposure times within the trainable range. It can be seen that using multiple ground-truth exposures with a model that is designed to output only a single one reduces the restoration quality for all the possible outputs.

We compare our approach with state-of-the-art adaptive methods. Table 2C presents the results for one of the most common use-cases: where the optimal exposure time is within the trainable range in runtime. We train the models to enhance input images with an exposure of 0.1 s and a ground truth exposure range of [1 s,10 s]. At inference time, the models can enhance an image to a range of exposures and the specific one is selected. We evaluate the models with input images of 0.1 s and optimal output exposure of 5 s. It can be seen that our approach outperforms all other methods.

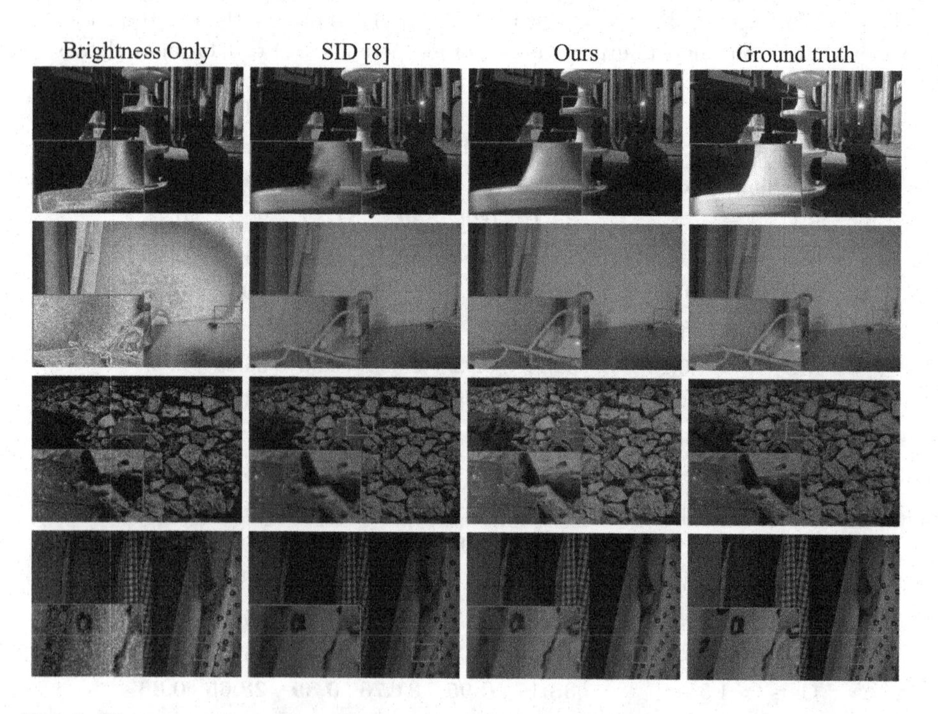

Fig. 5. The restoration effect of enhancing images to exposure level within the trained range. The first column is obtained by directly adjusting the brightness level to the optimal exposure by multiplication.

In real-world scenarios, the actual optimal exposure time of the enhanced image can be outside the trained range. We experiment with such cases, training the models for optimal exposure times of [1 s, 5 s], and testing with input images such that the ground truth exposure time is 10 s. The results are presented in Table 2D. As before, our approach achieves the best restoration accuracy.

4.2 Qualitative Comparisons

Figure 5 shows the effect of adjusting the exposure time for a value within the trained range, 5 s. The model is trained using input images with an exposure time of 0.1 s and ground truth images with exposure times of 1 s and 10 s. SID was trained on all possible output exposure times. The enhanced images after adjusting the brightness and enhancement parameters are shown. The left column shows the effect of multiplying the intensity of the input images by 50, which is the ratio between the ground truth exposure of the input (0.1 s) and the ground truth output (5 s). It can be seen that our model successfully removes the artifacts presented by the other approaches.

4.3 Ablation Study

Filter Size. We evaluate the sizes of different filters in the modulation module. We consider filter sizes of - 1×1, 3×3, 5×5, and 7×7. We train our base model with an exposure of 0.1 s and an output of 1 s, then fine-tune it to an output of 10 s. The test images are with an exposure of 5 s.

Table 3 shows our comparisons. It can be seen that the most significant gain is achieved when using a filter size of 3×3.

Table 3. Filter size comparisons. The model is trained from 0.1 s to 1 s and fine-tuned to 10 s, and tested for an unseen exposure level of 5 s.

1×1	3×3	5×5	7×7
31.87	32.35	32.39	32.48

Tuning Direction. We evaluate the optimal direction for the tuning. We compare two models. The first one is trained from 0.1 s to 1 s and fine-tuned for 10 s. The second one is trained from 0.1 s to 10 s and fine-tuned for 1 s. We compare the results with respect to unseen output images with an exposure time of 5 s. The forward direction from 0.1 s to 10 s achieved better results than the backward one, with a PSNR of 32.35 vs. 28.2.

5 Conclusion

Extreme low-light imaging is challenging and has recently gained growing interest. Current methods allow enhancement of dark images, assuming the input exposure and the optimal output exposure are known at inference time, which prevents its adaptation in practical scenarios. We collected a dataset of 1500 images with multiple exposure levels for extreme low-light imaging. We present an approach that enables continuously controlling of the optimal output exposure levels of the images at runtime, without the need to retrain the model and showed that our model presents promising results on a wide range of both indoor and outdoor images. We believe that our dataset as well as our model will support further research in the field of extreme low-light imaging, making a step forward towards its widespread adoption.

References

1. Abdelhamed, A., Lin, S., Brown, M.S.: A high-quality denoising dataset for smartphone cameras. In: IEEE Conference on Computer Vision and Pattern Recognition (CVPR) (2018)
2. Afifi, M., Derpanis, K.G., Ommer, B., Brown, M.S.: Learning multi-scale photo exposure correction. In: Proceedings of the IEEE/CVF Conference on Computer Vision and Pattern Recognition, pp. 9157–9167 (2021)
3. Anaya, J., Barbu, A.: Renoir-a dataset for real low-light image noise reduction. J. Vis. Commun. Image Represent. **51**, 144–154 (2018)
4. Cai, B., Xu, X., Guo, K., Jia, K., Hu, B., Tao, D.: A joint intrinsic-extrinsic prior model for retinex. In: Proceedings of the IEEE International Conference on Computer Vision, pp. 4000–4009 (2017)
5. Cai, J., Gu, S., Zhang, L.: Learning a deep single image contrast enhancer from multi-exposure images. IEEE Trans. Image Process. **27**(4), 2049–2062 (2018)
6. Celik, T., Tjahjadi, T.: Contextual and variational contrast enhancement. IEEE Trans. Image Process. **20**(12), 3431–3441 (2011)
7. Chen, C., Chen, Q., Xu, J., Koltun, V.: Learning to see in the dark. In: Proceedings of the IEEE Conference on Computer Vision and Pattern Recognition, pp. 3291–3300 (2018)
8. Wei, C., Wang, W., Yang, W., Liu, J.: Deep retinex decomposition for low-light enhancement. In: British Machine Vision Conference (2018)
9. Fu, X., Zeng, D., Huang, Y., Zhang, X.P., Ding, X.: A weighted variational model for simultaneous reflectance and illumination estimation. In: Proceedings of the IEEE Conference on Computer Vision and Pattern Recognition, pp. 2782–2790 (2016)
10. Gharbi, M., Chen, J., Barron, J.T., Hasinoff, S.W., Durand, F.: Deep bilateral learning for real-time image enhancement. ACM Trans. Graph. (TOG) **36**(4), 1–12 (2017)
11. Guo, X., Li, Y., Ling, H.: LIME: low-light image enhancement via illumination map estimation. IEEE Trans. Image Process. **26**(2), 982–993 (2016)
12. He, J., Dong, C., Qiao, Y.: Modulating image restoration with continual levels via adaptive feature modification layers. In: Proceedings of the IEEE Conference on Computer Vision and Pattern Recognition, pp. 11056–11064 (2019)

13. He, J., Dong, C., Qiao, Y.: Multi-dimension modulation for image restoration with dynamic controllable residual learning. arXiv preprint arXiv:1912.05293 (2019)
14. Hu, Z., Cho, S., Wang, J., Yang, M.H.: Deblurring low-light images with light streaks. In: Proceedings of the IEEE Conference on Computer Vision and Pattern Recognition, pp. 3382–3389 (2014)
15. Hwang, S.J., Kapoor, A., Kang, S.B.: Context-based automatic local image enhancement. In: Fitzgibbon, A., Lazebnik, S., Perona, P., Sato, Y., Schmid, C. (eds.) ECCV 2012. LNCS, vol. 7572, pp. 569–582. Springer, Heidelberg (2012). https://doi.org/10.1007/978-3-642-33718-5_41
16. Ignatov, A., Kobyshev, N., Timofte, R., Vanhoey, K., Van Gool, L.: DSLR-quality photos on mobile devices with deep convolutional networks. In: Proceedings of the IEEE International Conference on Computer Vision, pp. 3277–3285 (2017)
17. Jiang, Y., et al.: EnlightenGAN: deep light enhancement without paired supervision (2021)
18. He, J., Dong, C., Qiao, Yu.: Interactive multi-dimension modulation with dynamic controllable residual learning for image restoration. In: Vedaldi, A., Bischof, H., Brox, T., Frahm, J.-M. (eds.) ECCV 2020. LNCS, vol. 12365, pp. 53–68. Springer, Cham (2020). https://doi.org/10.1007/978-3-030-58565-5_4
19. Lee, C., Lee, C., Kim, C.S.: Contrast enhancement based on layered difference representation of 2D histograms. IEEE Trans. Image Process. **22**(12), 5372–5384 (2013)
20. Liu, J., Xu, D., Yang, W., Fan, M., Huang, H.: Benchmarking low-light image enhancement and beyond. Int. J. Comput. Vision **129**, 1153–1184 (2021)
21. Lore, K.G., Akintayo, A., Sarkar, S.: LLNet: a deep autoencoder approach to natural low-light image enhancement (2016)
22. Plotz, T., Roth, S.: Benchmarking denoising algorithms with real photographs. In: Proceedings of the IEEE Conference on Computer Vision and Pattern Recognition, pp. 1586–1595 (2017)
23. Remez, T., Litany, O., Giryes, R., Bronstein, A.M.: Deep convolutional denoising of low-light images. arXiv preprint arXiv:1701.01687 (2017)
24. Ronneberger, O., Fischer, P., Brox, T.: U-net: convolutional networks for biomedical image segmentation. In: Navab, N., Hornegger, J., Wells, W.M., Frangi, A.F. (eds.) MICCAI 2015. LNCS, vol. 9351, pp. 234–241. Springer, Cham (2015). https://doi.org/10.1007/978-3-319-24574-4_28
25. Shen, L., Yue, Z., Feng, F., Chen, Q., Liu, S., Ma, J.: MSR-net: low-light image enhancement using deep convolutional network (2017)
26. Shoshan, A., Mechrez, R., Zelnik-Manor, L.: Dynamic-net: tuning the objective without re-training for synthesis tasks. In: Proceedings of the IEEE International Conference on Computer Vision, pp. 3215–3223 (2019)
27. Wang, R., Zhang, Q., Fu, C.W., Shen, X., Zheng, W.S., Jia, J.: Underexposed photo enhancement using deep illumination estimation. In: Proceedings of the IEEE Conference on Computer Vision and Pattern Recognition, pp. 6849–6857 (2019)
28. Wang, W., Guo, R., Tian, Y., Yang, W.: CFSNet: toward a controllable feature space for image restoration. In: Proceedings of the IEEE International Conference on Computer Vision, pp. 4140–4149 (2019)
29. Wang, X., Yu, K., Dong, C., Tang, X., Loy, C.C.: Deep network interpolation for continuous imagery effect transition. In: Proceedings of the IEEE Conference on Computer Vision and Pattern Recognition, pp. 1692–1701 (2019)
30. Wei, C., Wang, W., Yang, W., Liu, J.: Deep retinex decomposition for low-light enhancement. arXiv preprint arXiv:1808.04560 (2018)

31. Xu, K., Yang, X., Yin, B., Lau, R.W.: Learning to restore low-light images via decomposition-and-enhancement (supplementary material) (2020)
32. Xu, K., Yang, X., Yin, B., Lau, R.W.: Learning to restore low-light images via decomposition-and-enhancement. In: Proceedings of the IEEE/CVF Conference on Computer Vision and Pattern Recognition, pp. 2281–2290 (2020)
33. Yan, Z., Zhang, H., Wang, B., Paris, S., Yu, Y.: Automatic photo adjustment using deep neural networks. ACM Trans. Graph. (TOG) **35**(2), 1–15 (2016)
34. Ying, Z., Li, G., Gao, W.: A bio-inspired multi-exposure fusion framework for low-light image enhancement. arXiv preprint arXiv:1711.00591 (2017)
35. Yuan, L., Sun, J.: Automatic exposure correction of consumer photographs. In: Fitzgibbon, A., Lazebnik, S., Perona, P., Sato, Y., Schmid, C. (eds.) ECCV 2012. LNCS, vol. 7575, pp. 771–785. Springer, Heidelberg (2012). https://doi.org/10.1007/978-3-642-33765-9_55
36. Zhang, B., et al.: DarkVision: a benchmark for low-light image/video perception. arXiv preprint arXiv:2301.06269 (2023)
37. Zhang, X., Shen, P., Luo, L., Zhang, L., Song, J.: Enhancement and noise reduction of very low light level images. In: Proceedings of the 21st International Conference on Pattern Recognition (ICPR2012), pp. 2034–2037. IEEE (2012)
38. Zhang, Y., Zhang, J., Guo, X.: Kindling the darkness: a practical low-light image enhancer. In: Proceedings of the 27th ACM International Conference on Multimedia, pp. 1632–1640 (2019)

Semi-supervised Classification and Segmentation of Forest Fire Using Autoencoders

Akash Koottungal, Shailesh Pandey, and Athira Nambiar(✉)

Department of Computational Intelligence, Faculty of Engineering and technology, SRM Institute of Science and Technology, Kattankulathur 603203, Tamil Nadu, India
{ak5749,sp5153,athiram}@srmist.edu.in

Abstract. Forests play a crucial role in sustaining life on earth by providing vital ecosystem services and supporting a wide range of species. The unprecedented increase in forest fires aka '*infernos*' due to global warming i.e. rising temperatures and changing weather patterns, is quite alarming. Recently, machine learning and computer vision-based techniques are leveraged to proactively analyze forest fire events. To this end, we propose novel semi-supervised classification and segmentation techniques using autoencoders to analyse forest fires, that require significantly less labelling effort in contrast to the fully-supervised methods. In particular, semi-supervised classification of forest fire using Convolutional autoencoders is proposed. Further, Class Activation Map-based techniques and patch-wise extraction methods are envisaged for the segmentation task. Extensive experiments are carried out on two publicly available large datasets i.e. FLAME and Corsican datasets. The proposed models are found to be outperforming the state-of-the-art approaches.

Keywords: Forest fire · Semi-supervised · Autoencoder

1 Introduction

Forests are one of the most important commodities in the world. Apart from providing natural habitat to numerous species, they also provide us with resources such as wood, resin, herbs etc. Forest fires are a huge threat to the vast expanses of forest cover. In recent years, we have seen uncontrollable forest fires around the world some of which are burning to date. Every year, hundred million hectares of land are destroyed by forest fires and over two lac fires happen every year over a total area of about 3.5–4.5 million km^2 [1,2]. The increase in forest fires in forest areas around the world has resulted in increased motivation for developing fire warning systems for the early detection of wildfires. Such early fire detection systems can act as deterrents and can prevent the excessive damage causing to flora and fauna due to wildfire.

Computer Vision and Machine Learning have been utilized to develop fire detection systems [3]. For instance, one of the classical approaches such as Support Vector Machines, has been used to classify image processing features in

J. Blanc-Talon et al. (Eds.): ACIVS 2023, LNCS 14124, pp. 27–39, 2023.
https://doi.org/10.1007/978-3-031-45382-3_3

detecting forest fire regions [4]. With the rise of deep learning, artificial neural network-based architectures, including Convolutional Neural Networks, Vision Transformers, and U-net, have also been employed [3]. Although supervised learning-based approaches have shown promise, they require laborious and time-consuming annotations. For instance, the task of semantic segmentation requires humans to provide strong pixel-level annotations for millions of images. To address this issue, semi-supervised learning has been used, which reduces the amount of human effort required for data labeling by incorporating information from a large set of unlabeled data [11,12]. Semi-supervised learning results in improved accuracy through the incorporation of additional information from unlabeled data thus providing enhanced generalization capabilities.

In this work, semi-supervised methods for forest fire classification and segmentation using autoencoder are proposed. In particular, a convolutional autoencoder (CAE) model is leveraged to reconstruct the input image in an unsupervised manner, by which a rich representation is learned at the latent space. Further, the encoder part will be used towards classification task in a semi-supervised manner using the annotations. On top of this, state-of-the-art weakly supervised methods viz., Class Activation Mapping (CAM) [5] is incorporated for the visualization of fire region. Further, for the semi-supervised segmentation, model training via patch-wise extraction of images and corresponding masks is facilitated.

Extensive analysis of classification and segmentation is carried out on the FLAME dataset and Corsican database, respectively. Both classification and segmentation models outperform the state-of-the-art semi-supervised approaches as well as are quite competitive with fully supervised approaches, even with a much lower annotation labeling effort. The key contributions of the paper are as follows:

- Proposal of a novel approach for semi-supervised segmentation and classification of forest fire using autoencoders, one of its first kind.
- Visualization of the classification results via Class Activation Mapping (CAM) saliency heat maps/ CAM binary masks to depict the relevant frame regions, thus making our approach interpretable (ExplainableAI).
- Semi-supervised learning for segmentation by training autoencoder, using patches drawn from limited forest fire images.

The remainder of the paper is organized as follows: Sect. 2 provides a literature review of the detection and segmentation of forest fires. Section 3 describes the methodology used for the classification and segmentation task, focusing on the use of autoencoders in a semi-supervised manner. Section 4 presents the experimental setup and result analysis. Finally, the paper concludes with a summary of the findings and suggestions for future research in Sect. 5.

2 Related Works

A variety of fire-sensing technologies were utilized in the early forest fire detection systems that involve gas, flame, heat, and smoke detectors. While these systems have managed to detect fire, they have faced some limitations related to coverage areas, false alarms, and slow time response [4]. Tlig et al. [10] proposed a new color image segmentation method based on principal component analysis(PCA) and Gabor filter responses on various color images and was found to be not sensitive to add noises. Using YCbCr color space, Mahmoud et al. [9] proposed a forest fire algorithm able to separate high-temperature fire centre pixels. The method had good detection rates and fewer false alarms.

Recently, various deep learning techniques are used for fire detection. Perrolas et al. [6] proposed a scalable method to detect and segment the region of fire and smoke in the image using a quad-tree search algorithm. The method was capable of localizing very small regions of fire incidence in the images. In another work, Ghali et al. [7], presented the use of Vision-based Transformers on visible spectrum images to perform segmentation of forest fire.

In addition to the supervised approaches, unsupervised and semi-supervised approaches are also reported. For example, Meenu et al. [8] developed an unsupervised segmentation system that can be used for early detection of fire in real-time using spatial, temporal and motion information from the video frames. The use of Class Activation Mapping (CAM) for weakly supervised fire and smoke segmentation are addressed by the recents works on semi-supervised approaches Amaral et al. [11] and Niknejad et al. [12]. In [11], Conditional Random Fields (CRF) are incorporated along with CAM to accurately detect fire/smoke masks at the pixel level.

In our proposed work, we leverage an Autoencoder-based semi-supervised approach for fire detection. Some application of autoencoder for semi-supervised learning is reported in the medical field for medical imaging and diagnostics. For example, Kucharski et al. [13] developed a semi-supervised convolutional autoencoder architecture for segmenting nests of melanocytes in two stages. Mousumi et al. [14] developed a similar architecture for segmenting viable tumour regions in liver whole-slide images. Varghese et al. [15] utilized stacked denoising autoencoders in a semi-supervised learning approach to develop a model capable of detecting and segmenting brain lesions using fewer data. To the best of our knowledge, our work represents one of the first studies of an autoencoder-based semi-supervised learning approach in the context of fire detection and segmentation tasks.

3 Methodology

This section will provide a comprehensive explanation of our autoencoder-based approach, which utilizes semi-supervised learning to achieve the classification and segmentation goals for forest fires. Our proposed Autoencoder (AE) model can effectively process aerial images captured from drones or other airborne vehicles, regardless of distance or the size of the fire area.

3.1 Autoencoders

Autoencoders are neural network architectures that comprises of an *Encoder* and *Decoder* part that are trained to learn efficient data representations or compress data by mapping it to a lower-dimensional space [16]. The encoder encodes the input data into a compressed representation known as the **latent space features** in the bottleneck layer or latent space, and then the decoder reconstructs the input data from the latent space features (see Fig. 1).

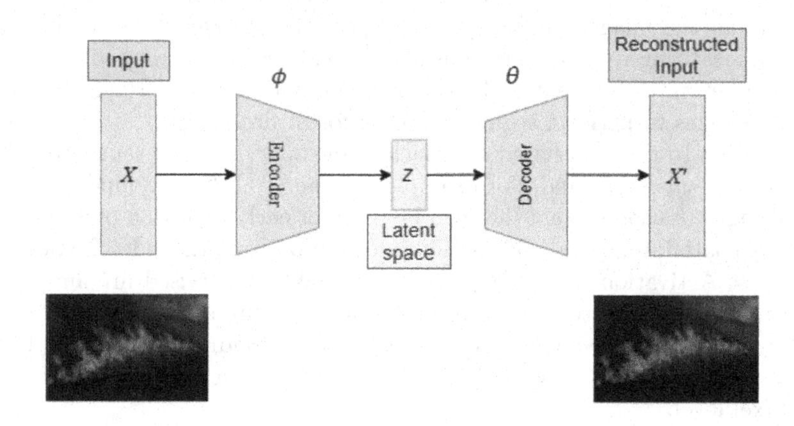

Fig. 1. Schema of a basic autoencoder

The encoder function (ϕ) maps the original data X to a latent space Z, which is present in the bottleneck layer(latent space). The decoder function (θ) maps the latent space Z to the output X', where the output is the same as the input function. Thus, the algorithm is trying to recreate the original image after some generalized nonlinear compression.

$$\phi : X \rightarrow Z$$

$$\theta : Z \rightarrow X'$$

$$\phi, \theta = \arg\min_{\phi\theta} ||X - (\phi.\theta)X||^2 \quad (1)$$

The characteristics of autoencoders have been rendered useful in various image processing tasks, such as image denoising and image restoration [17,18].

Convolutional Autoencoders (CAE): CAEs are the neural networks that use convolutional layers as feature extractors and exploit the correlation between neighbouring spatial features [16]. A CAE network uses convolutional layers to perform both encoding and decoding of input data. It is commonly used in computer vision applications for image data compression and reconstruction. Since CAEs perform convolutional operations on the input data, it is observed to extract more rich and more crisp features from the input data in the latent space compared to the ones by fully connected or dense layers in a normal autoencoder. The decoder then takes this lower-dimensional representation and reconstructs the original input image through a series of transposed convolutional operations. The objective of a CAE is to minimize the difference between the input image and the reconstructed image, which is typically measured using a loss function such as Mean Squared Error as in Eq. (3). By minimizing this loss, the network learns to encode and decode images in a way that preserves important features while compressing the input data into a smaller representation.

3.2 Semi-supervised Classification and Segmentation Using Autoencoder

Semi-supervised Approach for Forest Fire Classification: In the semi-supervised approach, we use a convolutional autoencoder(CAE) model and train it to perform the classification of input images into two classes namely *'Fire'* and *'Non_Fire'*. After training the model using unlabeled training images, it learns to reconstruct the input image from the latent space features as the output. On getting a good reconstruction score or minimum reconstruction loss, the encoder part from the models is extracted and a logistic layer is added to create a classification model that predicts whether an image belongs to *'Fire'* and *'Non_Fire'* class. This image classifier is trained using a semi-supervised learning approach, which leverages annotated data to increase the accuracy of the predictions.

Class Activation Mapping on the Classification Results: After the classification, we leverage Class Activation Mapping (CAM) [5] technique to visualize the fire in the classified image, thus creating an interpretable AI model. The CAM algorithm relies on a classification CNN's ability to localize objects despite being trained with only image-level labels. This is achieved by adding a global average pooling (GAP) layer after the last convolutional layer to obtain a single activation value for each feature map. These values are then multiplied by the class-specific weights of the output layer to generate the class scores. Feature maps are positively or negatively weighted based on their relevance to the output class. Formally, the method can be explained by the following equation:

$$M_{i,j} = \frac{1}{K} \sum_{k} B^{k}_{i,j} \tag{2}$$

where, K is the number of features in penultimate layer, B^k is its kth feature map. We obtain the activation map $M_{i,j}$ by averaging the feature maps at each position i, j in penultimate convolution layer.

To obtain the output mask in weakly-supervised segmentation, features used by the classification model are analyzed. However, this approach may not cover the entire object as it focuses on the discriminative part of the image. To address this limitation, we propose a patch-wise segmentation approach, where patches from training images are used for segmentation to obtain better results.

Semi-supervised Approach for Forest Fire Segmentation: Inspired by [13,15], we adopt an analogous unsupervised approach using an autoencoder, in the first stage of training. The objective is to train the autoencoder to achieve high accuracy in reconstructing input images while minimizing the reconstruction loss. This approach is similar to the initial training stage used for classification, but with a key difference: we begin by creating image patches and train the model using these patches.

During the pretraining phase, random sections of the image are extracted as patches. Here, we reduce the size of a whole image while retaining semantic information by shrinking its spatial dimensions, then each patch is compressed with a neural network mapping every image into a low-dimensional embedding vector. Further, each embedding is placed into an array that keeps the original spatial arrangement intact so that neighbor embeddings in the array represent neighbour patches in the original image. In our case, to achieve this low-dimensional embedding vector, we make use of autoencoders as discussed in Sect. 3.1. Later, the encoder is extracted and the weights are frozen and the network is made whole to generate the masks. The semi-supervised approach comes here when we use labels/ masks to train this network to output the segmented regions of fire. In the end, we recreate the mask for the whole image by unifying the generated masks of the patches. This will yield the segmented region of fire.

4 Experiments

4.1 Datasets

In this work, images from two forest fire datasets are extensively used i.e., Corsican Dataset and FLAME dataset.

(i) **Corsican Fire Dataset** [19] is an open fire database that includes pixel-level segmented images of wildfires and controlled fires. It consists of 1135 RGB images of forest fires captured at a close range, including some with sequences of frames of different fires. The database also contains 635 Near Infrared (NIR) images, but they are not used in this study. The images have dimensions of 1024×768.

(ii) **FLAME Dataset** [20] consists of 25018 images that contain fire and 14357 images that do not have fire. Each image is a frame from a video which covers different forest regions that contain fires. The images are shot from an aerial view with varying heights using a drone. The dimensions of the images are 256×256.

4.2 Evaluation Protocols

In the classification problem, the autoencoder is trained to reconstruct images, with the loss of reconstruction being monitored using Mean-Squared Error (MSE) loss. MSE is computed for n data points as follows:

$$MSE = \frac{1}{n}\sum_{i=1}^{n}(Y_i - \hat{Y}_i)^2 \tag{3}$$

where Y_i is the vector of actual values of the variable being predicted and $\hat{Y}_i$ is the predicted values. In the later stage of training, the encoder is extracted from the trained autoencoder for classification task. Here, the training loss for classification is calculated using Binary Cross Entropy Loss (BCE loss).

$$BCE = \frac{1}{N}\sum_{i=1}^{N} -(y_i * log(p_i) + (1 - y_i) * log(1 - p_i)) \tag{4}$$

here, p_i is the probability of class 1 ("fire"), and (1-p_i) is the probability of class 0 ("non-fire"). For the segmentation task, one of the most commonly used performance metrics is the IoU (Intersection over Union) score. It is otherwise known as the Jaccard index (or the Jaccard similarity coefficient) which has been widely used to measure the similarity between finite sample sets [21].

$$IoU(A, B) = \frac{A \cap B}{A \cup B} = \frac{A \cap B}{|A| + |B| - (A \cap B)} \tag{5}$$

where, A and B are two finite sample set and their IoU is defined as the intersection $(A \cap B)$ divided by the union $(A \cup B)$ of A and B.

4.3 Implementation Details

The neural network models are built in Python version 3.8 and trained on Google Colab and Kaggle platforms. For speeding up the training and working on images, the model used NVIDIA Tesla K80 GPU. Also, for building the neural network, Tensorflow API, version 2.0 and Pytorch API, version 1.13.1+cu116 is used.

For semi-supervised classification, we use a convolutional autoencoder built using Tensorflow API and train the model on images from FLAME dataset. The training-validation split is 80-20 and training data contains 31501 images whereas, validation data contains 7874 images of both *Fire* and *Non-Fire* class respectively. For semi-supervised segmentation, the autoencoder model is in-built with Pytorch API. For training, we used the Corsican database which contains fire images and their corresponding masks.

Model Training for Semi-supervised Classification: To classify forest fire images a convolutional autoencoder (CAE) with 16 convolutional layers, 4 Max-Pooling layers in the encoder part and 4 Upsampling layers in the decoder. The latent space contains a feature vector of size $(16 \times 16 \times 32)$. The CAE is trained on RGB images from the FLAME dataset that are resized into shape 256×256 and normalized in the preprocessing stage. The CAE first learns to reconstruct input images, with the model trained using the Adam optimizer and a learning rate of 0.001 for 50 epochs. We then appended the trained encoder part with additional dense layers, resulting in an output layer with a single neuron with

Sigmoid activation function. Refer to Table 1 for the CAE architecture used for semi-supervised classification. The entire model was then trained for an additional 30 epochs. We evaluate the fire classification model's performance not only through accuracy but also by examining the fire localization with Class Activation Mapping (CAM), as explained in Sect. 3.2. A global average pooling layer is employed after the last convolutional layer resulting in the activation mapping. The ADAM optimizer is used to train the model for 20 epochs on labeled data, achieving 0.97 accuracy.

Model Training for Semi-supervised Segmentation: The segmentation model is trained by feeding patches of images instead of whole images taken from Corsican Database. The model uses 500 images for training. Each image is divided into patches and was fed into the network to reconstruct the patches. As a part of the experiments, we experimented with variable patch sizes of 5×5, 9×9 and 16×16, respectively. We trained the model with 1000 patches from each image. The whole autoencoder is trained to reconstruct the RGB patches in unsupervised manner and is trained for 1000 epochs. In the later stage of training, the autoencoder is modeled to output the segmented image in the size of patches which later is rejoined to obtain the whole predicted mask. Here, the model is trained using binary masks that correspond to the input image. Here, the model is trained for 500 epochs.

Table 1. The architecture of the Convolutional Autoencoder (CAE) designed for semi-supervised classification

Layer (type)	Output Shape	Activation
Input	(None, 256, 256, 3)	
Conv2D	(None, 256, 256, 32)	ReLU
Conv2D	(None, 256, 256, 32)	ReLU
MaxPooling2D	(None, 128, 128, 64)	
Conv2D	(None, 128, 128, 64)	ReLU
Conv2D	(None, 128, 128, 128)	ReLU
MaxPooling2D	(None, 64, 64, 128)	
Conv2D	(None, 64, 64, 128)	ReLU
Conv2D	(None, 64, 64, 64)	ReLU
MaxPooling2D	(None, 32, 32, 64)	
Conv2D	(None, 32, 32, 64)	ReLU
Conv2D	(None, 32, 32, 32)	ReLU
MaxPooling2D	(None, 16, 16, 32)	
Flatten	(None, 8192)	
Dense	(None,512)	ReLU
Dense	(None, 64)	ReLU
Dense	(None, 1)	Sigmoid

4.4 Experimental Results

In this section, the performance of our semi-supervised classification and segmentation models are assessed and compared the state-of-the-art results.

Semi-supervised Classification: As explained in Sect. 3.2, for the classification task of the problem a Convolutional Autoencoder (CAE) is pre-trained using a large number of unlabeled images taken from FLAME dataset. The model is trained to perform reconstruction in order to learn rich latent representations. The trained encoder of CAE is further used for classification after semi-supervised fine-tuning. The reconstruction performance of the CAE model is depicted in Fig. 2. Similarly, the MSE loss and the classification results i.e. BCE loss & Accuracy are shown in Table 2. From Table 2, it can be observed that even with fewer data proportions (labeled vs. unlabeled data) i.e. 2/8 and 3/7, the CAE models are able to achieve good performance of 79.56% and 81.01% respectively. In our work, we used 2/8 ratio as our default setting.

To facilitate the localization of the fire in the *'Fire'* classes, Class Activation Mapping (CAM) is utilized. In addition to the good scores on the classification task, CAM saliency heat-map can highlight the fire regions in the image. Some examples of correct fire classification are visualized by means of the CAM model is shown in Fig. 3. Although it is possible to generate binary masks based on the CAM results, we found that it is suboptimal only imparting very little detail, in the weakly supervised setup. Hence semi-supervised patch-wise segmentation is performed.

Table 2. CAE performance for semi-supervised classification for FLAME dataset

CAE performance	Ratio of labeled and unlabeled (%)			
	2/8	3/7	5/5	Full (Supervised)
MSE loss	0.005	0.0052	0.0064	0.0049
BCE loss	0.972	0.947	1.0621	0.3565
Accuracy (in %)	79.56	81.01	84.85	97.27

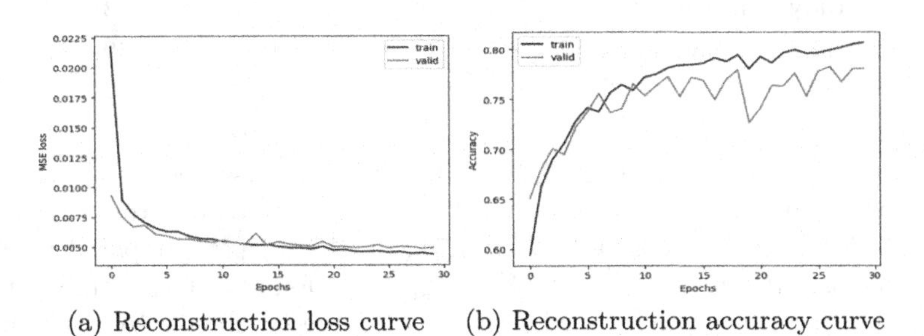

(a) Reconstruction loss curve (b) Reconstruction accuracy curve

Fig. 2. Reconstruction loss and accuracy curves of semi-supervised CAE

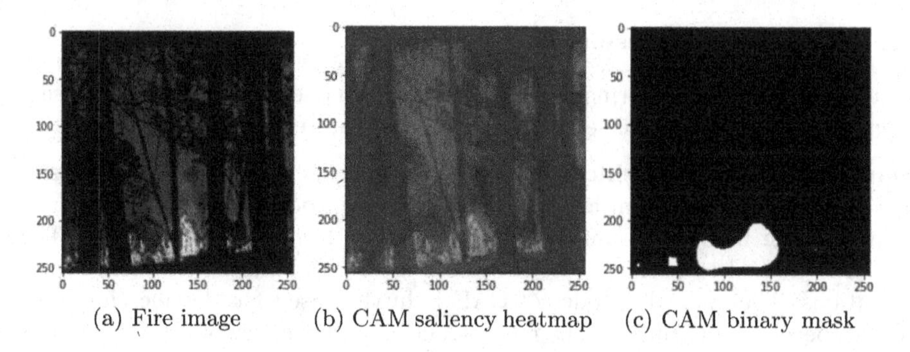

(a) Fire image (b) CAM saliency heatmap (c) CAM binary mask

Fig. 3. Visualization of the Class activation map (CAM) heatmap and binary mask, conducted upon the CAE-based classification results.

Semi-supervised Segmentation: Referring to Sect. 3.2, patch-wise extraction of the image is carried out for the segmentation. We generate segmentation masks by extracting 1000 random patches from each image of various patch sizes, as shown in Table 3. Different case study experiments by changing the patchsize are carried out. Following the segmentation process, a post-processing step is performed to obtain the predicted mask. This involves binarizing the segmentation images and applying morphological operations such as dilation to improve the quality of the predicted mask and reduce noise. The resulting images are displayed in Fig. 4. Additionally, the segmentation performance in terms of mean Intersection over Union (mIoU) is reported in Table 3.

Table 3. Performance of patch-wise approach for semi-supervised segmentation

Model experiments	Variants of patch-wise segmentation	
	Patch size	mIoU score
Case study 1: Small patch size	5×5	0.756
Case study 2: Medium patch size	9×9	0.742
Case study 3: Large patch size	16×16	0.723
Case study 4: Combined Model	$5 \times 5 + 9 \times 9 + 16 \times 16$	0.749

From the results (Case study 1–3) in Table 3, it is observed that the model trained with patches of size 5×5 and 9×9 are observed to produce a much better mIoU scores between 0.756 and 0.742 on the images compared to large patch size 16×16 (mIoU of 0.723). Further, it was found in Case study 4 that, while fusing the small patch based results with large patch size 16×16, mIoU is increased from 0.723 to 0.749.

State-of-the-Art Comparison: The proposed approach is compared against some of the recent state-of-the-art works, Amaral et al. [11] and Niknejad et al.

Table 4. Comparison of mean IoU (mIoU) on the test set for our proposed method compared to other state-of-the-art weakly-supervised segmentation methods.

Paper	Approach	mIoU score
Amaral et al. [11]	CAM and CRF	0.735
Niknejad et al. [12]	Via mid-layer of classification CNN	0.728
Ours	Classification+ CAM	0.547
Ours	**Patch-wise segmentation using Autoencoder**	**0.756**

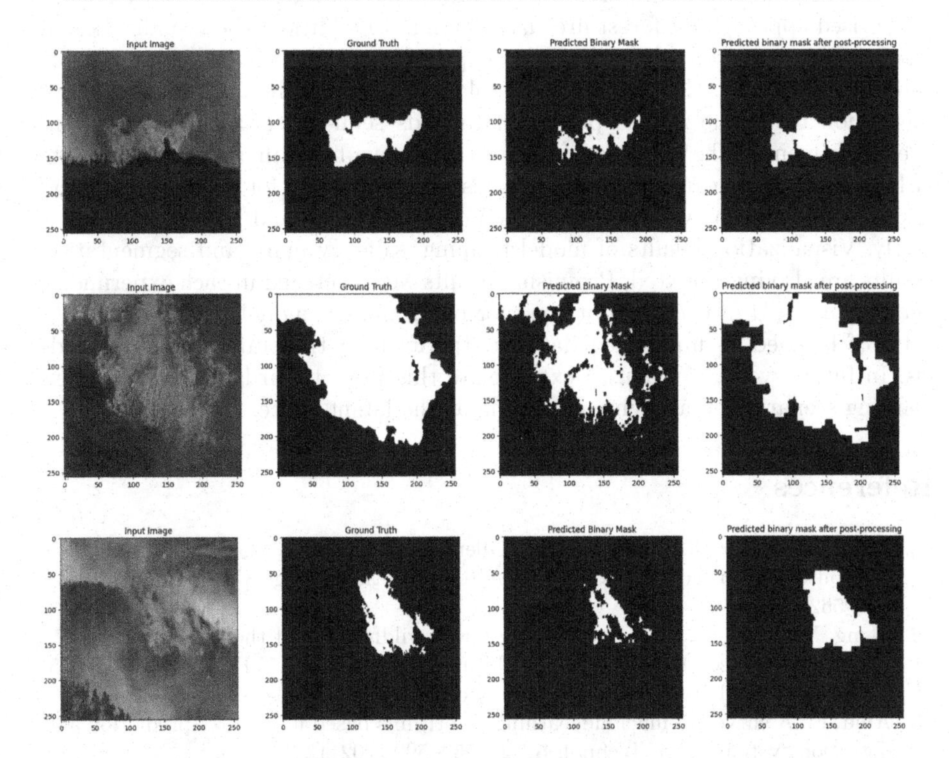

Fig. 4. Visualization of the semi-supervised segmentation results: Columnwise in the order as follows: Input image, Ground truth, Predicted mask and the binary mask after post-processing. First, second and third rows utilize patch sizes of dimensions 5 × 5, 9 × 9 and 16 × 16 respectively.

[12] that use weakly-supervised approaches to segment forest fire, as well as our CAM-based fire localization approach. Note that, since the ground truth segmentation images are unavailable in the FLAME dataset, the segmentation task and its comparison studies are carried out in the Corsican Dataset. The comparision results are reported in Table 4. From the result, it is observed that the CAM-based fire localization produced a mIoU of 0.547, which could be ascribable to the missing of non-discriminative part while training for classification. Whereas, our proposed CAE based patch-wise segmentation model (as in Case

study 1, Table 3) outperforms the state-of-the-art weakly-supervised segmentation methods (Amaral et al. [11] - 0.735 and Niknejad et al. [12] - 0.728) in forest fire segmentation with an mIoU of 0.756.

5 Conclusions and Future Work

We analysed the potential of exploiting autoencoders for semi-supervised learning for forest fire classification and segmentation. We leveraged CAE based Semi-supervised approach for forest fire classification. In addition, the visualization of the fire region is also realized with the help of Class Activation Mapping (CAM) saliency heat maps, thus facilitating explainable AI. Further, a patch-wise segmentation technique is also presented towards semi-supervised segmentation. We experimented the proposed frameworks on two benchmarking fire datasets (FLAME and Corsican). Extensive studies were conducted for both classification and segmentation by varying the labeled data split ratio and patch size, respectively. Visualization results of model training, saliency maps and segmentation results are also investigated. Promising results were reported in each experiment, showing a mIoU of 0.756 is obtained for our proposed model (even with a 2/8 ratio of labelled vs unlabelled data), outperforming the state-of-the-art models. In future work, we plan to extrapolate this work towards an unsupervised learning scenarios by applying clustering at the latent space.

References

1. Martinez-de Dios, J.R., Arrue, B.C., Ollero, A., Merino, L., Gómez-Rodríguez, F.: Computer vision techniques for forest fire perception. Image Vis. Comput. **26**(4), 550–562 (2008)
2. Meng, Y., Deng, Y., Shi, P.: Mapping forest wildfire risk of the world. In: Shi, P., Kasperson, R. (eds.) World Atlas of Natural Disaster Risk. IERGPS, pp. 261–275. Springer, Heidelberg (2015). https://doi.org/10.1007/978-3-662-45430-5_14
3. Abid, F.: A survey of machine learning algorithms based forest fires prediction and detection systems. Fire Technol. **57**(2), 559–590 (2021)
4. Chen, T.H., Wu, P.H., Chiou, Y.C.: An early fire-detection method based on image processing. In: 2004 International Conference on Image Processing, ICIP 2004, Singapore, vol. 3, pp. 1707–1710 (2004)
5. Zhou, B., Khosla, A., Lapedriza, A., Oliva, A., Torralba, A.: Learning deep features for discriminative localization. In: 2016 IEEE conference on Computer Vision and Pattern Recognition (CVPR), Las Vegas, NV, USA, pp. 2921–2929 (2016)
6. Perrolas, G., Niknejad, M., Ribeiro, R., Bernardino, A.: Scalable fire and smoke segmentation from aerial images using convolutional neural networks and quad-tree search. Sensors **22**(5), 1701 (2022)
7. Ghali, R., Akhloufi, M.A., Jmal, M., Souidene Mseddi, W., Attia, R.: Wildfire segmentation using deep vision transformers. Remote Sens. **13**(17), 3527 (2021)
8. Ajith, M., Martínez-Ramón, M.: Unsupervised segmentation of fire and smoke from infra-red videos. IEEE Access **7**, 182381–182394 (2019)
9. Mahmoud, M.A., Ren, H.: Forest fire detection using a rule-based image processing algorithm and temporal variation. Math. Probl. Eng. (2018)

10. Tlig, L., Bouchouicha, M., Tlig, M., Sayadi, M., Moreau, E.: A fast segmentation method for fire forest images based on multiscale transform and PCA. Sensors **20**(22), 6429 (2020)
11. Amaral, B., Niknejad, M., Barata, C., Bernardino, A.: Weakly supervised fire and smoke segmentation in forest images with CAM and CRF. In: 26th International Conference on Pattern Recognition (ICPR), Montreal, QC, Canada, pp. 442–448 (2022)
12. Niknejad, M., Bernardino, A.: Weakly-supervised fire segmentation by visualizing intermediate CNN layers. arXiv, abs/2111.08401 (2021)
13. Kucharski, D., Kleczek, P., Jaworek-Korjakowska, J., Dyduch, G., Gorgon, M.: Semi-supervised nests of melanocytes segmentation method using convolutional autoencoders. Sensors **20**(6), 1546 (2020)
14. Roy, M., et al.: Convolutional autoencoder based model HistoCAE for segmentation of viable tumor regions in liver whole-slide images. Sci. Rep. **11**, 139 (2021)
15. Alex, V., Vaidhya, K., Thirunavukkarasu, S., Kesavadas, C., Krishnamurthi, G.: Semisupervised learning using denoising autoencoders for brain lesion detection and segmentation. J. Med. Imaging **4**(4), 041311 (2017)
16. Hinton, G.E., Salakhutdinov, R.R.: Reducing the dimensionality of data with neural networks. Science **313**(5786), 504–507 (2006)
17. Gondara, L.: Medical image denoising using convolutional denoising autoencoders. In: IEEE 16th International Conference on Data Mining Workshops (ICDMW), Barcelona, Spain, pp. 241–246 (2016)
18. Liu, Y., Li, C., Zhao, Y., Xu, J.: Unified image restoration with convolutional autoencoder. In: 2022 2nd International Conference on Networking, Communications and Information Technology (NetCIT), pp. 143–146 (2022)
19. Toulouse, T., Rossi, L., Campana, A., Celik, T., Akhloufi, M.A.: Computer vision for wildfire research: an evolving image dataset for processing and analysis. Fire Saf. J. **92**, 188–194 (2017)
20. Alireza, S., Fatemeh, A., Abolfazl, R., Liming, Z., Peter, F., Erik, B.: The FLAME dataset: aerial imagery pile burn detection using drones (UAVs). IEEE Dataport (2020)
21. Zhou, D., Fang, J., Song, X., Guan, C., Yin, J., Dai, Y., Yang, R.: IoU loss for 2D/3D object detection. In: 2019 International Conference on 3D Vision (3DV), pp. 85–94 (2019)

Descriptive and Coherent Paragraph Generation for Image Paragraph Captioning Using Vision Transformer and Post-processing

Naveen Vakada(✉) and C. Chandra Sekhar

Indian Institute of Technology, Madras, Chennai, India
vakadanaveen@gmail.com, chandra@cse.iitm.ac.in

Abstract. The task of visual paragraph generation involves generating a descriptive and coherent paragraph based on an input image. The current state-of-the-art approaches use Regions of Interest (RoI) identification to generate paragraphs. The proposed approach eliminates the need for RoI identification. A transformer-based encoder-decoder model is used for paragraph generation. A post-processing step is introduced to enhance the semantic relevance of the generated paragraphs. This is achieved by incorporating the image-text similarity scores and related-classes similarity scores. The results of our studies demonstrate that the proposed model generates paragraphs with improved coherence and a higher Flesch reading ease score.

1 Introduction

Image captioning is the task of generating a textual description of an image. Early methods for image captioning used the encoder-decoder models, where the encoder extracted features from the image and the decoder generated the caption. These methods have difficulty in capturing the nuances and complexities of the image content.

The advent of transformer-based models revolutionized the field of image captioning. The transformer architecture [1], originally developed for natural language processing tasks, allows for capturing the long-range dependencies and relationships in the image and text. Transformer-based models have significantly improved the quality of generated captions, allowing for more descriptive and coherent descriptions of the image content.

Image paragraph captioning involves generating a paragraph of descriptive text for an input image. Image paragraph captioning is a more challenging task as it requires the model to not only generate a description of the image content but also ensure coherence and consistency within the paragraph.

To address this challenge, the current approaches have leveraged the techniques used in image captioning, such as transformer-based encoder-decoder models and post-processing steps, to generate the paragraph. However, there

J. Blanc-Talon et al. (Eds.): ACIVS 2023, LNCS 14124, pp. 40–52, 2023.
https://doi.org/10.1007/978-3-031-45382-3_4

is scope for improvement in terms of coherence and descriptive power of the generated paragraph.

In this paper, we present a novel approach to image paragraph captioning that uses transformer-based models and incorporates a post-processing step to enhance the semantic relevance of the generated paragraph. Our method outperforms the existing state-of-the-art methods in terms of coherence, diversity, and semantic relevance on the Stanford image paragraph dataset.

The rest of the paper is organized as follows: Sect. 2 provides a background and related work on image captioning and visual paragraph generation. Section 3 describes the proposed method in detail. Section 4 presents the dataset details and experimental settings. In Sect. 5, results on evaluation metrics and analysis of the results are given. Finally, in Sect. 6, we conclude the paper.

2 Background and Related Work

Image captioning involves generating a textual description of an image. The goal of image captioning is to generate a description that accurately conveys the content of the image, and is easily understood by human readers. Image captioning models typically consist of two main components: an encoder and a decoder. The encoder is responsible for encoding the visual features of an image into a compact representation. The decoder is responsible for generating a textual description based on the encoded features. The encoder is typically a convolutional neural network (CNN) trained on a large-size dataset of images. The decoder is typically a recurrent neural network (RNN) that is trained to generate a textual description based on the encoded features [2].

The attention mechanism has been shown to be effective in capturing the attention of the decoder towards the most relevant parts of an image during the caption generation process [3]. Additionally, the integration of transformer-based models has shown promising results [4].

Image paragraph captioning, also known as visual paragraph generation, involves generating a coherent and informative paragraph for a given image. This task has potential applications in various domains such as image retrieval, visual question answering, and visual story-telling.

Region-based methods have been used for image paragraph captioning. These methods typically involve identifying regions of interest (RoIs) in the image and use these RoIs for paragraph generation. The main advantage of this approach is the ability to generate paragraphs that are coherent and semantically relevant, as the RoIs provide a strong cue for semantic content.

[5] proposed a hierarchical approach for generating descriptive paragraphs. They introduced a hierarchical RNN model and evaluated it on the Stanford dataset. The model consists of two RNNs: a sentence-RNN and a word-RNN. The sentence-RNN takes the region of interest (RoI) features as input and generates sentence topic vectors, which are then used as input to the word-RNN to generate the words in a sentence. It is shown that the hierarchical approach outperformed models that generated paragraphs using a single RNN.

In [6], the authors proposed an adversarial framework for paragraph generation, using a paragraph generator and a discriminator to enhance the plausibility of sentences and the coherence of the paragraph, respectively. The plausibility of a sentence refers to the likelihood or probability that a sentence makes sense and is credible or acceptable within a given context. In other words, it refers to how reasonable or believable a sentence is based on the context in which it is used. The plausibility of a sentence can depend on various factors such as the vocabulary used, grammar, context, and coherence with other related sentences or information. They incorporated language and region-based visual attention in the paragraph generator and showed the effectiveness of Recurrent-Topic Transition Generative Adversarial Network (RTT-GAN) framework. This approach aims to generate visually and semantically coherent paragraphs, taking into account the transitions between sentences. The authors demonstrated the superiority of RTT-GAN approach.

In [7], the limitations of traditional image captioning methods and the naive approach of concatenating multiple short sentences to generate paragraphs are highlighted. To overcome these challenges, they propose to use coherence vectors and global topic vectors to ensure coherence and consistency in the generated paragraphs. They also propose a modified approach using sentence-RNN and word-RNN. The global topic vector is the mean of all the topic vectors generated by the sentence-RNN. The coherence vectors are the context vectors taken from the word-RNN of the previous sentence generated. Coherence Vectors ensure cross-sentence topic smoothness and global topic vector captures the summarizing information about the image.

The approach in [8] is inspired by the human process of composing a mental script before generating a paragraph and uses a hierarchical scene graph. The Hierarchical Scene Graph Encoder-Decoder (HSGED) uses the image scene graph as a script to incorporate rich semantic knowledge and hierarchical constraints. The model consists of two RNNs, the Sentence Scene Graph RNN and the Word Scene Graph RNN, both of which help generate sub-graph level topics and sentences. The authors also propose a sentence-level loss to encourage the sequence of generated sentences to be similar to that of the ground-truth paragraphs.

The related work in image paragraph captioning highlights the need for more sophisticated models and techniques that can generate paragraphs that are both coherent and semantically relevant. Our approach addresses these challenges by leveraging the power of transformer-based models and incorporating a post-processing step to enhance the semantic relevance of the generated paragraphs.

3 Architecture of the Proposed Model

The architecture of the proposed model is shown in Fig. 1. It consists of an encoder and a decoder. After the paragraph is generated by the decoder, post-processing techniques are used to enhance the caption.

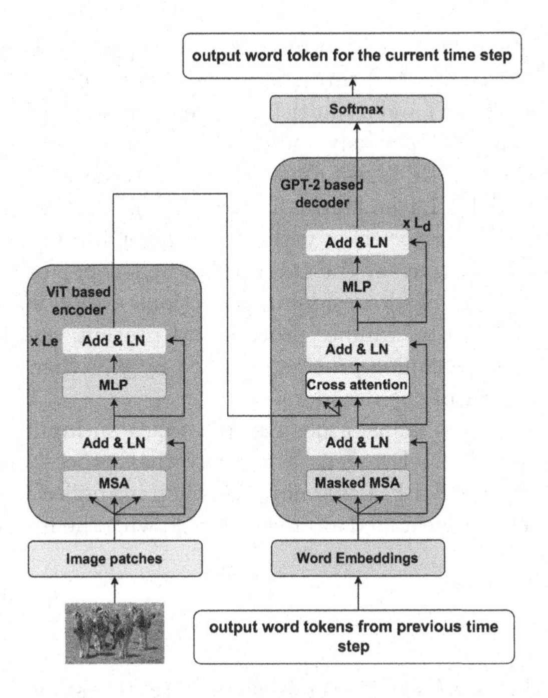

Fig. 1. Architecture of the proposed model for image paragraph captioning

3.1 Encoder

As proposed by [9] in vision transformers (ViT), we use patches of images as input to the encoder. The input image is denoted by $\mathbf{I} \in \mathbb{R}^{H*W*C}$, where H represents the number of pixels across the height of the image, W represents the number of pixels along the width of the image and C represents the number of channels. First, the image is resized into a fixed size (H, W) and then it is converted into N patches of size (k, k), where $N = \frac{HW}{k^2}$. The 2-D patches are flattened to 1-D patches denoted by $(\mathbf{x}_1, \mathbf{x}_2, ..., \mathbf{x}_N)$, where $\mathbf{x_p} \in \mathbb{R}^{k^2C\times 1}$. They are converted into embeddings $(\mathbf{e}_1, \mathbf{e}_2, .., \mathbf{e}_N)$ by using the embedding matrix $\mathbf{E} \in \mathbb{R}^{d\times k^2C}$, where $\mathbf{e}_p = \mathbf{E}\mathbf{x}_p$. The 1-D positional embeddings are added to these embeddings to retain the position information. The position embedding matrix [1] is denoted by $\mathbf{E}_{pos} \in \mathbb{R}^{d\times N}$. The following is the sequence of operations performed in the encoder:

$$\mathbf{z}_0 = [\mathbf{e}_1; \mathbf{e}_2; ...; \mathbf{e}_N] + \mathbf{E}_{pos} \quad (1)$$

$$\mathbf{z}_l^{'} = \mathrm{LN}(\mathrm{MSA}(\mathbf{z}_{l-1}^{'})) + \mathbf{z}_{l-1}^{'} \quad (2)$$

$$\mathbf{z}_l = \mathrm{LN}(\mathrm{MLP}(\mathbf{z}_l^{'})) + \mathbf{z}_l^{'} \quad (3)$$

$$y = \mathbf{z}_L \quad (4)$$

where, $l = 1, 2, .., L$, and L denotes the number of layers of the encoder.

The input image is first transformed into a sequence of one-dimensional tokens, which are then processed through a series of alternating layers of Multi-Head Self-Attention (MSA) and a Multi-Layer Perceptron (MLP) network. Each position of the sequence is separately and identically processed through the MLP network. The MSA and MLP layers are applied in alternating blocks, with each block being preceded by Layer Normalization (LN). This normalization step helps to ensure that the activations within each layer have similar means and variances, which is important for the stability and efficiency of the network.

After each processing block, residual connections are added to the output, which helps to stabilize the training process and prevent vanishing gradients. The final output of the encoding process, denoted by y, is then used as input to the cross-attention layer of the decoder.

In summary, the vision transformer encoder takes an input image and converts it into a sequence of 1-D tokens, which are then processed through alternating layers of MSA and MLP. Layer normalization is applied before each block and residual connections are added after each block, with the final output serving as input to the decoder.

3.2 Decoder

The decoder in the proposed model is similar to the standard transformer decoder. The output of the encoder is used as input to the cross-attention block of the decoder. The cross-attention layer of the decoder is the same as the MSA block in the encoder except that we get the key K and value V from the output of the encoder block and Q from the masked MSA layer of the decoder. Masked multi-head self-attention [1] is used to prevent the model from using future tokens for prediction. The output of the transformer decoder block is given to softmax layer whose size is equal to the vocabulary size. The final layer of the decoder is used to predict the next word of the caption given the words upto the previous time step.

Since GPT-2 is also a transformer decoder but without the cross-attention layer, we use GPT-2 as the decoder by adding a cross-attention layer initialized with random weights. Fine-tuning GPT-2 for image paragraph captioning is a better approach due to several key advantages.

First, GPT-2 is a large and powerful language model, trained on a diverse range of text sources. This gives it a strong foundation in natural language processing and the ability to generate coherent and semantically meaningful text. When fine-tuned for image captioning, this language generation capability can be adapted to the specific task of generating captions for images.

Second, fine-tuning is computationally efficient compared to training a new model from scratch. This is because the pre-trained GPT-2 model serves as a strong starting point, allowing the model to converge faster and with less data than if it were trained from scratch.

In summary, fine-tuning GPT-2 for image paragraph captioning is a better approach compared to training a new model from scratch, its ability to be adapted to the specific task, its ability to fine-tune to the target data distribution, and its computational efficiency.

3.3 Post-processing

In [10] a post-processing technique is used to generate multiple paragraphs for a given image and then combine them into a single paragraph. The final paragraph caption generated includes sentences that are semantically not related to the image given. To generate more semantically related sentences in the final paragraph, we use a modified method to pick the sentences for the final paragraph in the post-processing phase. We propose to use the image-text similarity score and the related-classes similarity score along with the language score and the dissimilarity score proposed by [10] to generate paragraphs that are more semantically related to the input image.

For a given input image, we generate 10 paragraph captions. We take all the sentences from these 10 paragraphs and use it as a list of candidate sentences denoted by S. From these captions, we choose the most suitable sentences based on a combination of their language score, dissimilarity score, image-text similarity score, related-classes score and length-penalty. Each of these factors is explained in detail below. We will also present the post-processing method used to determine the final captions.

Language Score: As suggested by [10], we construct a dataset of two classes: the positive class is made up of grammatically correct sentences, while the negative class consists of sentences with repeated words or jumbled word pairs. An LSTM is trained for binary classification to predict whether a sentence is positive or negative. The last layer's sigmoid probability is used to obtain a score between 0 and 1, which represents the language score, denoted by $L(s)$ for a candidate sentence s.

Dissimilarity Score: For the dissimilarity score, we use the word mover's distance (WMD), a measure of the dissimilarity between two text documents that takes into account both the semantic meaning and the distance between words in a vector space representation. Let F_i denote the i-th sentence in the final caption generated denoted by F. Let T be the number of sentences in the final caption. Then the dissimilarity score for a sentence s is calculated as shown in Eq. (5)

$$D(s,F) = \frac{\sum_{i=1}^{T} WMD(s,F_i)}{t}, t > 0 \tag{5}$$

Length Penalty: To avoid selecting overly brief sentences for the final caption, a length penalty has been applied to sentences of short length. The length of a candidate sentence, s, is represented by $|s|$. The median length of all candidate sentences (S) is represented as $ML(S)$, while MML is the minimum median length (set to 3 as suggested by [10]). The length penalty is calculated as shown in Eq. (6).

$$LP(s, S) = min(1, \frac{|s|}{max(MML, ML(S))}) \quad (6)$$

Image-Text Similarity Score: The image-text similarity score is represented by $IT(I, s)$, where I is the input image and s is a candidate sentence for generating the paragraph. For calculating the image-text similarity score, we use the pre-trained vision and language model CLIP, proposed by [11]. The CLIP is pre-trained for the task of image-text matching. We pass the candidate sentence and the input image to CLIP and obtain its output which consists of a score representing the relatedness of the sentence to the given input image.

Related-Classes Similarity Score: A pre-trained vision transformer, as introduced in [9], for image classification trained on the Imagenet dataset is used to compute this score. The image is processed through the ViT model, and the top 10 classes with the highest probabilities, denoted as $K = (k_1, k_2, .., k_{10})$, are selected as the related-classes for the image. Then we convert each of these ten labels in the text format to a text embedding by using the word2vec model [12]. Word2Vec is a neural network-based model that learns vector representations (embeddings) of words in a high-dimensional space, preserving the semantic meaning and relationships between words in the process. The mean of these 10 class label embeddings represents the related-classes information of the image. The words in a candidate sentence s are represented by $W_s = (w_1, w_2, .., w_{|s|})$, where $|s|$ is the number of words in sentence s. Again, we use word2vec model to obtain the text embedding for each of the words in the sentence. Then we take the mean of all the word embeddings to represent the sentence. We compare the mean embedding of related classes and the mean embedding of the candidate sentence using cosine similarity to obtain the related-classes similarity score as shown in Eq. (7)

$$RC(K, W_s) = \text{cosine similarity}(\frac{\sum_{i=1}^{10} word2vec(k_i)}{10}, \frac{\sum_{i=1}^{|s|} word2vec(w_i)}{|s|}) \quad (7)$$

The process to generate the final caption based on the scores described is given below:

For the first sentence of the final paragraph, we pick the candidate sentence with the highest language score and the image-text similarity score, as in Eq. 8.

$$F_1 = \arg\max_s (L(s) + IT(I, s)) \quad (8)$$

The next sentence (NS) is picked from the list of available candidate sentences(S) as shown in Eq. (9). After the next sentence is chosen, it is concatenated to the final caption.

Table 1. Ablation study for the proposed model

Model	METEOR
Ours (Base model)	17.59
Ours + post-processing by [10]	18.86
Ours + post-processing with image-text score only	19.02
Ours + post-processing with related-classes similarity score only	19.07
Ours + post-processing using all scores combined	19.16

$$NS = \arg\max_{s} ((L(s) + D(F_{t-1}, s) + IT(I, s) + RC(K, W_s)) * LP(s, S)) \quad (9)$$

$$F_t = concatenate(F_{t-1}, NS) \quad (10)$$

where, $t = 2, 3, ..., T$

It is important to note that once a sentence is chosen for the final caption it is removed from the candidate sentences list. We stop adding new sentences to the final caption when there are no sentences with a score above 0.5 as suggested in [10].

4 Experiments

4.1 Dataset

Stanford dataset introduced by [5] was used in the experiments. It consists of 19551 images with their corresponding paragraphs. Out of these 19551 images, 14575 images are present in the training set. Of the remaining images, 2487 images are in the validation set and the rest of the 2489 images are used as the test set.

4.2 Experimental Settings and Implementation Details

The number of layers in the encoder and decoder is denoted by L and it is set to 12. The dimension of the embeddings in both the encoder and decoder, denoted by d, is set to 762. The dropout parameter is set to 0.1 for the decoder. The model is trained for 24 epochs with cross-entropy loss function. The AdamW optimizer is used with learning rate of $5e-6$. Beam search is used to sample the captions during inference. Beam search with beam size 3 is used. The encoder is initialized with the weights of the vision transformer and the decoder is initialized with the weights of GPT-2 model.

In our experiments, we use METEOR [13] score metric to evaluate our model. METEOR score can be used to evaluate if the hypothesis and the candidate are semantically closer. A higher METEOR score indicates a high similarity between the hypothesis and candidate paragraphs. METEOR score has a significantly higher correlation to human judgments. Hence, we consider METEOR score for evaluating our model.

Experiments for finding the best beam size were conducted and beam size 3 gave the best METEOR score of 16.83. Fixing the beam size as 3, we perform experiments to study the effect of the n-gram penalty, where we block the n-gram repetitions in paragraphs by setting their probability scores to zero once they already appeared in the generated text. This approach is proposed by [14] to generate text without redundant phrases, thereby increasing the diversity of the text. For $n = 2$, we get the best METEOR score of 17.62.

The effect of using the image text similarity score and the related-classes similarity score is studied and the results are given in Table 1. Individually, the image-text similarity score and the related-classes similarity score improved the METEOR score of the base model from 17.59 to 19.02 and 19.07 respectively. When both of these are used in post-processing, we get the best METEOR score of 19.16.

Table 2. Comparison of performance with other state-of-the-art models

Model name	METEOR
SCST+dissimilarity score+length penalty	19.01
CAE-LSTM	18.82
Diverse and Coherent paragraph generation (VAE)	18.62
RTT-GAN	18.39
HSGED	18.33
SCST+trigram penality	17.86
Curiosity driven RL	17.71
Visual Relationship Embedding network	17.40
IMAP	17.36
Meshed transformers + off policy	17.20
Depth Supervised Hierarchical Policy value Network	17.02
Para CNN	17.00
Regions - Hierarchical (LSTMs)	15.95
Dual-CNN	15.80
Meshed transformers + MLE	15.40
Template	14.30
Depth aware Attention model (DAM)	13.90
SCST	13.60
Top down attention	12.9
Image flat	12.80
Sentence concat	12.10
IMG+LNG	11.30
Patch based vision encoder + image-text similarity score +related-classes similarity score	**19.16**

5 Results and Analysis

5.1 METEOR Score

Our base model gives a METEOR score of 17.59. The proposed model that uses the post-processing techniques proposed by [10] along with the image-text similarity score and the related-classes similarity score gives a score of 19.16, which is the best METEOR score achieved till now. The comparison with other models is given in Table 2.

5.2 Anaphora Analysis

An anaphora consists of an antecedent and an anaphor. Consider the example shown in Fig. 2. The phrases highlighted in blue color are antecedents. Antecedents can be nouns or noun phrases followed by indefinite articles. The phrases highlighted in purple color are anaphors. An anaphor can be a noun followed by a definite article or a personal pronoun. For example, in the paragraph generated by our model, the noun phrase 'A man' is the antecedent which is referred to by the anaphor 'He' in the next sentence.

Anaphoras connect different sentences of a discourse. They improve the coherence of the paragraphs by referring to the antecedents. Hence, anaphoras can be used to measure the coherence of the paragraphs generated. We define the anaphora rate as the average number of anaphoras that can be resolved between a pair of sentences in the discourse. The comparison of the anaphora rate with [5] can be seen in Table 3. It shows that our model generates more anaphoras, which indicates the higher coherence of paragraphs by the proposed approach.

Image	Proposed apporach	HRNN	Ground Truth
	A man is skateboarding outside on a sunny day. He is wearing a black t-shirt, blue jeans, and black sneakers. The skateboard is black with white wheels on it. There are tall green trees behind the man. Part of a building can be seen in the background of the photo. A large blue sky is seen above the trees.	the man is wearing a black helmet on his head . he is wearing a white shirt and black pants . the man has dark hair and is wearing dark sunglasses . the man is wearing a green t shirt and black pants . the man on the right is wearing a gray jacket and black pants . there are other people standing behind the fence .	A young man wearing a green graphic t-shirt, a black skull cap, and gray pants is seen riding a skateboard on a small wall in front of a tall glass skyscraper on a clear day. The man has one arm in the air as he steadies himself on the skateboard. The sky scraper in the distance is white with blue glass windows, and there is a construction crane on top of the building.

Fig. 2. Anaphora examples

Table 3. Anaphora rate and Flesch reading ease score for the proposed approach, HRNN and humans

Model	Proposed approach without post-processing	Proposed approach with post-processing	HRNN	Human
Anaphora rate	0.44	0.55	0.45	0.48
Flesch reading ease score	91	95	41	82

5.3 Readability Analysis

Ease of readability and comprehensibility are indicators of a good discourse. We evaluate the ease of readability of the paragraphs generated using the Flesch reading ease score given in Eq. 11.

$$RE = 206.835 - (1.015 * ASL) - (84.6 * ASW) \tag{11}$$

Here, ASL stands for the Average Sentence Length (i.e., the number of words divided by the number of sentences) and ASW stands for the Average number of syllables per word (i.e., the number of syllables divided by the number of words). The paragraphs generated by our model are in the category of 'very easy' and those generated by HRNN [5] are in the category of 'difficult'.

5.4 Qualitative Analysis

Ground Truth: A man holds onto a white frisbee. He wears a white shirt and black shorts. He is standing in the park. There is nobody next to him. There is a lot of trees behind him. The trees have many leaves. There is a lot of grass on the ground.

Proposed approach: A man is standing on a grassy field. He is wearing a short sleeve shirt and blue shorts. The man has short brown hair. Part of a blue frisbee can be seen flying in the air in front of the man. There is a chain link fence surrounding the field, and a large green tree trunk behind the fence.

HSGED: A man is standing on a field. The field is full of grass. The man is wearing a white shirt. he holds a white frisbee. There are some trees behind him.

(a) Comparison of paragraphs generated by HSGED, the proposed approach, and the ground truth

Ground Truth: This is a group of soccer players walking off of the field. One of the players has the soccer ball tucked under his arm. There are many referee's and coaches standing around watching them.

Proposed approach: A group of people are playing soccer on a soccer field. There is a red and white soccer ball in front of the people. A man in a white shirt and blue shorts is standing near the ball. Another man can be seen standing next to him wearing a blue shirt, blue pants and black shoes. Tall green trees are standing in the background behind the soccer players.

HRNN: A group of people are standing in the background watching a baseball game . he is wearing a white shirt and blue pants . they are all holding umbrellas . there are also two people standing on the ground watching the snowboarder . there are several people standing in the background . in the distance there are three men wearing jackets .

(b) Comparison of paragraphs generated by HRNN, the proposed approach, and the ground truth

Fig. 3. Qualitative results

We compare our model with HSGED [8]. From Fig. 3, we can see that our model tends to be more descriptive of the image when compared to HSGED model.

Our model gives a more detailed description about the person's shirt or his hair. This shows that our model tends to give better descriptions of the objects, than the other model.

Figure 3b shows the output paragraphs generated by our model and compares it with [5] and the ground truth. The example shows that our model uses anaphoras very well to maintain the coherence of the paragraphs. The global attention of the transformer models is the reason for the high coherence of the paragraphs. The examples also indicate that the paragraphs generated by the proposed model are easy to read and understand when compared to those generated by HRNN.

6 Conclusion

In conclusion, the proposed approach for visual paragraph generation presents a novel solution to the challenge of generating descriptive and coherent paragraphs based on input images. By eliminating the need for Regions of Interest (RoI) supervision and utilizing a transformer-based encoder-decoder model, the proposed approach effectively generates paragraphs that exhibited improved coherence and a high Flesch reading ease score. It outperforms the existing state-of-the-art models as determined by the METEOR score metric. The addition of a post-processing step, which incorporates both the image-text similarity scores and the related-classes similarity score, further enhances the semantic relevance of the generated paragraphs.

References

1. Vaswani, A., et al.: Attention is all you need. In: Advances in Neural Information Processing Systems, vol. 30 (2017)
2. Vinyals, O., Toshev, A., Bengio, S., Erhan, D.: Show and tell: a neural image caption generator. In: Proceedings of the IEEE Conference on Computer Vision and Pattern Recognition, pp. 3156–3164 (2015)
3. Anderson, P., et al.: Bottom-up and top-down attention for image captioning and visual question answering. In: Proceedings of the IEEE conference on Computer Vision and Pattern Recognition, pp. 6077–6086 (2018)
4. Cornia, M., Stefanini, M., Baraldi, L., Cucchiara, R.: Meshed-memory transformer for image captioning. In: Proceedings of the IEEE/CVF Conference on Computer Vision and Pattern Recognition, pp. 10578–10587 (2020)
5. Krause, J., Johnson, J., Krishna, R., Fei-Fei, L.: A hierarchical approach for generating descriptive image paragraphs. In: Proceedings of the IEEE Conference on Computer Vision and Pattern Recognition, pp. 317–325 (2017)
6. Liang, X., Hu, Z., Zhang, H., Gan, C., Xing, P.: Recurrent topic-transition GAN for visual paragraph generation. In: Proceedings of the IEEE International Conference on Computer Vision, pp. 3362–3371 (2017)
7. M. Chatterjee and G. Schwing. Diverse and coherent paragraph generation from images. In: Proceedings of the European Conference on Computer Vision (ECCV), pp. 729–744 (2018)

8. Yang, X., Gao, C., Zhang, H., Cai, J.: Hierarchical scene graph encoder-decoder for image paragraph captioning. In: Proceedings of the 28th ACM International Conference on Multimedia, pp. 4181–4189 (2020)
9. Dosovitskiy, A., et al.: An image is worth 16×16 words: transformers for image recognition at scale. arXiv preprint arXiv:2010.11929 (2020)
10. Kanani, S., Saha, S., Bhattacharyya, P.: Improving diversity and reducing redundancy in paragraph captions. In: 2020 International Joint Conference on Neural Networks (IJCNN), pp. 1–8. IEEE (2020)
11. Radford, A., et al.: Learning transferable visual models from natural language supervision. In: International Conference on Machine Learning, ICML, pp. 8748–8763 (2021)
12. Mikolov, T., Chen, K., Corrado, G., Dean, J.: Efficient estimation of word representations in vector space. arXiv preprint arXiv:1301.3781 (2013)
13. Banerjee, S., Lavie, A.: METEOR: an automatic metric for MT evaluation with improved correlation with human judgments. In: Proceedings of the ACL Workshop on Intrinsic and Extrinsic Evaluation Measures for Machine Translation and/or Summarization, pp. 65–72 (2005)
14. Paulus, R., Xiong, C., Socher, R.: A deep reinforced model for abstractive summarization. arXiv preprint arXiv:1705.04304 (2017)

Pyramid Swin Transformer for Multi-task: Expanding to More Computer Vision Tasks

Chenyu Wang[1,2](✉), Toshio Endo[1], Takahiro Hirofuchi[2], and Tsutomu Ikegami[2]

[1] Tokyo Institute of Technology, Tokyo, Japan
wang.c.ao@m.titech.ac.jp
[2] National Institute of Advanced Industrial Science and Technology (AIST), Tokyo, Japan

Abstract. We presented the Pyramid Swin Transformer, a versatile and efficient architecture tailored for object detection and image classification. This time we applied it to a wider range of tasks, such as object detection, image classification, semantic segmentation, and video recognition tasks. Our architecture adeptly captures local and global contextual information by employing more shift window operations and integrating diverse window sizes. The Pyramid Swin Transformer for Multi-task is structured in four stages, each consisting of layers with varying window sizes, facilitating a robust hierarchical representation. Different numbers of layers with distinct windows and window sizes are utilized at the same scale. Our architecture has been extensively evaluated on multiple benchmarks, including achieving 85.4% top-1 accuracy on ImageNet for image classification, 51.6 AP^{box} with Mask R-CNN and 54.3 AP^{box} with Cascade Mask R-CNN on COCO for object detection, 49.0 mIoU on ADE20K for semantic segmentation, and 83.4% top-1 accuracy on Kinetics-400 for video recognition. The Pyramid Swin Transformer for Multi-task outperforms state-of-the-art models in all tasks, demonstrating its effectiveness, adaptability, and scalability across various vision tasks. This breakthrough in multi-task learning architecture opens the door to new research and applications in the field.

Keywords: Computer Vision · Transformer Vision · Swin Transformer

1 Introduction

Transformers have gained considerable attention in the computer vision community due to their ability to capture long-range dependencies and model complex interactions between visual elements more effectively than CNNs. The Vision Transformer (ViT) by Dosovitskiy et al. [8] was one of the first successful adaptations of transformers for vision tasks, demonstrating competitive performance in image classification tasks. Following the ViT, several other transformer-based

J. Blanc-Talon et al. (Eds.): ACIVS 2023, LNCS 14124, pp. 53–65, 2023.
https://doi.org/10.1007/978-3-031-45382-3_5

architectures have been proposed for various computer vision tasks. Despite their remarkable results, these architectures often suffer from scalability issues and high computational complexity, which has led to the development of more efficient vision transformer architectures. One such architecture is the Swin Transformer, introduced by Liu et al. [16]. It is a hierarchical vision transformer that addresses the limitations of traditional transformers by incorporating a local window-based self-attention mechanism and a shifted window strategy. This innovative approach has resulted in state-of-the-art performance in various computer vision benchmarks, including image classification, object detection, semantic segmentation, and video recognition. Furthermore, the Swin Transformer has demonstrated strong scalability and adaptability to a wide range of vision tasks, making it a highly promising architecture for future research and applications.

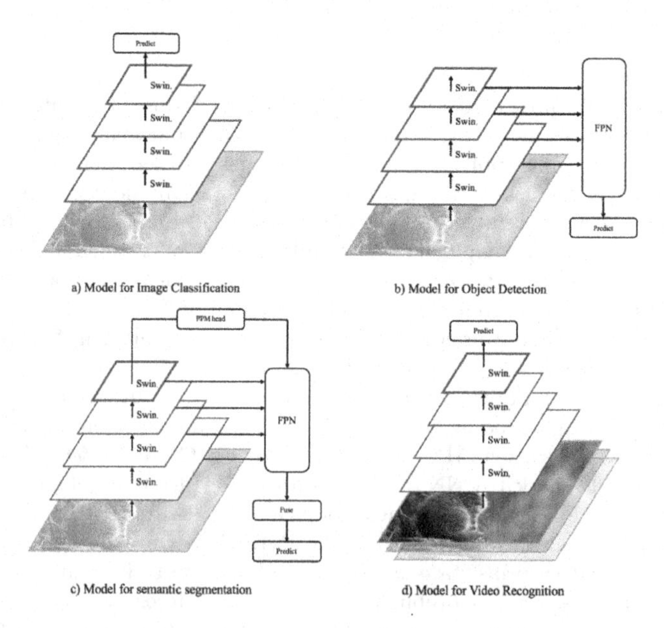

Fig. 1. Pyramid Swin Transformer. a) model is used for image classification, b) model is used for object detection, c) model is used for semantic segmentation, d) model is used for video recognition

In our previous work [21], we proposed the Pyramid Swin Transformer, which employed different-size windows in the Swin Transformer architecture to enhance its performance in image classification and object detection tasks. Building upon the success of our original Pyramid Swin Transformer, we extend its capabilities in this paper to address two additional vision tasks: semantic segmentation and video recognition. Our Pyramid Swin Transformer solves the problem of the lack of connections among windows on a large-scale in Swin Transformer. By implementing multiple windows of varying sizes on an extensive feature map, we construct a layered hierarchy of windows. In this arrangement, a single window in the upper layer encapsulates the features of four windows in the immediate lower layer. This progression significantly strengthens the interconnections among the windows, thereby enhancing the overall representational capability of the model.

This paper presents the Pyramid Swin Transformer for Multi-task, detailing our changes and their rationale. We also provide extensive experimental results, demonstrating the improved performance achieved by our new architecture across all four vision tasks: image classification on ImageNet [19], object detection on COCO [14], semantic segmentation on ADE20K [26], and video recognition on Kinetics-400 [11], as shown in Fig. 1. In this paper, our objective is to further test our improved Pyramid Swin Transformer on a wider range of computer vision tasks beyond image classification and object detection, demonstrating its potential for various vision applications and its superior performance. We aim to provide valuable resources for researchers and practitioners interested in leveraging the enhanced Swin Transformer, as well as to encourage further exploration and innovation in the field of computer vision.

2 Related Work

In this section, we will briefly describe the studies that are relevant to our research.

Swin Transformer: The Swin Transformer [16] is a hierarchical vision transformer that addresses traditional transformers' limitations in computer vision tasks. It introduces a local window-based self-attention mechanism and a shifted window strategy, significantly improving the model's efficiency and scalability. The Swin Transformer has achieved state-of-the-art performance on various computer vision benchmarks, including image classification, object detection, and semantic segmentation.

PVT v2: The Pyramid Vision Transformer v2 [23] introduces improvements to the original Pyramid Vision Transformer [22], creating a new baseline known as PVT v2. The enhancements include a linear complexity attention layer, overlapping patch embedding, and a convolutional feed-forward network. These modifications reduce the computational complexity of the model to a linear scale and yield significant improvements in fundamental vision tasks such as classification, detection, and segmentation.

3 Pyramid Swin Transformer Detailed Architecture

The Pyramid Swin Transformer [21], an extension of the Swin Transformer, is designed to address the limitations of the original model, particularly the issue of insufficient information exchange between windows in the window-based window-based multi-head self-attention mechanism on large-size feature map, as shown in the Fig. 2. This limitation is mitigated by incorporating multi-scale windows

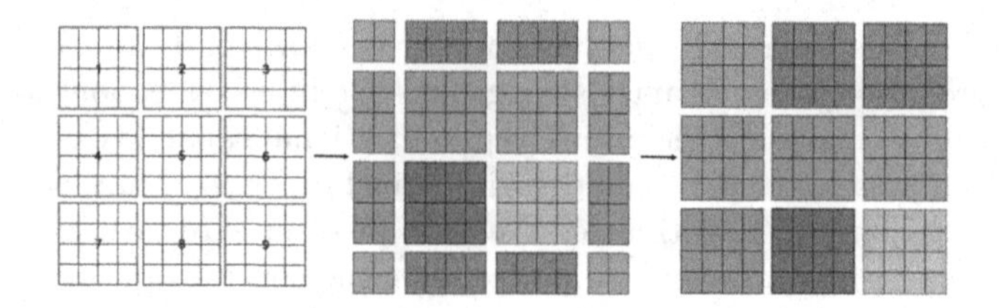

Fig. 2. Swin Transformer architecture problem

on the same scale, thereby enhancing the connections among large-scale windows and improving the model's ability to capture both local and global information.

The Pyramid Swin Transformer for Multi-task is based on the Swin Transformer architecture [16] and incorporates multi-scale windows on the same scale to address the lack of connections among large-scale windows. The architecture consists of four stages, each with multiple layers of Swin Transformer blocks. The input to the architecture is divided into non-overlapping patches, which are then linearly embedded and processed through the hierarchical transformer layers. Each layer consists of two steps: one for window multi-head self-attention and the other for shift window multi-head self-attention. We split it into smaller blocks, and the number of different-sized windows in each layer follows a hierarchical progression from more to less, facilitating global connections. In each stage, except for the fourth stage, the last layer has a 2×2 window, enhancing window-to-window information interaction and increasing global relevance (Fig. 3).

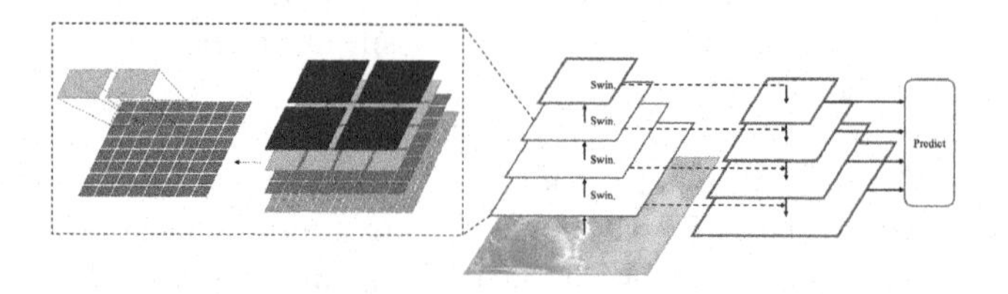

Fig. 3. Pyramid Swin Transformer architecture

The Pyramid Swin Transformer is a variant of the Swin Transformer architecture that aims to improve the efficiency and scalability of self-attention in computer vision multi-tasks. The overall architecture detail is shown in Table 1. The input image size is 256×256. In contrast to our prior work, we have incorporated two additional scale models in the current study. These models employ varying channel sizes, numbers of headers, window sizes, and quantities of windows. This enhancement is intended to provide a more comprehensive evaluation of our framework's performance and to facilitate a more robust comparison with alternative frameworks.

Table 1. Pyramid Swin Transformer Detailed architecture specifications

P. Swin-S	Output Size	Layers	Channel	Windows	Window size	Heads
Stage 1	64^2	3	96	$8^2, 4^2, 2^2$	$8^2, 16^2, 32^2$	3
Stage 2	32^2	2	192	$4^2, 2^2$	$8^2, 16^2$	6
Stage 3	16^2	2	384	$4^2, 2^2$	$4^2, 8^2$	12
Stage 4	8^2	1	768	1^2	8^2	24
P. Swin-R	Output Size	Layers	Channel	Windows	Window size	Heads
Stage 1	64^2	4	96	$16^2, 8^2, 4^2, 2^2$	$4^2, 8^2, 16^2, 32^2$	3
Stage 2	32^2	3	192	$8^2, 4^2, 2^2$	$4^2, 8^2, 16^2$	6
Stage 3	16^2	2	384	$4^2, 2^2$	$4^2, 8^2$	12
Stage 4	8^2	2	768	$2^2, 1^2$	$4^2, 8^2$	24
P. Swin-B	Output Size	Layers	Channel	Windows	Window size	Heads
Stage 1	64^2	4	128	$16^2, 8^2, 4^2, 2^2$	$4^2, 8^2, 16^2, 32^2$	4
Stage 2	32^2	3	256	$8^2, 4^2, 2^2$	$4^2, 8^2, 16^2$	8
Stage 3	16^2	3	512	$8^2, 4^2, 2^2$	$2^2, 4^2, 8^2$	16
Stage 4	8^2	2	1024	$2^2, 1^2$	$4^2, 8^2$	32
P. Swin-L	Output Size	Layers	Channel	Windows	Window size	Heads
Stage 1	64^2	4	192	$16^2, 8^2, 4^2, 2^2$	$4^2, 8^2, 16^2, 32^2$	6
Stage 2	32^2	3	384	$8^2, 4^2, 2^2$	$4^2, 8^2, 16^2$	12
Stage 3	16^2	3	768	$8^2, 4^2, 2^2$	$2^2, 4^2, 8^2$	24
Stage 4	8^2	2	1536	$2^2, 1^2$	$4^2, 8^2$	48

3.1 Pyramid Swin Transformer for New Tasks

In this section, we describe the proposed Pyramid Swin Transformer for Multi-task, which extends our previous Pyramid Swin Transformer [21] to semantic segmentation and video recognition tasks.

Architecture Overview. We make several modifications to the original architecture to adapt the Pyramid Swin Transformer to semantic segmentation and video recognition tasks. For semantic segmentation, we introduce a head pyramid pooling module referred to [24] that predicts per-pixel class probabilities and adopts a multi-scale feature fusion strategy to improve the segmentation accuracy at object boundaries. For video recognition, we extend the architecture to process space-time video volumes by incorporating temporal information into the transformer layers and modifying the attention mechanism to consider temporal dependencies.

Semantic Segmentation Head. To address the semantic segmentation task, we integrate the Pyramid Swin Transformer with UperNet [24], and add a pyramid pooling module in our network, which serves to aggregate context from varying regions of an image by applying diverse scales of pooling, thereby encapsulating information that might be more apparent or impactful at different scales, a unified perceptual parsing network incorporating multi-scale features for high-resolution segmentation maps, as shown in Fig. 1 (c model). The UperNet-based semantic segmentation head is designed to predict per-pixel class probabilities from the feature maps generated by the Multi-task Pyramid Swin Transformer. UperNet employs a pyramid pooling module that combines feature maps from different stages of the architecture with varying levels of abstraction. This strategy enables the model to leverage high-level semantic information and low-level spatial details, resulting in more accurate and detailed segmentation outputs. The fused feature maps are then processed through a series of convolutional layers and activation functions to generate the final segmentation predictions. To integrate the Pyramid Swin Transformer with UperNet, we replace the backbone of the original UperNet architecture with our Pyramid Swin Transformer, feeding the multi-scale feature maps generated by the transformer layers into the pyramid pooling module. This combination allows us to harness the advantages of the Pyramid Swin Transformer and UperNet, resulting in a powerful and efficient architecture for semantic segmentation. By integrating the Pyramid Swin Transformer with UperNet, we demonstrate the potential of our proposed architecture for semantic segmentation tasks, achieving state-of-the-art performance on the ADE20K dataset [26].

Video Recognition Adaptations. To adapt the Pyramid Swin Transformer to video recognition tasks, as shown in Fig. 1 (d model), we incorporate temporal information into the architecture by extending the input patches to include both spatial and temporal dimensions. Specifically, the input patches are formed by stacking consecutive frames from the input video along the temporal dimension, creating space-time video volumes. The attention mechanism in the transformer layers is further altered to account for temporal dependencies. This is accomplished by introducing temporal positional encodings into the self-attention computation, similar to how the spatial positional encodings are used in the original Swin Transformer. These temporal encodings could be learned or precomputed and allow the model to learn and exploit temporal relationships between video frames effectively. The output of this video recognition model is a sequence of class probabilities corresponding to each input frame. These probabilities can represent the likelihood of specific actions or events at any given frame. The model can thus generate fine-grained, frame-level predictions, which can be aggregated over the entire video to generate a comprehensive video-level prediction. This aggregation could be a simple average or a more complex operation based on the specific task requirements. By integrating temporal information, the adapted Pyramid Swin Transformer for Multi-task has the potential to offer superior performance on video recognition tasks by capturing complex spatiotemporal patterns within the video data.

4 Result

We conduct experiments on ImageNet-1K image classification [7], COCO object detection [14], ADE20K semantic segmentation [26] and Kinetics-400 video recognition [11]. In the following sections, we will compare the suggested Pyramid Swin Transformer architecture to the prior state-of-the-art on these tasks. We use 4 pieces of NVIDIA Tesla V100 to make training and test.

4.1 Image Classification

To ensure fairness in our benchmarks, we evaluate the proposed Pyramid Swin Transformer on ImageNet-1K, which consists of 1.28M training images and 50K validation images across 1,000 classes. We report the top-1 accuracy, using the same training method of Swin Transformer [16]. Our training setting mostly follows [20], where we employ an AdamW [12] optimizer for 300 epochs with a cosine decay learning rate scheduler, similar to the Swin Transformer (Table 2).

Table 2. Results of image classification on Imagenet

Method	Resolution	Params	FLOPs	Top-1 Acc.
MViTv2-T	224^2	24M	5G	82.3
MViTv2-S	224^2	35M	7G	83.6
MViTv2-B	224^2	52M	11G	84.4
Swin-T	224^2	28M	5G	81.3
Swin-S	224^2	50M	9G	83.0
Swin-B	384^2	88M	47G	84.5
SwinV2-T	256^2	28M	7G	82.8
SwinV2-S	266^2	50M	13G	84.1
SwinV2-B	256^2	88M	22G	84.6
P. Swin-S	256^2	64M	11G	**83.9**
P. Swin-R	256^2	77M	14G	**84.6**
P. Swin-B	256^2	123M	27G	**85.1**
P. Swin-L	256^2	164M	39G	**85.4**

Our proposed design, the Pyramid Swin Transformer, has been shown to outperform several Transformer systems, even when utilizing a small model and regular model (Pyramid Swin-S and Pyramid Swin-R) on ImageNet. However, our design does not exhibit significant advantages over Transformer systems in image classification. Compared to the original Swin Transformer and Swin Transformer V2 [15], our improved version achieves greater accuracy while utilizing fewer parameters. For example, Pyramid Swin-R achieves the same accuracy as SwinV2-B while utilizing fewer parameters. For bigger models, Pyramid Swin-B and Pyramid Swin-L perform better than SwinV2 [15]. On the regular-size

Table 3. Results of object detection on COCO with Mask R-CNN.

Model	AP^{box}	FLOPs	Params
Res50	46.3	739G	82M
Res101	47.7	819G	101M
ViL-S-RPB	47.1	277G	45M
ViL-M-RPB	48.9	352G	60M
ViL-B-RPB	49.6	384G	76M
Swin-T	46.0	264G	48M
Swin-S	48.5	354G	69M
Swin-B	48.5	496G	107M
P. Swin-S	**49.9**	402G	78M
P. Swin-R	**50.3**	463G	94M
P. Swin-B	**51.1**	657G	137M
P. Swin-L	**51.6**	864G	193M

Table 4. Results of object detection on COCO with Cascade Mask R-CNN.

Model	AP^{box}	FLOPs	Params
Res50	46.3	739G	82M
Res101	47.7	819G	101M
MViTv2-T	52.2	701G	76M
MViTv2-S	53.2	748G	87M
MViTv2-B	54.1	814G	103M
Swin-T	50.5	745G	86M
Swin-S	51.8	838G	107M
Swin-B	51.9	982G	145M
P. Swin-S	**53.1**	812G	114M
P. Swin-R	**53.6**	902G	136M
P. Swin-B	**54.0**	1247G	201M
P. Swin-L	**54.3**	1567G	273M

model, Pyramid Swin-R outperforms Swin-B by +0.1% in accuracy, while on the large-size model, Pyramid Swin-B achieves an improvement of +0.5% over SwinV2-B. Moreover, when compared to MVit with a (320×320) image size, our big model(Pyramid Swin-B) and large model (Pyramid Swin-L) achieves higher accuracy but at a higher computational cost.

4.2 Object Detection

We use the Microsoft COCO dataset [14] for our object detection experiments. We conduct an ablation study on the validation set and report system-level comparisons on the test-dev set to evaluate our proposed architecture. To compare the backbones of various models in our object detection experiments, we only utilized the Mask R-CNN [9] and Cascade Mask R-CNN [4] framework. The backbone networks of the objects we compared are Resnet [10], PVT-S [22], ViL-S-RPB [25] and Swin [16]. We follow the same settings as the Swin Transformer [16] to ensure a fair comparison. For the four frameworks, we use multi-scale training [5], with input sizes set to $[64, 32, 16, 8]$ for multi-scale four stages, consistent with the self-attention size used in Imagenet-1K pre-training. For the Pyramid Swin architecture, we also use the backbone pre-trained from Imagenet-1K (Tables 3 and 4).

When utilizing the Mask R-CNN framework, our Pyramid Swin Transformer achieved relatively high accuracy on the regular-size model, outperforming other models in the comparison. Specifically, Pyramid Swin-R achieved a AP^{box} of 50.3, improving by +1.8 AP^{box} over Swin-B [16], while utilizing fewer parameters. Compared to Vit-B-RPB, our Pyramid Swin architecture demonstrated an advantage with a +0.7 box AP improvement. On the big-size and large-size model, Pyramid Swin-L achieved a AP^{box} of 51.1 and AP^{box} of 51.6, achieving better performance than Swin-B, despite utilizing more parameters. These results demonstrate the superior performance of our Pyramid Swin Transformer architecture when utilized with the Mask R-CNN framework for object detection tasks. In the Cascade Mask R-CNN framework, our Pyramid Swin Transformer also achieved exceptional results, surpassing other models in the comparison. In particular, our Pyramid Swin-S outperformed all other models with a AP^{box} of 53.1, which is +1.3 AP^{box} higher than Swin-S [16] however, MViTv2-S [13] achieves higher scores. Pyramid Swin-R achieved a AP^{box} of 53.6, surpassing Swin-B by +1.7 AP^{box} and MViTv2-B by +0.5 AP^{box}, while utilizing fewer parameters. In the large-size model, Pyramid Swin-L achieved a AP^{box} of 54.3. Our Pyramid Swin Transformer architecture outperformed other models when used in the Cascade Mask R-CNN framework for object detection tasks.

4.3 Semantic Segmentation

We adopt ADE20K [26] as our semantic segmentation dataset, which is widely used and covers a broad range of 150 semantic categories. The dataset contains a total of 25,000 images, with 20,000 for training, 2,000 for validation, and another 3,000 for testing. We use UperNet [24] in mmseg [6] as our base framework for its high efficiency. Pyramid Swin-S and Pyramid Swin-R are trained with the same standard setting as previous approaches, using an input size of 512 × 512.

Table 5. Results of ADE20K samantic segmentation with UperNet.

Backbone	Method	mIoU	FLOPs	Params
ResNet-101	DANet	47.1	1119G	69M
ResNet-101	OCRNet	46.0	1249G	69M
ResNet-101	DNL	49.6	384G	76M
ResNet-101	UperNet	44.9	1029G	86M
XCiT-S24/8	UperNet	47.1	–	74M
XCiT-M24/16	UperNet	45.9	–	109M
XCiT-M24/8	UperNet	46.9	–	109M
Swin-T	UperNet	44.5	945G	60M
Swin-S	UperNet	47.6	1038G	81M
Swin-B	UperNet	48.1	1188G	121M
P. Swin-S	UperNet	**47.9**	926G	92M
P. Swin-R	UperNet	**48.5**	1091G	113M
P. Swin-B	UperNet	**48.8**	1452G	161M
P. Swin-L	UperNet	**49.0**	2036G	237M

The Transformer model exhibits superior performance characteristics compared to traditional Convolutional Neural Networks (CNNs) on accuracy, as shown in Table 5. Our design achieves higher accuracy than Swin-B and XCiT-S24/8 [1] with comparable numbers of parameters for the regular-size model, with a +0.4 mIoU improvement over Swin-B and a +1.4 mIoU improvement over XCiT-S24/8. Furthermore, increasing the number of parameters can lead to even better results with our architecture. Specifically, our big-size model achieves 48.8 mIoU, and our large-size model achieves 49.0 mIoU, demonstrating the effectiveness of our approach in improving semantic segmentation performance.

4.4 Video Recognition

We evaluate the performance of our Pyramid Swin Transformer on the Kinetics-400 dataset [11] (K400), which comprises approximately 240k training videos and 20k validation videos spanning 400 human action categories. Our training methodology follows that of [17]. Specifically, we employ an AdamW [12] optimizer for 30 epochs using a cosine decay learning rate scheduler and 2.5 epochs of linear warm-up, with a batch size of 64. As the backbone is initialized from a pre-trained model while the head is randomly initialized, we multiply the backbone learning rate by 0.1 to improve performance. The initial learning rates for the ImageNet pre-trained backbone and the randomly initialized head are set to 3e−5. To compute the final score, we take the average score overall views.

Table 6. Results of Kinetics-400 video recognition.

Model	Pre-train	Top-1 Acc	FLOPs × views	Params
SlowFast 16 × 8	–	79.8	234 × 3 × 10	60M
X3D-XL	–	79.1	48 × 3 × 10	11M
MoViNet-A6	–	81.5	386 × 1 × 1	31M
ViT-B-TimeSformer	ImageNet-21K	80.7	2380 × 1 × 3	121M
ViT-B-VTN	ImageNet-21K	78.6	4218 × 1 × 1	11M
ViViT-L/16 × 2	ImageNet-21K	80.6	1446 × 4 × 3	311M
ViViT-L/16 × 2 320	ImageNet-21K	81.3	3992 × 4 × 3	311M
Swin-T	ImageNet-1K	78.8	88 × 4 × 3	28M
Swin-S	ImageNet-1K	80.6	166 × 4 × 3	50M
Swin-B	ImageNet-1K	80.6	282 × 4 × 3	88M
Swin-B	ImageNet-21K	82.7	282 × 4 × 3	88M
P. Swin-S	ImageNet-1K	**80.8**	206 × 4 × 3	64M
P. Swin-R	ImageNet-1K	**81.2**	261 × 4 × 3	77M
P. Swin-R	ImageNet-21K	**83.4**	261 × 4 × 3	77M

Table 6 presents the results of Kinetics-400 video recognition, where we compare our Pyramid Swin Transformer models with other state-of-the-art models.

TimeSformer-L [3], ViT-B-VTN [18], ViViT-L/16×2 [2], and ViViT-L/16×2 320 [2]are pre-trained on ImageNet-21K. Swin-T, Swin-S, and Swin-B are pre-trained on ImageNet-1K or ImageNet-21K. Our Pyramid Swin Transformer models, including Pyramid Swin-S, Pyramid Swin-R, and Pyramid Swin-L, outperform Swin-B and other models in terms of accuracy with comparable or fewer parameters. Specifically, our Pyramid Swin-S achieves a top-1 accuracy of 80.8%, and Pyramid Swin-R achieves a top-1 accuracy of 81.2%, outperforming the other models. Moreover, Pyramid Swin-R pre-trained on ImageNet-21K achieves the highest accuracy of 83.4% among all the models.

5 Conclusion

The Multi-Task Pyramid Swin Transformer is a versatile and efficient architecture for object detection, image classification, semantic segmentation, and video recognition tasks. The architecture adeptly captures local and global contextual information by employing more shift window operations and integrating diverse window sizes on the same scale. The structure of the Multi-Task Pyramid Swin Transformer is divided into four stages, each consisting of layers with varying window sizes, facilitating a robust hierarchical representation. Different numbers of layers with distinct windows and window sizes are utilized at the same scale. Extensive evaluations across various tasks, including image classification on ImageNet, object detection on COCO, semantic segmentation on ADE20K, and video recognition on Kinetics-400, demonstrate that the Multi-Task Pyramid Swin Transformer exhibits exceptional detection and recognition performance. This demonstrates its effectiveness, adaptability, and scalability across various vision tasks. In conclusion, the Multi-Task Pyramid Swin Transformer is a promising architecture for multi-task learning in computer vision. Its ability to capture local and global contextual information through diverse window sizes makes it highly adaptable to various tasks. The exceptional performance demonstrated in various benchmarks paves the way for future research and applications in this field.

Acknowledgements. This work was partly supported by JSPS KAKENHI Grant Number 20H04165.

References

1. Ali, A., et al.: XCiT: cross-covariance image transformers. Adv. Neural. Inf. Process. Syst. **34**, 20014–20027 (2021)
2. Arnab, A., Dehghani, M., Heigold, G., Sun, C., Lučić, M., Schmid, C.: ViViT: a video vision transformer. In: Proceedings of the IEEE/CVF International Conference on Computer Vision, pp. 6836–6846 (2021)
3. Bertasius, G., Wang, H., Torresani, L.: Is space-time attention all you need for video understanding? In: ICML, vol. 2, p. 4 (2021)

4. Cai, Z., Vasconcelos, N.: Cascade R-CNN: delving into high quality object detection. In: Proceedings of the IEEE Conference on Computer Vision and Pattern Recognition, pp. 6154–6162 (2018)
5. Carion, N., Massa, F., Synnaeve, G., Usunier, N., Kirillov, A., Zagoruyko, S.: End-to-end object detection with transformers. In: Vedaldi, A., Bischof, H., Brox, T., Frahm, J.-M. (eds.) ECCV 2020. LNCS, vol. 12346, pp. 213–229. Springer, Cham (2020). https://doi.org/10.1007/978-3-030-58452-8_13
6. Contributors, M.: MMSegmentation: OpenMMlab semantic segmentation toolbox and benchmark (2020). https://github.com/open-mmlab/mmsegmentation
7. Deng, J., Dong, W., Socher, R., Li, L.J., Li, K., Fei-Fei, L.: ImageNet: a large-scale hierarchical image database. In: 2009 IEEE Conference on Computer Vision and Pattern Recognition, pp. 248–255. IEEE (2009)
8. Dosovitskiy, A., et al.: An image is worth 16×16 words: transformers for image recognition at scale. arXiv preprint arXiv:2010.11929 (2020)
9. He, K., Gkioxari, G., Dollár, P., Girshick, R.: Mask R-CNN. In: Proceedings of the IEEE International Conference on Computer Vision, pp. 2961–2969 (2017)
10. He, K., Zhang, X., Ren, S., Sun, J.: Deep residual learning for image recognition. In: Proceedings of the IEEE Conference on Computer Vision and Pattern Recognition, pp. 770–778 (2016)
11. Kay, W., et al.: The kinetics human action video dataset. arXiv preprint arXiv:1705.06950 (2017)
12. Kingma, D.P., Ba, J.: Adam: a method for stochastic optimization. arXiv preprint arXiv:1412.6980 (2014)
13. Li, Y., et al.: MViTv 2: improved multiscale vision transformers for classification and detection. In: Proceedings of the IEEE/CVF Conference on Computer Vision and Pattern Recognition, pp. 4804–4814 (2022)
14. Lin, T.-Y., et al.: Microsoft COCO: common objects in context. In: Fleet, D., Pajdla, T., Schiele, B., Tuytelaars, T. (eds.) ECCV 2014. LNCS, vol. 8693, pp. 740–755. Springer, Cham (2014). https://doi.org/10.1007/978-3-319-10602-1_48
15. Liu, Z., et al.: Swin transformer v2: scaling up capacity and resolution. In: Proceedings of the IEEE/CVF Conference on Computer Vision and Pattern Recognition, pp. 12009–12019 (2022)
16. Liu, Z., et al.: Swin transformer: hierarchical vision transformer using shifted windows. arXiv preprint arXiv:2103.14030 (2021)
17. Liu, Z., et al.: Video swin transformer. In: Proceedings of the IEEE/CVF Conference on Computer Vision and Pattern Recognition, pp. 3202–3211 (2022)
18. Neimark, D., Bar, O., Zohar, M., Asselmann, D.: Video transformer network. In: Proceedings of the IEEE/CVF International Conference on Computer Vision, pp. 3163–3172 (2021)
19. Russakovsky, O., et al.: ImageNet large scale visual recognition challenge. Int. J. Comput. Vision **115**, 211–252 (2015)
20. Touvron, H., Cord, M., Douze, M., Massa, F., Sablayrolles, A., Jégou, H.: Training data-efficient image transformers & distillation through attention. In: International conference on machine learning, pp. 10347–10357. PMLR (2021)
21. Wang, C., Endo, T., Hirofuchi, T., Ikegami, T.: Pyramid swin transformer: different-size windows swin transformer for image classification and object detection. In: Proceedings of the 18th International Joint Conference on Computer Vision, Imaging and Computer Graphics Theory and Applications, pp. 583–590 (2023). https://doi.org/10.5220/0011675800003417

22. Wang, W., et al.: Pyramid vision transformer: A versatile backbone for dense prediction without convolutions. In: Proceedings of the IEEE/CVF International Conference on Computer Vision, pp. 568–578 (2021)
23. Wang, W., et al.: PVT v2: improved baselines with pyramid vision transformer. Comput. Vis. Media **8**(3), 415–424 (2022)
24. Xiao, T., Liu, Y., Zhou, B., Jiang, Y., Sun, J.: Unified perceptual parsing for scene understanding. In: Proceedings of the European Conference on Computer Vision (ECCV), pp. 418–434 (2018)
25. Zhang, P., et al.: Multi-scale vision longformer: a new vision transformer for high-resolution image encoding. In: Proceedings of the IEEE/CVF International Conference on Computer Vision, pp. 2998–3008 (2021)
26. Zhou, B., et al.: Semantic understanding of scenes through the ADE20K dataset. Int. J. Comput. Vision **127**, 302–321 (2019)

Person Activity Classification from an Aerial Sensor Based on a Multi-level Deep Features

Fatma Bouhlel[1], Hazar Mliki[2,3](✉), and Mohamed Hammami[1]

[1] MIRACL-FS, Faculty of Sciences of Sfax, University of Sfax, Road Sokra Km 3, 3018 Sfax, Tunisia
{fatma.bouhlel,mohamed.hammami}@fss.usf.tn
[2] MIRACL Laboratory, University of Sfax, Sfax, Tunisia
[3] National Institute of Applied Science and Technology, University of Carthage, Tunis, Tunisia
mliki.hazar@gmail.com

Abstract. The intelligent surveillance system is considered as an important research field since it provides continuous personal security solution. In fact, this surveillance system support security guards by warning them in suspicious situations such as recognize the abnormal person activity. In this context, we introduced a new method for person activity classification that includes offline and inference phases. Based on convolutional neural networks, the offline phase aims to generate the person activity model. Whereas, the inference phase relies on the generate model to classify the person's activity. The main contribution of the proposed method is to introduce a multi-level deep features to handle inter- and intra-class variation. Through a comparative study, performed on the UCF-ARG dataset, we assessed the performance of our method compared to the state-of-the-art works.

Keywords: Person activity classification · multi-level deep features · aerial sensor · CNN

1 Introduction

Nowadays, surveillance is seen as an essential step to preserve personal security. In fact, monitoring is mandatory in a number of locations, including malls banks, airports, and protests. As a result, there was a greater need to automate surveillance tasks so as to support security guards and enable them to do their work more effectively. Indeed, abnormal person activity analysis is a practical illustration of an intelligent surveillance system. In fact, it entails identifying the achieved activity label and thereafter recognize the abnormal activity. Hence, a significant interest has so been awarded to person activity classification by the scientific community [8]. Particularly, person activity classification from an aerial sensor allows monitoring wide regions and surveilling limited access areas

J. Blanc-Talon et al. (Eds.): ACIVS 2023, LNCS 14124, pp. 66–75, 2023.
https://doi.org/10.1007/978-3-031-45382-3_6

[3,5,8]. However, person activity classification from an aerial sensor is considered as a difficult task since the aerial sensor affects scale and point of view variations. Withal, the inter- and intra-person activity class variations delineate a foremost constraint in the person activity classification field. In this paper, we introduce a new method for person activity classification from an aerial sensor. The introduced method involves an offline phase that aims to generate the person activity classification model and an inference phase that allows classifying the person activity. The main contribution of our method is to propose a multi-level deep features, in order to generate a performant and robust feature vector that is able to handle the aerial sensor constraints and the inter as well as intra-class variations. The remaining parts of this paper are structured as follows. Section 2 reveals some related works on person activity classification. Section 3 details the proposed method. Section 4 presents the experimental results and the discussion. The last section emphasizes the main contribution of the proposed method and outlines the future perspectives.

2 Related Works

Referring to the literature [6], person activity classification methods can be categorized into handcraft methods and deep learning methods.

In the context of handcraft methods, Moussa et al. [9] introduced the spatial features person activity classification method. Indeed, the SIFT (Scale Invariant Feature Transform) technique was applied in order to extract the points of interest. Then, the K-means technique was used to generate a BoW (Bag of Word), which match each obtained feature vector to the closest visual word. Subsequently, the visual word frequency histogram was calculated. At last, the person activity class was determined through an SVM (Support Vector Machine) classifier. In contrast to [9], Sabri et al. [11] proposed the spatio-temporal features person activity classification method. Hence, the HOG (Histograms of Oriented Gradients) and the HOF (Histograms of the Optical Flow) were combined to generate the spatio-temporal features vectors. Finally, each feature vector was assigned to the closest visual word using the K-means algorithm. Within the same framework, Burghouts et al. [4] used STIP (Spatio-Temporal Interest Points) technique since it exhibited higher discrimination in the context of person activity classification. In [6], the authors introduced a method based on the skeleton representation of spatio-temporal features. Indeed, they extracted relative and temporal derivatives of joint positions features which refer to spatial and temporal features. The temporal features include joint acceleration and joint velocity. Despite the fact that these methods asserted their performance in the person activity classification field, they rely heavily on the choice of the features extraction technique [2].

With regard to deep learning methods, Baccouche et al. [2] introduced a two-part deep learning model for person activity classification. The first part consists of a 3D-CNN (Convolutional Neural Networks) network that aims to extract and learn spatio-temporal features. In fact, the 3D-CNN is an extension of the CNN where the input layer is of size $34 \times 54 \times 9$, corresponding to 9 consecutive images of size 34×54 pixels. The second part, uses the 3D-CNN generated features to train an LSTM (Long Short Term Memory networks) so as to take into account the long-term dependencies. In [14], Wang et al. proposed to extract spatial features relative to each frame using CNN. Then, in order to extract temporal features from consecutive frames two types of LSTM and an attention model were applied. Indeed, the two types of LSTM allow to study both of the features generated by the convolution layer which produces spatial information as well as the fully connected layer which provides semantic information. Subsequently, they used an optimization layer to combine the two types of information. The temporal attention model has been integrated to determine the relevant frames in a video sequence for person activity classification. Sargano et al. [12] showed that deep learning-based activity classification methods which rely on 3D-CNN, RNN, or LSTM require a large dataset. Thus, it is inadequate to use one of these models in the context of a small dataset, since, they fail to circumvent the overfitting problem. Hence, the authors applied the transfer learning and used the pre-trained CNN AlexNet. In order to suit the pre-trained CNN architecture to their study context, they replaced the softmax layer through a hybrid SVM-KNN classifier. Based on the archived experimental study, the authors demonstrated that using pre-trained CNN to extract spatial features is better than learning spatio-temporal features from scratch.

The study of the state-of-the-art methods has incited us to propose person activity classification method based on pre-trained deep neural network. Different from reference methods, we introduced a multi-level deep features, in order to handle the inter- and the intra-class variation.

3 Proposed Method

In the context of person activity classification from an aerial sensor, we introduce a method based on a multi-level deep features. The proposed method involves an offline phase that aims to generate the person activity classification model and an inference phase that allows classifying the person activity according to two scenarios: an instant classification of the video frames and an entire classification of the video sequence.

3.1 Offline Phase: Person Activity Model Generation

To generate the model person activity, we automatically extract multi-level spatial features using a pre-trained model. Relying on the comparative study performed by Kaiming et al. [7], we opt for the pre-trained GoogLeNet CNN [13]. Our choice is based on two fundamental criteria which are the number of layers and the classification error rate. Moreover, the CNN GoogLeNet has shown a good compromise among the classification error rate and the computation time. Indeed, GoogLeNet includes 9 inception modules namely: inception 3a, inception

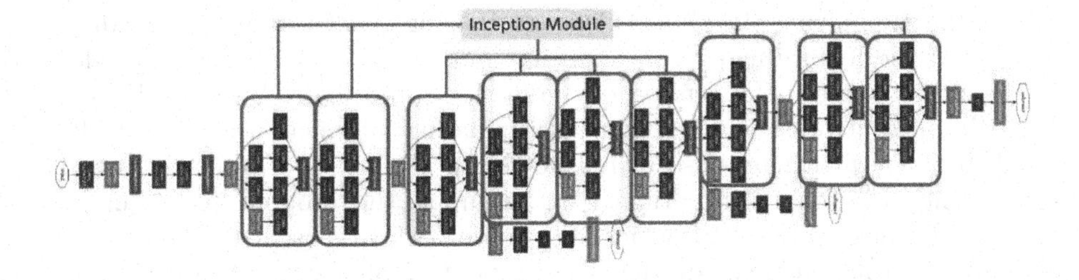

Fig. 1. GoogLeNet's architecture

3b, inception 4a, inception 4b, inception 4c, inception 4d, inception 4e, inception 5a, and inception 5b (cf. Fig. 1).

These inception modules integrate different-sized convolutions, enabling the learning of features at varying scales. Figure 2 shows an illustrative example of an inception module.

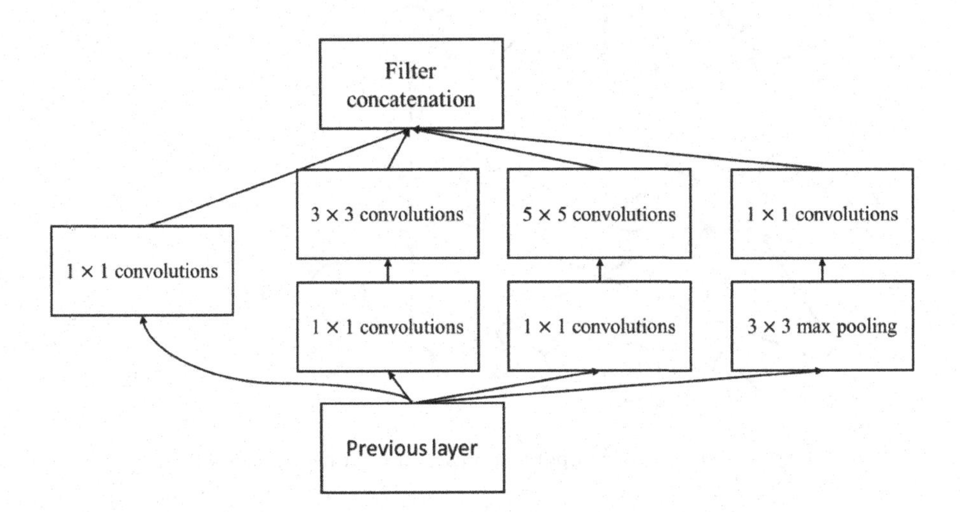

Fig. 2. Example of GoogLeNet's Inception module

Thereafter, we adapt GoogLeNet to obtain a multi-level deep features vector that is more efficient and robust against inter- and intra-class variation. For this purpose, we keep the GoogLeNet architecture until the last layer of the 5b inception module. Then, we concatenate the feature maps of the inception module 5a of size $7 \times 7 \times 832$ with the feature maps of the 5b inception module of size $7 \times 7 \times 1024$. Thus, we obtain a multi-level feature map of size $7 \times 7 \times 1856$ which results from the of the features at the level of the third dimension. Such a multi-level representation makes it possible to encode the activity classes in a more discriminating way. These feature maps constitute the input of a pooling layer applying the Global Average Pooling (GAP) strategy to reduce the size of the feature maps from $(n \times n \times nc)$ to $(1 \times 1 \times nc)$, with nc refer to the size of the third dimension of the feature maps. Indeed, the GAP does not only make it possible to extract discriminating information and avoid the problem of overfitting but also to reduce the total number of parameters and consequently the computation time. Next, we apply the 'dropout' technique on the Fully-Connected Layer FC which follows the GAP to consolidate the process of the overfitting problem. In fact, the dropout technique is a regularization technique that solves the over-learning constraint by temporarily deactivating some neurons (cf. Fig. 3).

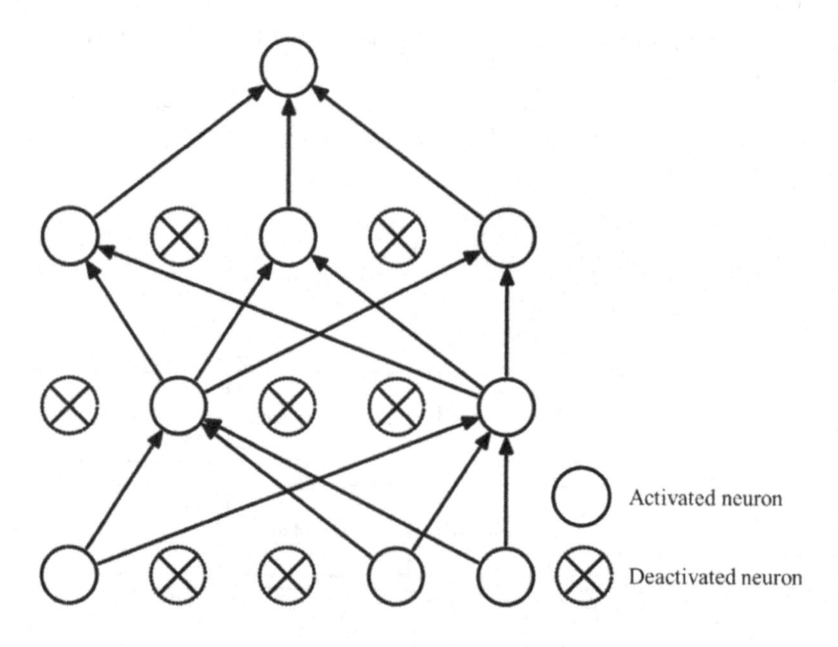

Fig. 3. Example of Dropout application on a neural network

Finally, we add a softmax layer which includes a number of neurons referring to the number of activities to classify. The softmax layer designates a loss layer which makes it possible to determine the class of an image based on the probabilities resulting from the softmax activation function. This function normalizes the output of the last layer of the CNN to produce a vector of size n indicating the number of activity classes.

3.2 Inference Phase: Person Activity Classification

In order to classify the person activity, we use the pre-generated person activity model resulting from the offline phase. Thus, two person activity classification scenarios are proposed: an instant classification of the video frames and an entire classification of the video sequence. The instant classification of the video frames allows recognizing the activity in every frame of the sequence. This scenario of classification is sought in the context of intelligent video surveillance systems since an instant alert is set off in the case of distrustful activities. Figure 4 illustrates the proposed instant classification of a person's activity.

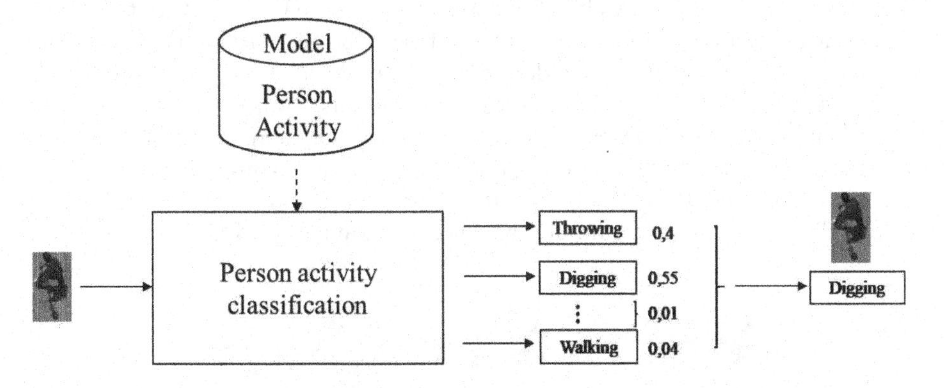

Fig. 4. Instant person activity classification

As for the entire classification, it provides an activity label to the entire video sequence. Thus, by resorting to instant classification scenario, we determine the activity label. The average of each activity label throughout the video sequence is then determined. Finally, we assign to the video sequence the label of the activity class which has the highest average (cf. Fig. 5).

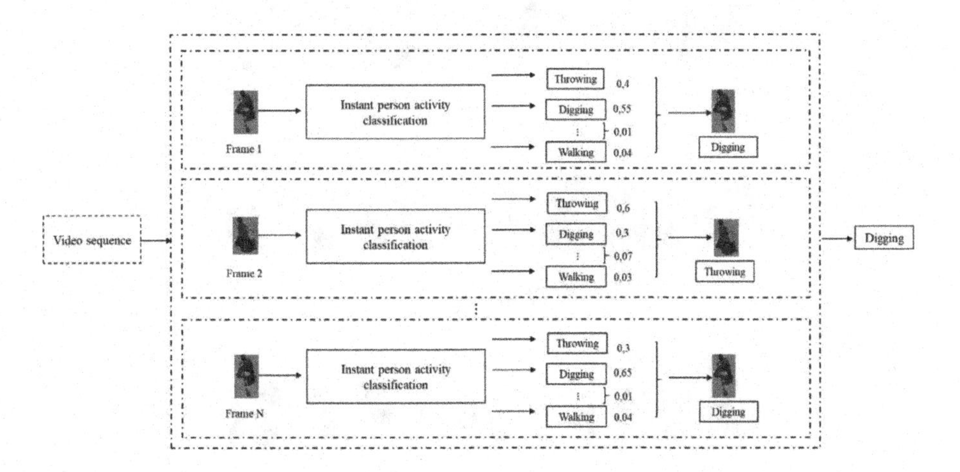

Fig. 5. Entire person activity classification

4 Experimental Study

In order to evaluate the proposed method, we first describe the used dataset. Subsequently, we compare and discuss the obtained results.

4.1 Dataset Description

We assess our person activity classification method on the UCF-ARG dataset [10] which is acquired by an aerial sensor. This dataset represents 10 human activities; each activity is displayed in 48 video sequences performed by 12 persons. This dataset deals with a multitude of constraints such as: point of view variation, scale variation, occlusion, inter- and intra-class variation, and lighting conditions variation. Therefore, it is considered in the literature as a complex dataset [1,5, 8]. For fair comparison to [5], we are interested in the following five activities, namely: digging, throwing, running, walking and waving (cf. Fig. 6).

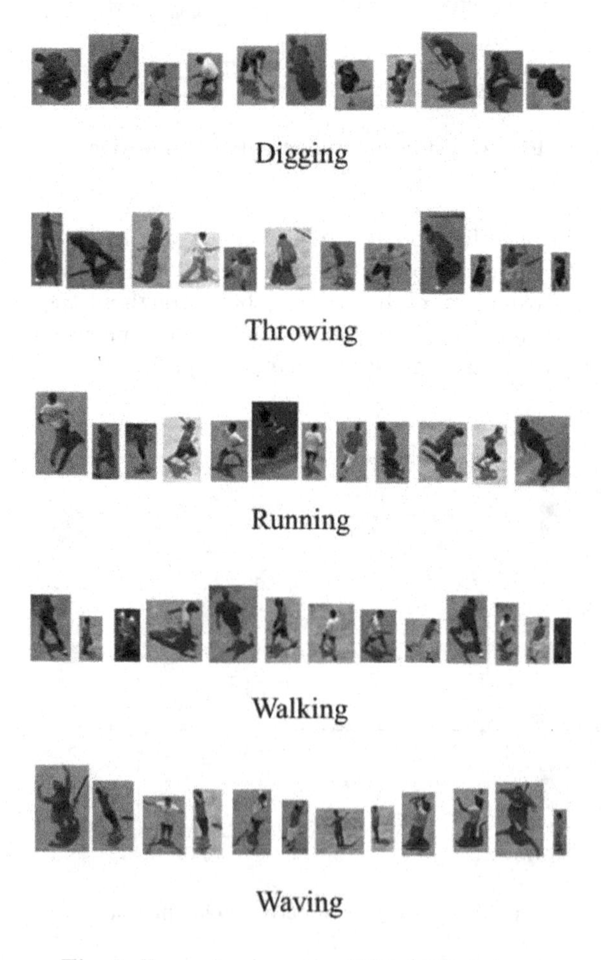

Fig. 6. Examples from the UCF-ARG dataset

4.2 Experimental Results

In order to validate the introduced person activity classification method, two series of experiments were performed. The first series aims to assess the multi-level deep features contribution. Whereas, the second compares the performance of the proposed instant classification as well as an entire classification with Burghouts et al. [5]. For a fair evaluation with [5], a LOO CV (Leave-One-Out Cross-Validation) was applied, with p = 1.

First Series of Experiments. In order to evaluate the multi-level deep features contribution, we calculate the accuracy rates and we compare the obtained rates without multi-level deep features (1), and with multi-level deep features (2). Tables 1 reports the obtained results achieved on the UCF-ARG dataset.

Table 1. Comparative results of the proposed multi-level deep features in terms of accuracy rate on the UCF-ARG dataset

Activity	Proposed Method	
	(1)	(2)
Throwing	79 %	84 %
Digging	69 %	80 %
Runing	67 %	82 %
Walking	67 %	57 %
Waving	56 %	67 %
	68 %	74 %

Relying to this evaluation, we validate the multi-level deep features contribution. Indeed, this made it possible to record a gain of 6.5%. Such contribution, addressed the constraint of inter- and intra-class variation.

Second Series of Experiments. Through this series we compare the performance of the proposed instant classification as well as the entire classification with the state-of-the-art method [5]. Table 2 illustrate this comparative study.

Relying to this evaluation, we noted that the accuracy rate of our method exceeds that obtained by Burghouts et al. [5]. This is justified by the fact that the pre-trained GoogLeNet CNN, suitable for multi-level feature extraction, handle the constraint of inter- and intra-class variation. Even though, the instant classification doesn't perform well compared to the entire classification, it exceeds the Burghouts et al. [5] method. Moreover, this instant classification is more suitable for intelligent video surveillance system.

Table 2. Proposed person activity classification method performance comparison in terms of accuracy on the UCF-ARG dataset

Activity	Proposed method		Burghouts et al. [5] Entire classification
	Instant classification	Entire classification	
Throwing	67 %	84 %	50 %
Digging	60 %	80 %	33 %
Runing	64 %	82 %	91 %
Walking	50 %	57 %	75 %
Waving	56 %	67 %	33 %
	59 %	74 %	57 %

5 Conclusion

The person activity classification from an aerial sensor is a prosperous research axis. In this context, we proposed a method that involves offline and inference phases. The offline phase uses convolutional neural networks to generate person activity model. The inference phase makes use the generated model to perform, the person activity classification. A multi-level deep features highlights the main contribution of the proposed method, which aims to deal with the inter- and the intra-class variation. Furthermore, we introduced two person activity classification scenarios an instant and an entire classification. By means of a comparative study, achieved on the UCF-ARG dataset, we demonstrated the performance and the contribution of our method compared to the state-of-the-art works. As future perspectives, we foresee introducing a bimodal person re-identification method which combines the person appearance with its activity to enrich the semantic description of persons in intelligent video surveillance system context.

References

1. AlDahoul, N., Md Sabri, A.Q., Mansoor, A.M.: Real-time human detection for aerial captured video sequences via deep models. Comput. Intell. Neurosci. **2018** (2018)
2. Baccouche, M., Mamalet, F., Wolf, C., Garcia, C., Baskurt, A.: Sequential deep learning for human action recognition. In: Salah, A.A., Lepri, B. (eds.) HBU 2011. LNCS, vol. 7065, pp. 29–39. Springer, Heidelberg (2011). https://doi.org/10.1007/978-3-642-25446-8_4
3. Bouhlel, F., Mliki, H., Hammami, M.: Crowd behavior analysis based on convolutional neural network: Social distancing control COVID-19. In: VISIGRAPP (5: VISAPP), pp. 273–280 (2021)
4. Burghouts, G.J., Schutte, K.: Spatio-temporal layout of human actions for improved bag-of-words action detection. Pattern Recogn. Lett. **34**(15), 1861–1869 (2013)
5. Burghouts, G., van Eekeren, A., Dijk, J.: Focus-of-attention for human activity recognition from UAVs. In: Electro-Optical and Infrared Systems: Technology and Applications XI, vol. 9249, pp. 256–267. SPIE (2014)

6. Dang, L.M., Min, K., Wang, H., Piran, M.J., Lee, C.H., Moon, H.: Sensor-based and vision-based human activity recognition: a comprehensive survey. Pattern Recogn. **108**, 107561 (2020)
7. He, K., Zhang, X., Ren, S., Sun, J.: Deep residual learning for image recognition. In: Proceedings of the IEEE Conference on Computer Vision and Pattern Recognition, pp. 770–778 (2016)
8. Mliki, H., Bouhlel, F., Hammami, M.: Human activity recognition from UAV-captured video sequences. Pattern Recogn. **100**, 107140 (2020)
9. Moussa, M.M., Hamayed, E., Fayek, M.B., El Nemr, H.A.: An enhanced method for human action recognition. J. Adv. Res. **6**(2), 163–169 (2015)
10. Nagendran, A., Harper, D., Shah, M.: UCF-ARG dataset, university of central Florida (2010). http://crcv.ucf.edu/data/UCF-ARG.php
11. Sabri, A., Boonaert, J., Lecoeuche, S., Mouaddib, E.: Caractérisation spatio-temporelle des co-occurrences par acp à noyau pour la classification des actions humaines. In: GRETSI 2013 (2012)
12. Sargano, A.B., Wang, X., Angelov, P., Habib, Z.: Human action recognition using transfer learning with deep representations. In: 2017 International Joint Conference on Neural Networks (IJCNN), pp. 463–469. IEEE (2017)
13. Szegedy, C., et al.: Going deeper with convolutions. In: Proceedings of the IEEE Conference on Computer Vision and Pattern Recognition, pp. 1–9 (2015)
14. Wang, L., Xu, Y., Cheng, J., Xia, H., Yin, J., Wu, J.: Human action recognition by learning spatio-temporal features with deep neural networks. IEEE Access **6**, 17913–17922 (2018)

Person Quick-Search Approach Based on a Facial Semantic Attributes Description

Sahar Dammak[1], Hazar Mliki[2,3](✉), and Emna Fendri[1]

[1] MIRACL-FS, Faculty of Sciences of Sfax, University of Sfax, Sfax, Tunisia
sahardammak@fsegs.u-sfax.tn, fendri.msf@gnet.tn

[2] MIRACL Laboratory, University of Sfax, Sfax, Tunisia
mliki.hazar@gmail.com

[3] National Institute of Applied Science and Technology, University of Carthage, Tunis, Tunisia

Abstract. Person search based on semantic attributes description presents an interest task for intelligent video surveillance applications. The main objective is to locate a suspect or to find a missing person in public areas using a semantic description (*e.g.* a 40-year-old asian woman) provided by an eyewitness. Such a description provides the facial soft biometric related to the facial semantic attributes (*i.e.* age, gender and ethnicity). In this paper, we introduced a new approach for person search named "Quick-Search" based on a facial semantic attributes description to enhance the person search task in an unconstrained environment. The main contribution of the paper is to introduce a multi-attributes score fusion method which relies on soft biometric features (age, gender, ethnicity) to improve the person search in a large dataset. An experimental study was conducted on the challenging FairFace dataset to validate the effectiveness of the proposed person search approach.

Keywords: Person search · Facial semantic attributes detection · Gender classification · Age group classification · Ethnicity classification · CNN

1 Introduction

Person search in real world scenarios is a challenging computer vision task that attract the interest of several researchers. The main objective is to locate a suspect or to find a missing person in public areas such as airports, shopping malls, parks, among others. Traditional person search methods [9,15] are based on appearance characteristics covering the whole human body which may present limited capacity for automated surveillance solutions. Since the face remains the most informative and accessible source about a person, the recent advancement of research in facial semantic attributes detection has provided the way for person search using a facial semantic attributes description (*i.e.* gender, age and ethnicity).

J. Blanc-Talon et al. (Eds.): ACIVS 2023, LNCS 14124, pp. 76–87, 2023.
https://doi.org/10.1007/978-3-031-45382-3_7

In the literature, several studies were devoted to address the facial semantic attributes detection. Early proposed methods [8,12,17], known as handcraft methods, are based on a standard descriptor followed by a statistical classifier. The recent success of the Convolutional Neural Network (CNN) based methods, has encouraged the research community to adapt CNNs to the facial semantic attributes detection. For the gender classification, Aslam et al. [3] applied the Cascaded Deformable Shape model to extract the facial feature regions namely: the mouth, the nose, the eyes and the foggy faces. Then, a four-dimensional (4-D) representation is constructed based on these facial feature regions. Finally, the VGG-16 pre-trained model [21] is fine-tuned using the 4-D array for a final gender decision. In addition, Serna et al. [20] proposed a gender classification method that provided a preliminary investigation on how biased data impact the deep neural network architectures learning process in terms of activation level. Two gender detection models based on VGG16 and ResNet are used to assess the impact of bias on the learning process.

Regarding the age groups classification, Chen et al. [4] proposed to fine-tune the VGG-Face model and adapt it to the age classification task. Next, the activation of the penultimate layer and the last layer of the VGG-Face are used as a separate local feature to feed to the maximum joint probability classifier (MJPC). In the same context, a survey of different CNN architectures for facial age classification is presented in [1]. In addition, several studies proposed a joint gender and age classification methods. In this context, Duan et al. [7] combined an Extreme Learning Machine (ELM) [11] with a Convolutional Neural Networks (CNN) in one network and used the interaction of two classifiers to deal with gender and age groups classification. The convolutional layer of the CNN was applied to extract features. Then, the ELM structure was combined with the fully connected layers of the CNN model, to generate a hybrid gender and age groups classification model.

As for the ethnicity classification, Ahmed et al. [2] proposed a new CNN architecture with nine layers called "RNet". The first six layers constitue the convolution layers while the final three layers represent the fully connected FC layers followed by the softmax loss function. In addition, DropOut regularization technique is used to avoid the overfitting problem. Finally, the trained model is optimized using the stochastic gradient descent. In the same context, Luong et al. [16] improved the face ethnicity, age and gender classification performance by using a contrastive loss based metric learning. In fact, a multi-output supervised contrastive loss was applied on the ResNet-50 model. Kärkkäinen et al. [13] used the ResNet model based on the softmax loss to perform the face ethnicity, age and gender classification. The proposed model was optimized using the Adaptive Moment estimation (ADAM) algorithm.

Referring to the studied related works, we noticed that the CNN-based methods have been shown to achieve significantly results thanks to their capability to capture complex visual variations by applying a large amount of training data, without the need for any specific descriptor. Therefore, to tackle the facial semantic attributes problem in an unconstrained environment, we have opted for a CNN-based methods.

In this paper, we address an interesting and challenging person search problems based on facial semantic attributes description provided by an eyewitness. Our main objective is to highlight the contribution of using facial semantic attributes description in the person search methods. The proposed approach for person search named "Quick-Search", does not seek to provide the person identity, but, it aims to reduce the subject of interest list. Thus, the output may contain multiple retrievals matching the query. The main contribution of the paper is to introduce a multi-attributes score fusion method which relies on facial semantic attributes description (age, gender, ethnicity) to improve the person search in a large dataset.

The remainder of this paper is organized as follows. Section 2 introduce and detail the proposed method. The experiments and results are given in Sect. 3 Finally, Sect. 4 provided the conclusion and some perspectives.

2 Proposed Approach

The proposed approach introduces a new person search solution based on a facial semantic attributes description provided by an eyewitness in terms of a query. The main objective is to determine a list of people that matches the query using a facial semantic attributes prediction models. As illustrated in Fig. 1 the proposed approach consists of two main steps: (1) Facial semantic attributes detection, and (2) Multi-attributes score fusion.

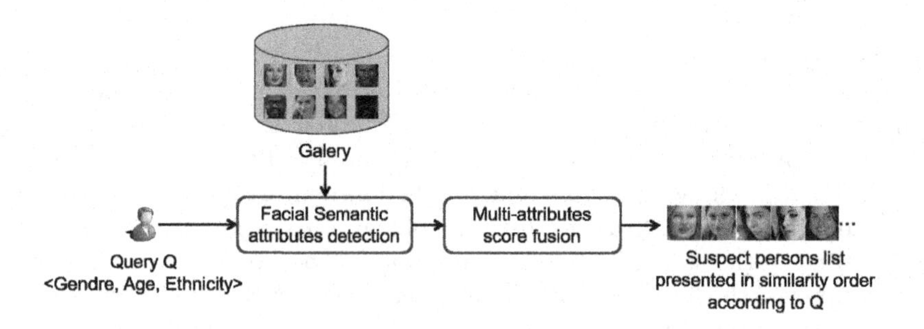

Fig. 1. Proposed approach for person search based on facial semantic attributes description.

Formally, the query Q of the person to be re-identified is provided in terms of a vector of combinations (A_i, S_i) (*cf.* Eq. 1).

$$Q = \{(A_1, S_1), \ldots, (A_p, S_p)\} \quad (1)$$

Where A_i correspond to the set of semantic attributes gender, age and ethnicity and S_i presents its status. The S_i values depend on the attribute. For the gender attribute, we deal with two values (male/female). As for the age attribute, we handle four values corresponding to the four age groups: 0–9, 10–29, 30–49 and more than 50. Regarding to the ethnicity attribute, we address four values of ethnicity groups: Asian, Black, Indian and White. A query may not specify the values of the three attributes. In fact, unselected attributes are not considered during the search process.

2.1 Facial Semantic Attributes Detection

In order to learn a set of facial semantic attributes classifiers, we proposed to build a prediction model for each semantic attribute (*i.e.* gender, age and ethnicity). Unlike methods that propose a single model to classify the facial semantic attributes, we propose to treat each attribute separately in order to deal with the different constraints of each attribute. Therefore, we generated three prediction models: gender classification model, age groups classification model and ethnicity classification model. In the classification phase, we assume that the gallery is fed by a set of suspect person faces acquired by the different cameras of the video surveillance system. For each person face $p \in SP$, each prediction model of an attribute $a \in Q$ predicts the probability of the presence of the concerned attribute from the face image. This probability varies between 0 and 1. If the value is close to 1, it indicates the presence of a facial attribute. Otherwise, it is considered unavailable.

In the following sub-sections, we described each of the proposed semantic attributes detection methods.

Gender Classification. For the gender classification, we introduced a facial gender classification method based on a hybrid architecture which combines deep learned features and handcrafted features by performing information fusion at the score-level [6]. As illustrated in Fig. 2, two models were generated: the deep learned features based model and the handcrafted features based model.

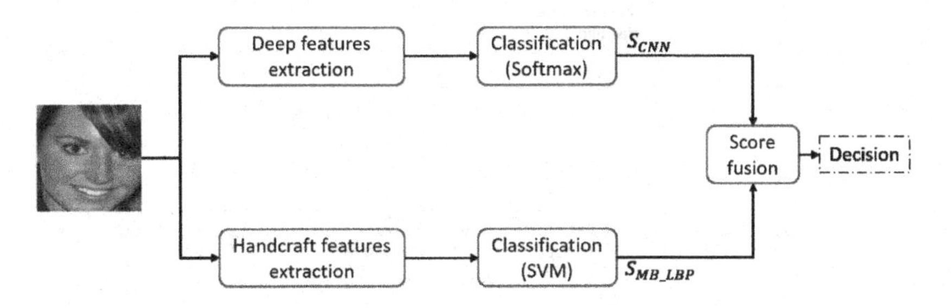

Fig. 2. Proposed method for the gender classification.

For the first model, the deep features are extracted using the fine-tuned VGG-16 [21] model and the classification is achieved using the softmax classifier. As for the second model, the handcrafted features are computed using the MB-LBP (multi block LBP) [26] which is an extension of the LBP descriptor [18] and the classification is performed using the SVM classifier.

The scores output from the softmax and the SVM classifiers indicate the probabilities of the gender class. In the testing phase, we fused the extracted scores from the softmax and the SVM classifiers to obtain a final score using the maximum rule defined as follows:

$$S = max(S_{CNN}, S_{MB-LBP}) \tag{2}$$

where S_{CNN} is the obtained score from the deep learned features based model and $S_{(MB-LBP)}$ is the obtained score from the handcrafted features based model.

Age Groups Classification. Several studies [22,23] prove that the aging process differs from one gender to another. Therefore, we proposed to improve the age groups classification by exploring the correlation between age and gender information [5]. In addition, in order to reduce the confusion between the intra- and inter-age groups, we proceed with a two-level age classification strategy. This two-level age classification is based on deep learning and consists of the age models generation and the age classification process. The proposed method for age groups classification is illustrated in Fig. 3.

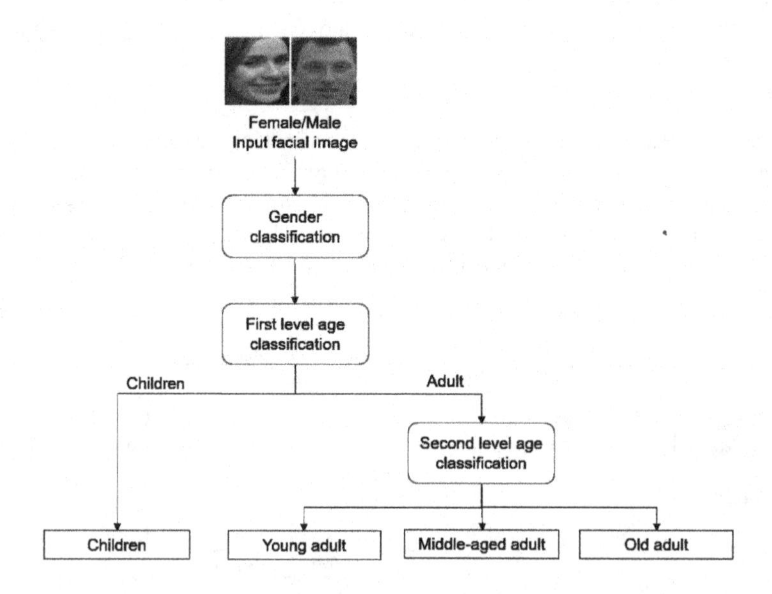

Fig. 3. Proposed method for the age groups classification.

For the different age groups prediction models, we adapted the FaceNet model [19] to the context of each classification level. First, we use the dedicated first level CNN model to classify the age groups into two categories: children and adults. Second, we use the second level model of the detected class to refine the classification. Thus, a face classified as "adult" will be classified among three age groups "young adult", "middle-aged adult" and "old adult".

Ethnicity Classification. For the ethnicity classification, we adapted the ResNet model [10] to the ethnicity classification task. We proposed to jointly optimize the softmax loss function with the center loss function [25] which aim to learn the centers for deep features of each class. Then we use these centers to penalize the distance between deep features and their corresponding class centers in the euclidean space (*cf.* Fig. 4). In fact, the center loss adds the square sum of the distance between samples and similar samples into the original softmax loss function as a penalty term, as shown in Eq. 3:

$$L = L_s + \mu L_c \tag{3}$$

where, L_s is the traditional Softmax Loss, μ is the penalty term in the Center Loss [24], and L_c is defined in Eq. 4:

$$L_c = \frac{1}{2} \sum_{i=1}^{N} \|x_i - c_{y_i}\|_2^2 \tag{4}$$

where x_i is the deep feature of the ith sample, c_{y_i} denotes the $y_{i^{th}}$ class center of the deep features.

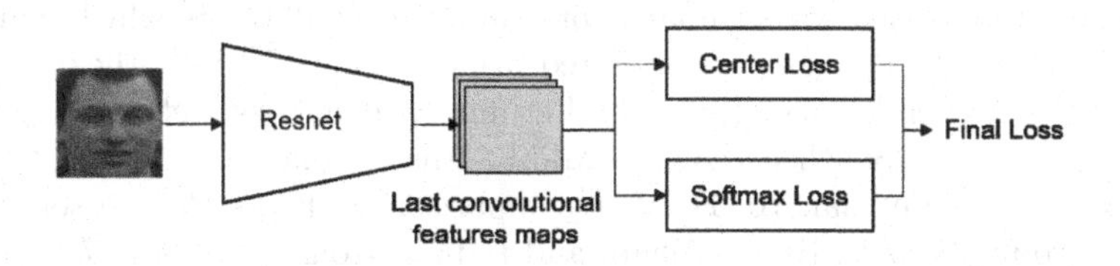

Fig. 4. Proposed method for the ethnicity classification.

2.2 Multi-attributes Score Fusion

In this step, we compute the final score that describe the similarity of the suspect person matching the query. This involves a function F that fuses the scores of the semantic attributes prediction models involved in the query Q to obtain a final score FS, $F : Q \longrightarrow FS$. Formally, we have a set of N attributes $A = A_1, \ldots, A_N$ and their prediction scores are presented as probabilities $P = P_1, \ldots, P_N$, where $P_i = [0..1]$.

To merge the scores, we applied the sum of the probabilities (*cf.* Eqs. 5) [14].

$$FS_i = \sum_{i=1}^{N} P_i \tag{5}$$

Then, the faces are sorted according to these scores in a descending order. The first rank is given to the most similar person to the query. This system helps security agent to locate the suspect person in a video surveillance network by reducing the list of possible suspects. Thus, the identification of the target person from a reduced list provides a gain in terms of time execution and increases the possibility of finding the requested person.

3 Experimental Study

To study the performance of the proposed method, three series of experiments were carried out on FairFace dataset [13].

3.1 Dataset Description

The proposed approach was evaluated on the FairFace datasets which contains 108,501 images. The images were gathered from the YFCC-100M Flickr dataset and labeled with gender, age groups and ethnicity. It defines 7 ethnicity groups (White, East Asian, Southeast Asian, Middle East, Indian, Black, and Latino) aged from 0 to more than 70. For fair comparison with [13,16], we selected 4 ethnicity groups (White, Black, Asian, and Indian) from among the 7 ethnicity groups. The images were collected under real world conditions (*e.g.* low resolution, pose variation, occlusion, and illumination variation).

3.2 Experimental Results

In order to evaluate the performance of the proposed "Quick-Search" approach, we carried out three series of experiments. The two first series of experiments assessed the performance of each of the proposed facial semantic attributes classification methods. The last one evaluated the performance of the proposed approach for person search based on facial semantic attributes description.

First Series of Experiments. This series of experiments highlights the importance of the proposed contributions for each attribute classification. For the gender classification, we proved the importance of the hybrid architecture. The fine-tuned VGG-16 and MB-LBP models were evaluated apart and then the performance of the score-level fusion strategy was studied. As for the age groups classification, we evaluated the age classification performance with and without applying the two-level classification strategy as well as with and without proceeding with the gender estimation. Regarding to the ethnicity classification, we highlight the importance of the joint supervision of the softmax loss and the center loss. In fact, we compared the performance of the CNN model using the softmax loss and the joint supervision of the softmax loss with the center loss. Table 1 reports the classification accuracy rates for each of the generated models on the FairFace datasets.

Table 1. Comparative study of the generated facial semantic attributes classification models on the FairFace dataset in terms of the classification accuracy rate.

Attributes classification	Methods	Accuracy
Gender classification	Handcraft based method (MB-LBP)	66.84%
	CNN based method (VGG16)	89.00%
	Score fusion based method	**93.70%**
Age classification	Independent on gender (Without level strategy)	61.49%
	Independent on gender (With level strategy)	66.56%
	Dependent on gender (With level strategy)	**68.45%**
Ethnicity classification	Softmax Loss	84.30%
	Softmax Loss +Center Loss	**85.50%**

Results in Table 1 confirm the efficiency of the proposed contributions for the facial semantic attributes classification task. For the gender classification, the obtained results reveal the important gain while combining the deep learned and handcrafted features. In fact, the proposed hybrid architecture produces more discriminant information. As for the age groups classification, the recorded results highlights the gain of using the two-level classification strategy. In fact, dividing the classification problem into subclasses contributed to reducing confusion between classes. Furthermore, the two-level age classification was significantly improved by the use of gender information. Regarding to the ethnicity classification, the joint supervision of the softmax loss with the center loss based method achieved an important gain compared to the softmax loss based method. Such results prove the effectiveness of using the center loss to enhance the discriminative ability of the deeply learned features.

Second Series of Experiments. To validate the performance of the generated facial semantic attributes classification models, we carried out a comparative study of the proposed method with the state-of-the art methods [13,16]. The authors in [13] used the ResNet model based on the softmax loss. In [16], the authors proposed to apply a multi-output supervised contrastive loss on the ResNet model. Table 2 represents the obtained results in terms of the classification accuracy rate.

Table 2. Comparative study of the generated facial semantic attributes classification models on the FairFace dataset in terms of the classification accuracy rate.

Methods	Gender	Age	Ethnicity
Kärkkäinen et al. [13]	95.70%	53.60%	81.50%
Luong et al. [16]	92.69%	57.81%	78.81%
Proposed method	93.70%	68.45%	85.50%

Referring to Table 2, the proposed method for age groups and ethnicty classification recorded, respectively, an accuracy rate of 68.45% and 85.50% on the FairFace dataset, which outperforms those obtained by the state-of-the art methods. Such performance confirms the effectiveness of the proposed methods. As for the gender classification, the proposed method achieved a competitive results. Compared to Luong et al. [16], we achieved a gain of 1.01% which validates the efficiency of the combination of the deep learned and handcrafted features. However, the proposed method in [13], which is based on the ResNet model, slightly outperforms our method. Although their method provides good results, its deep architecture yields to complex computing time, making it an expensive option in real-time applications.

Third Series of Experiments. To examine the performance of the proposed people search approach based on facial semantic attributes description, we proceed with a qualitative experimental evaluation on the FairFace dataset. Table 3 shows the qualitative results of ten queries in terms of the first five images returned by each query.The obtained results validate the contribution of using the facial semantic attributes description in the person search context. In fact, these attributes allow to filter a gallery based on a textual description provided by an eyewitness. This filtering leads to a reduced list of candidate persons, which improves the performance of re-identification in terms of time and efficiency.

Table 3. Qualitative results of the proposed person search approach based on a semantic description in terms of attribute queries.

#	Queries	First five images
1	Q={(Gender,Male),(Age,30-49),(Ethnicity,Black)}	
2	Q={(Gender,Female),(Age,10-29),(Ethnicity,White)}	
3	Q={(Gender,Male),(Age,0-9),(Ethnicity,Asian)}	
4	Q={(Gender,Female),(Age,30-49),(Ethnicity,Asian)}	
5	Q={(Gender,Female),(Age,more than50),(Ethnicity,Black)}	
6	Q={(Gender,Male),(Ethnicity,White)}	
7	Q={(Gender,Female),(Age,30-49)}	
8	Q={(Gender,Male),(Age,0-9)}	
9	Q={(Ethnicity,Indian)}	
10	Q={(Age,more than 50)}	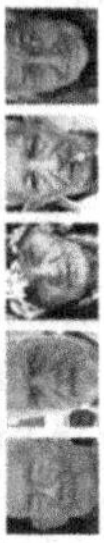

4 Conclusion

In this paper, we introduced a new people search approach "Quick-Search" based on facial semantic attributes description provided by eyewitnesses. Firstly, we proposed a facial semantic attributes detection based methods. In fact, we generated three prediction models: gender classification model, age groups classification model and ethnicity classification model. The gender classification method is based on a hybrid architecture which combines deep-learned and handcrafted features by performing information fusion at the score-level. As for the age groups classification, we proposed to explore the correlation between age and gender information. In addition, in order to reduce the confusion between the intra- and inter-age groups, we proceed with a two-level age classification strategy. Regarding to the ethnicity classification, we proposed to jointly optimize the softmax loss function with the center loss function. After that, we evaluated the performance of the multi-attributes search approach based on the learned attributes classifiers. Experimental results illustrate the effectiveness of the proposed approach. The present study presents novel insights on integrating soft facial semantic attributes description with person appearance features to enhance the effectiveness of person search in the context of video surveillance.

References

1. Agbo-Ajala, O., Viriri, S.: Deep learning approach for facial age classification: a survey of the state-of-the-art. Artif. Intell. Rev. **54**, 1–35 (2020)
2. Ahmed, M.A., Choudhury, R.D., Kashyap, K.: Race estimation with deep networks. J. King Saud Univ. Comput. Inf. Sci. **34**, 4579–4591 (2020)
3. Aslam, A., Hussain, B., Cetin, A.E., Umar, A.I., Ansari, R.: Gender classification based on isolated facial features and foggy faces using jointly trained deep convolutional neural network. J. Electron. Imaging **27**(5), 053–023 (2018)
4. Chen, L., Fan, C., Yang, H., Hu, S., Zou, L., Deng, D.: Face age classification based on a deep hybrid model. SIViP **12**(8), 1531–1539 (2018)
5. Dammak, S., Mliki, H., Fendri, E.: Gender effect on age classification in an unconstrained environment. Multimedia Tools Appl. **80**(18), 28001–28014 (2021)
6. Dammak, S., Mliki, H., Fendri, E.: Gender estimation based on deep learned and handcrafted features in an uncontrolled environment. Multimedia Syst. **29**, 1–13 (2022)
7. Duan, M., Li, K., Yang, C., Li, K.: A hybrid deep learning CNN-ELM for age and gender classification. Neurocomputing **275**, 448–461 (2018)
8. Eidinger, E., Enbar, R., Hassner, T.: Age and gender estimation of unfiltered faces. IEEE Trans. Inf. Forensics Secur. **9**(12), 2170–2179 (2014)
9. Frikha, M., Fendri, E., Hammami, M.: People search based on attributes description provided by an eyewitness for video surveillance applications. Multimedia Tools Appl. **78**, 2045–2072 (2019)
10. He, K., Zhang, X., Ren, S., Sun, J.: Deep residual learning for image recognition. In: Proceedings of the IEEE Conference on Computer Vision and Pattern Recognition, pp. 770–778 (2016)
11. Huang, G.B., Zhu, Q.Y., Siew, C.K.: Extreme learning machine: theory and applications. Neurocomputing **70**(1–3), 489–501 (2006)

12. Jagtap, J., Kokare, M.: Human age classification using facial skin aging features and artificial neural network. Cogn. Syst. Res. **40**, 116–128 (2016)
13. Kärkkäinen, K., Joo, J.: Fairface: Face attribute dataset for balanced race, gender, and age. arXiv preprint arXiv:1908.04913 pp. 1–11 (2019)
14. Kittler, J., Hatef, M., Duin, R.P., Matas, J.: On combining classifiers. IEEE Trans. Pattern Anal. Mach. Intell. **20**(3), 226–239 (1998)
15. Li, S., Xiao, T., Li, H., Zhou, B., Yue, D., Wang, X.: Person search with natural language description. In: Proceedings of the IEEE Conference on Computer Vision and Pattern Recognition, pp. 1970–1979 (2017)
16. Luong, T.K., Hsiung, P.A., Han, Y.T.: Improve gender, race, and age classification with supervised contrastive learning (2021). https://doi.org/10.13140/RG.2.2.14680.01286
17. Mohamed, S., Nour, N., Viriri, S.: Gender identification from facial images using global features. In: Conference on Information Communications Technology and Society (ICTAS), pp. 1–6. IEEE (2018)
18. Ojala, T., Pietikäinen, M., Harwood, D.: A comparative study of texture measures with classification based on featured distributions. Pattern Recogn. **29**(1), 51–59 (1996)
19. Schroff, F., Kalenichenko, D., Philbin, J.: FaceNet: a unified embedding for face recognition and clustering. In: Proceedings of the IEEE Conference on Computer Vision and Pattern Recognition, pp. 815–823 (2015)
20. Serna, I., Pena, A., Morales, A., Fierrez, J.: InsideBias: measuring bias in deep networks and application to face gender biometrics. In: 2020 25th International Conference on Pattern Recognition (ICPR), pp. 3720–3727. IEEE (2021)
21. Simonyan, K., Zisserman, A.: Very deep convolutional networks for large-scale image recognition. In: arXiv preprint arXiv:1409.1556, pp. 1–14 (2014)
22. Smulyan, H., Asmar, R.G., Rudnicki, A., London, G.M., Safar, M.E.: Comparative effects of aging in men and women on the properties of the arterial tree. J. Am. Coll. Cardiol. **37**(5), 1374–1380 (2001)
23. Sveikata, K., Balciuniene, I., Tutkuviene, J.: Factors influencing face aging. Lit. Revi. Stomatologija **13**(4), 113–116 (2011)
24. Wang, J., Feng, S., Cheng, Y., Al-Nabhan, N.: Survey on the loss function of deep learning in face recognition. J. Inf. Hiding Priv. Prot. **3**(1), 29–47 (2021)
25. Wen, Y., Zhang, K., Li, Z., Qiao, Yu.: A discriminative feature learning approach for deep face recognition. In: Leibe, B., Matas, J., Sebe, N., Welling, M. (eds.) ECCV 2016. LNCS, vol. 9911, pp. 499–515. Springer, Cham (2016). https://doi.org/10.1007/978-3-319-46478-7_31
26. Zhang, L., Chu, R., Xiang, S., Liao, S., Li, S.Z.: Face detection based on multi-block LBP representation. In: Lee, S.-W., Li, S.Z. (eds.) ICB 2007. LNCS, vol. 4642, pp. 11–18. Springer, Heidelberg (2007). https://doi.org/10.1007/978-3-540-74549-5_2

Age-Invariant Face Recognition Using Face Feature Vectors and Embedded Prototype Subspace Classifiers

Anders Hast(✉)

Uppsala University, 751 05 Uppsala, Sweden
anders.hast@it.uu.se
http://www.andershast.com

Abstract. One of the major difficulties in face recognition while comparing photographs of individuals of different ages is the influence of age progression on their facial features. As a person ages, the face undergoes many changes, such as geometrical changes, changes in facial hair, and the presence of glasses, among others. Although biometric markers like computed face feature vectors should ideally remain unchanged by such factors, face recognition becomes less reliable as the age range increases. Therefore, this investigation was carried out to examine how the use of Embedded Prototype Subspace Classifiers could improve face recognition accuracy when dealing with age-related variations using face feature vectors only.

Keywords: Face Recognition · Embedded Prototype Subspace Classifier · Biometric Markers

1 Introduction

The study explores various models for face recognition (FR) and their performance based on Face Feature Vectors (FFV). It also examines whether *Embedded Prototype Subspace Classification* (EPSC) can improve the accuracy in FR with age variations. The overall aim is to achieve age-invariant face recognition (AIFR).

The reliability of face recognition (FR) performed by Intelligent Vision Systems is constantly improving. Facial features are captured through images or videos and used as biometric markers. Face recognition methods involve analyzing these features and encoding them as Face Feature Vectors (FFV). These vectors enable the identification and verification of a face by matching it with a known identity in a face database. Hence, FFV serves as a valuable tool for identification and verification purposes.

However, the impact of age progression on facial features poses a considerable challenge for FR systems. As individuals age, their faces undergo changes that include geometric alterations, changes in facial hair, and the use of glasses, among

J. Blanc-Talon et al. (Eds.): ACIVS 2023, LNCS 14124, pp. 88–99, 2023.
https://doi.org/10.1007/978-3-031-45382-3_8

other factors. Although biometric markers, such as computed FFVs, are intended to be unaffected by such factors, FR systems become less reliable as the age range expands. This study investigates how well different models for FR and their respective FFV perform and whether EPSC could enhance the accuracy of FR in the presence of such age variations, with the goal of achieving AIFR. EPSC has already proven to be able to classify datasets of various kinds, such as digits, words and objects [9–12].

2 Previous Work

In their paper, Sawant and Bhurchandi [33] examine the difficulties encountered by practical AIFR systems concerning appearance variations and resemblances between subjects. They classify AIFR techniques into three categories: generative, discriminative, and deep learning, with each approach addressing the problem from a distinct standpoint.

The performance of generative approaches depends on aging models and age estimation. The main idea is to transform the recognition problem into general FR by simulating the aging process and synthesise face images [2,5,6,21,31,32], or extracted features [27], of the target age into the same age group. Enhancements in AIFR performance have been achieved using deep learning techniques; however, these methods necessitate considerable training data, posing a challenge when utilising smaller databases.

Discriminative techniques, on the other hand, depend on learning methods and local features. They strive to identify robust features for identity recognition that remain unaffected by age. The goal of both discriminative and generative approaches is to reduce the impact of age variation on FR systems. This study concentrates on discriminative methods that employ FFVs obtained from any FR pipeline. However, there has been relatively little written about discriminative methods for AIFR compared to other areas of FR. Nevertheless, an overview of the more recent paper will be given before showing how EPSC can improve classification.

Several works propose to decompose the features into age- and identity-related features. For an example Gong et al. [7] proposed Hidden Factor Analysis, using a probabilistic model on these two different features. An Expectation Maximisation learning algorithm was used to estimate the latent factors and model parameters. They reported improvement over then state-of-the-art algorithms on two public face aging datasets. Similarily, Huang et al. [13,14] introduced MTLFace, a multi-task learning framework for AIFR and identity-level face age synthesis. They showed how improved performance can be achieved by a selective fine-tune strategy. Experiments demonstrate the superiority of MTLFace, and the authors suggest a newly collected dataset could further advance development in this area. Xu and Ye [40] introduced a solution to the AIFR using coupled auto-encoder networks (CAN) and nonlinear factor analysis. By utilising CAN, the identity feature can be separated non-linearly to become age invariant in a given face image.

Zheng et al. [43] developed a novel deep face recognition network called AE-CNN, which uses age estimation to separate age-related variations from stable person-specific features. The CNN model learns age-invariant features for FR and the proposed approach was tested on two public datasets, showing good results. Wang et al. [37] introduced similar approach that enhances AIFR by separating deep face features into two components: age-related and identity-related. This approach enabled the extraction of highly discriminative age-invariant features using a multi-task deep CNN model.

Zhifeng et al. [23] proposed a discriminative model for age-invariant face recognition. The approach uses a patch-based local feature representation scheme and a multi-feature discriminant analysis (MFDA) method to refine the feature space for enhanced recognition performance. Ling et al. [24,25] proposed a robust face descriptor called the gradient orientation pyramid, which captures hierarchical facial information. The new approach was compared to several techniques and demonstrated promising results on challenging passport databases, even with large age differences. Yan et al. [41] propose an AIFR approach called MFD, which combines feature decomposition with fusion based on the face time series to effectively represent identity information that is not sensitive to the aging process. Self-attention is also employed to capture global information from facial feature series, and a weighted cross-entropy loss is adopted to address imbalanced age distribution in training data. Gong et al. [8] proposed a new approach for AIFR using a maximum entropy feature descriptor and identity factor analysis. The maximum entropy feature descriptor encodes the microstructure of facial images and extracts discriminatory and expressive information. Li et al. [22] proposed a two-level hierarchical learning model for AIFR. The first level involves extracting features by selecting local patterns that optimize common information. The second level involves refining these features using a scalable high-level feature refinement framework to create a powerful face representation. Experiments demonstrated improvement over existing methods on one public face aging dataset.

This work takes the Discriminative approach, previously described, and therefore the FFV resulting from face detection on face images and subsequent feature extraction will be used as a biometric marker. Several FR models are investigated and compared.

3 Face Recognition Methods

In this work, two freely available major pipelines for FR were used and compared, which are described below.

3.1 Insightface

The *InsightFace* [15] pipeline is an integrated Python library for 2D and 3D face analysis, mainly based on PyTorch and MXNet. *InsightFace* efficiently implements a rich variety of state of the art algorithms for both face detection, face alignment and face recognition, such as *RetinaFace* [4]. It allows for automatic extraction of highly discriminative FFVs for each face, based on the Additive Angular Margin Loss (ArcFace) approach [3]. The models used are $buffalo\_l$, *antvelope2*, $buffalo\_m$ and $buffalo\_s$, which are all provided by *InsightFace*.

Deepface. https://github.com/serengil/deepface *Deepface* [34–36] is a python-based, lightweight framework for FR as well as facial attribute analysis, including age, gender, emotion, and race. It incorporates cutting-edge models in a hybrid FR framework. In this work, the *RetinaFace* backend was used for face detection, while different models for creating the FFV were used. These models used herein are $Facenet512$ and $ArcFace$, which both produce 512 long FFVs just like *InsightFace* does.

4 Datasets

In the automatic FR process, images where faces were not properly detected or images with more than one face were removed. Furthermore, it was required that only person identities having several face images, covering several ages (decades) were selected. Some mislabeled faces and corrupt FFVs were removed.

4.1 AgeDB

The *AgeDB* dataset [29] contains 16,516 images. Of those, 9826 face images were extracted so that each person depicted had about 36 face images on average covering at least three different age decades, i.e. (0–10, 11–20, 21–30, 31–40, 41–50, 51–60, 61–70 and 70+). In addition, it was required that each person included should have at least 30 face images each, in order to ensure that there are several face images of the same person at different ages and decades.

4.2 CASIA

The original *CASIA-WebFace* [42] is rather large, containing around $500k$ images. However, the selection process resulted in a much smaller and feasible dataset containing 65579 face images. Nonetheless, it is still quite a lot larger than the *AgeDB* dataset. A similar approach was used here, like the one used for the *AgeDB* dataset, resulting in more than 50 face images per person on average.

5 Embedded Prototype Subspace Classification

Subspace Classification in pattern recognition was introduced by Watanabe et al. [38] in 1967 and was later developed further, especially by Kohonen and other Finnish researchers associated with him [16–18,30,39]. The following mathematical derivation follows from Kohonen and Oja [30] and Laaksonen [20].

EPSC, a shallow model that utilizes PCA and dimensionality reduction techniques such as t-SNE, UMAP, or SOM [12], has advantages over various deep learning methods, including not requiring powerful GPU resources for training due to the absence of hidden layers. EPSC creates subspaces that specialize in identifying class variations from feature vectors. Although EPSC may not always achieve higher accuracy than deep learning methods, it offers faster learning and inference, interpretability, explainability, and ease of visualisation.

Every detected face to be classified is represented by a FFV $\mathbf{x}$ with m real-valued elements $\mathbf{x}_j = \{x_1, z_2...x_m\}, \in \mathbb{R}$, such that the operations take place in a m-dimensional vector space $\mathbb{R}^m$. Here, m is equal to the FFV length, which depends on the model, usually 128 or 512. Any set of n linearly independent basis vectors $\{\mathbf{u}_1, \mathbf{u}_2, ...\mathbf{u}_n\}$, where $\mathbf{u}_i = \{w_{1,j}, w_{2,j}...w_{m,j}\}, w_{i,j} \in \mathbb{R}$, which can be combined into an $m \times n$ matrix $\mathbf{U} \in \mathbb{R}^{m \times n}$, span a subspace $\mathcal{L}_{\mathbf{U}}$

$$\mathcal{L}_{\mathbf{U}} = \{\hat{\mathbf{x}} | \hat{\mathbf{x}} = \sum_{i=1}^{n} \rho_i \mathbf{u}_i, \rho_i \in \mathbb{R}\} \tag{1}$$

where,

$$\rho_i = \mathbf{x}^T \mathbf{u}_i = \sum_{j=1}^{m} x_j w_{i,j} \tag{2}$$

Classification of a feature vector can be performed by projecting $\mathbf{x}$ onto each subspace $\mathcal{L}_{\mathbf{U}_k}$. The vector $\hat{\mathbf{x}}$ will therefore be a reconstruction of $\mathbf{x}$, using n vectors in the subspace through

$$\hat{\mathbf{x}} = \sum_{i=1}^{n} (\mathbf{x}^T \mathbf{u}_i) \mathbf{u}_i \tag{3}$$

$$= \sum_{i=1}^{n} \rho_i \mathbf{u}_i \tag{4}$$

$$= \mathbf{U}^T \mathbf{U} x^T \tag{5}$$

By normalising all the vectors in $\mathbf{U}$, the norm of the projected vector can be simplified as

$$||\hat{\mathbf{x}}||^2 = (\mathbf{U}x^T) \cdot (\mathbf{U}x^T) \tag{6}$$

$$= (\mathbf{U}x^T)^2 \tag{7}$$

$$= \sum_{i=1}^{n} {\rho_i}^2 \tag{8}$$

Therefore, the feature vector $\mathbf{x}$, which is most similar to the feature vectors that were used to construct the subspace in question $\mathcal{L}_{\mathbf{U}_k}$, will have the largest norm $||\hat{\mathbf{x}}||^2$.

The previously explained Subspace classification can be regarded as a two layer neural network [11,20,30], where the weights are all mathematically defined through Principal Component Analysis (PCA) [20], instead of the time consuming backpropagation.

Table 1. Embedded Prototype Subspace Classification are compared to Nearest Neighbor Classification. Accuracy as well as the Bhattacharyya distance and coefficient for different datasets and methods are given. (Better values in green and best value for AgeDB in bold).

Dataset	Model	Size	EPSC Acc.	NNC Acc.	EPSC Similarity	FFV Similarity	EPSC Coeff.	FFV Coeff.
AgeDB	buffalo_l	9753	**0.9975**	**0.9973**	**0.4117**	0.0412	0.0198	0.0768
	antelopev2	9755	0.9969	0.9966	0.3561	0.0246	0.0244	**0.0656**
	buffalo_m	9751	0.9973	0.9967	0.4115	**0.0415**	**0.0191**	0.0777
	buffalo_s	9766	0.9940	0.9922	0.3824	0.0338	0.0373	0.1550
	Facenet512	9548	0.9234	0.9095	0.0664	0.0000	0.1703	0.4155
	ArcFace	9525	0.9548	0.9357	0.2252	0.0270	0.1259	0.4099
CASIA	buffalo_l	64047	0.9954	0.9955	0.4213	0.0446	0.0378	0.0855

Generally, a group of prototypes within each class need to be chosen for the construction of each subspace, which is done by the embedding obtained from some dimensionality reduction method such as t-SNE [26], UMAP [28] or SOM [19]. However, here a subspace for each person was constructed, as explained later, regardless of age since relatively few FFVs are at hand per person.

6 Experiments

In order to verify the efficiency of the EPSC, experiments were conducted on each data set as follows. Each and every FFV was classified using the EPSC approach. Subspaces were created for all class labels, using all the FFVs for each class, except for the FFV to be tested, which was temporarily removed and a subspace was created for that class using the remaining FFVs. In this way, the FFV to be classified was never used for creating the subspaces itself. Therefore, the class to be classified will always have an disadvantage compared to the others. However, the result will indicate how well AIFR can be performed using EPSC.

This approach is compared to what a nearest neighbor classifier (NNC) would give. Here each face image, with its corresponding FFV is given the same label as its closest neighbour in FFV space, using the cosine similarity, which is simply the dot product between two normalised FFVs.

The projection depth variable n is set to 3, which was experimentally found to be a good value. This basically means that for each class, a dot product is computed three times. While for NNC it is computed as many times as the size of the dataset. NNC generally gives a very good classification, but is simply too costly to be used in practice as it has quadratic time complexity, even if there are methods to reduce the search space. EPSC on the other hand is very fast and efficient as a classifier as it has linear time complexity.

Histograms are shown in Fig. 1, where both the distribution of the intra similarities, i.e. cosine similarity between different FFVs of the same class (blue

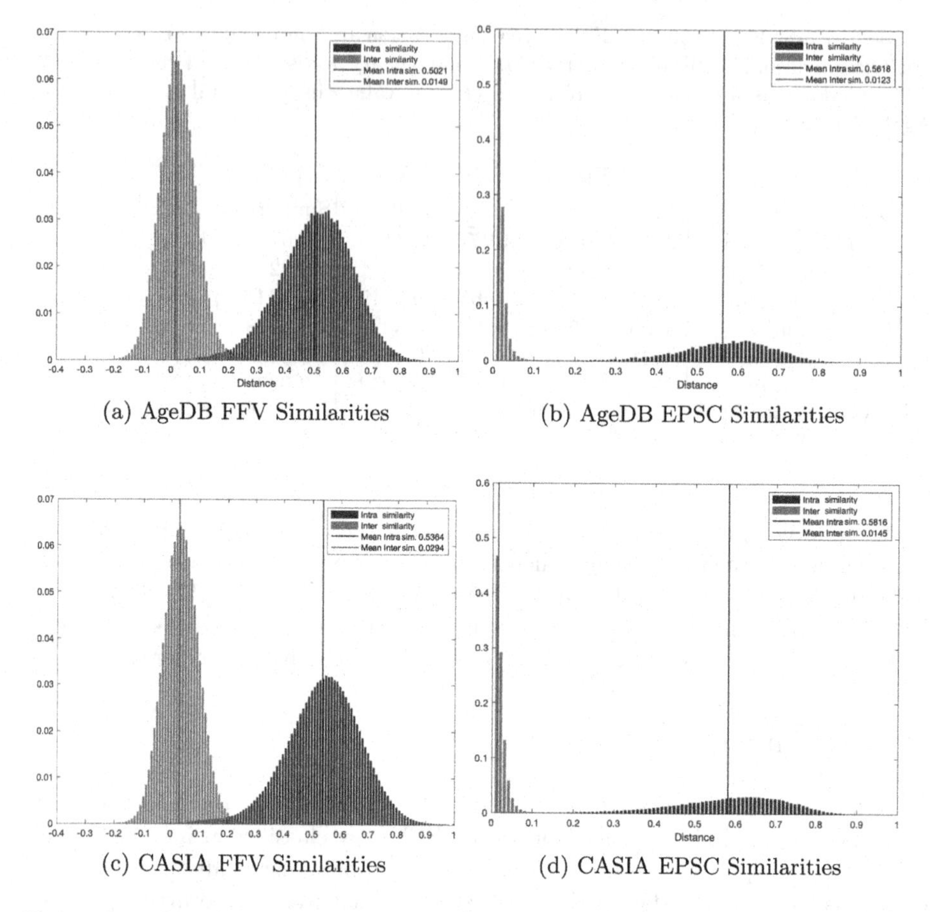

(a) AgeDB FFV Similarities

(b) AgeDB EPSC Similarities

(c) CASIA FFV Similarities

(d) CASIA EPSC Similarities

Fig. 1. Bar plot of a histograms of intra similarities (blue bars = same class label) and inter similarities (red bars = different class labels) between FFV (left) and EPSC (right). The mean of both similarities are indicated with vertical lines. (Color figure online)

bars) and the distribution of the inter similarities, i.e. the similarity between FFVs of one class and FFVs of all other classes (red bars) are computed. There is an overlap between the both distributions in Fig. 1a, which shows where misclassifications generally occur. When using EPSC the overlap becomes smaller. The Bhattacharyya distance and the related Bhattacharyya coefficients [1] are reported in Table 1.

The Bhattacharyya distance and coefficient have various applications in statistics, pattern recognition, and image processing. They are used to compare image similarity and cluster data points based on their probability distributions. Machine learning algorithms like neural networks and support vector machines also use them. Both measures are bounded by 0 and 1 and are symmetric.

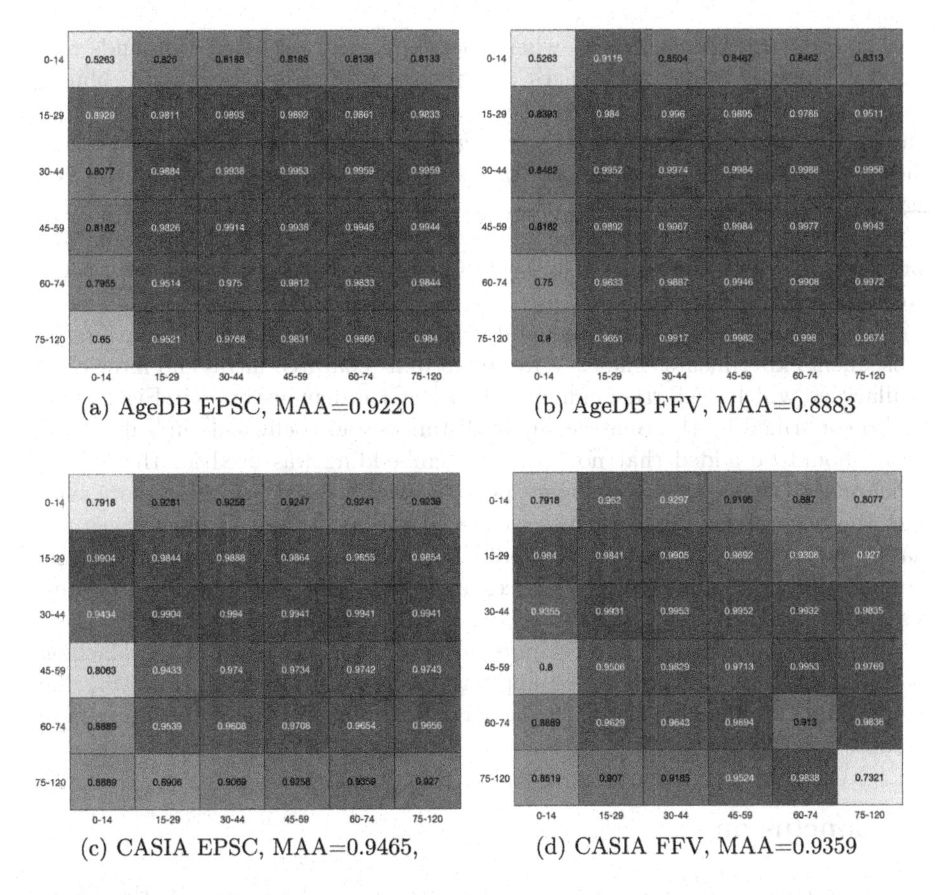

(a) AgeDB EPSC, MAA=0.9220

(b) AgeDB FFV, MAA=0.8883

(c) CASIA EPSC, MAA=0.9465,

(d) CASIA FFV, MAA=0.9359

Fig. 2. Similarity matrices showing the accuracy for classifying a person from one age group (per column) using images from another age group (per row). EPSC are used on the left and the FFV on the right. The Macro Average Arithmetic is given and shows that EPSC yields an overall improvement.

The Bhattacharyya distance is a measure of the similarity between two probability distributions. Specifically, it quantifies the closeness between two probability distributions. The Bhattacharyya distance is zero if and only if the two distributions are identical. Hence, it becomes larger the more separated they are.

The Bhattacharyya coefficient is a similarity measure between two probability distributions, which is closely related to the Bhattacharyya distance. Unlike the distance measure, 0 indicates no overlap between the distributions, and 1 indicates that the two distributions are identical.

7 Discussion

The selected images from both AgeDB and CASIA contains a rather large age span per person, which makes it quite challenging for FR. It also contains some very hard to recognise images, where faces are semi occluded by hands and glasses. Furthermore, some images are of low resolution or generally blurry. Others are exhibiting strange grimaces etc. Nevertheless, it could be noted from Table 1 that the models provided by *InsightFace* generally performed better than those of *DeepFace*. The reason most probably lies in the fact that the latter is a lightweight FR framework, which employs fast and simple alignment. Therefore, it will fail more often for such hard cases, and especially it fails to capture profile faces and to create good FFV's for such difficult face postures. Hence, the fever produced FFV's.

Since the model $buffalo\_l$ generally performed best it was chosen for the subsequent experiments. The FFV results in well-separated intra- and inter-class similarities, which are further enhanced by EPSC, as demonstrated in Fig. 2. This is also confirmed by the Bhattacharyya distances and coefficients in Table 1.

It should be added that no prototype embedding was used for the EPSC, which means that only one susbpace per class was created. By using more than one, the accurracies might increase, especially when having many images per class. Another way to go about, would be to have smaller age spans in the computation of the similarity matrices in Fig. 2. This is proposed for future research.

It is not claimed that EPSC is more accurate than any other of the Machine Learning or Deep Learning based approaches covered in Sect. 2. However, due to its simplicity and that it use the FFVs only, it should be a viable alternative to more complex methods.

8 Conclusion

It can be concluded that, among the two tested pipelines for FR, the FFVs from *InsightFace* using the model $buffalo\_l$ produced the best results for the goal of achieving AIFR, using the discriminative approach, previously explained. NNC performs very well, but is impractical due to its quadratic time complexity. EPSC, with its linear time complexity, increased the accuracies and is therefore both faster and more efficient.

Acknowledgments. This work has been partially supported by the Swedish Research Council (Dnr 2020-04652; Dnr 2022-02056) in the projects *The City's Faces. Visual culture and social structure in Stockholm 1880-1930* and *The International Centre for Evidence-Based Criminal Law (EB-CRIME)*. The computations were performed on resources provided by SNIC through UPPMAX under project SNIC 2021/22-918.

References

1. Bhattacharyya, A.: On a measure of divergence between two statistical populations defined by their probability distribution. Bull. Calcutta Math. Soc. **35**, 99–110 (1943)
2. Deb, D., Aggarwal, D., Jain, A.K.: Identifying missing children: Face age-progression via deep feature aging. In: 2020 25th International Conference on Pattern Recognition (ICPR), pp. 10540–10547 (2021). https://doi.org/10.1109/ICPR48806.2021.9411913
3. Deng, J., Guo, J., Xue, N., Zafeiriou, S.: Arcface: Additive angular margin loss for deep face recognition. In: Proceedings of the IEEE/CVF Conference on Computer Vision and Pattern Recognition, pp. 4690–4699 (2019)
4. Deng, J., Guo, J., Zhou, Y., Yu, J., Kotsia, I., Zafeiriou, S.: Retinaface: single-stage dense face localisation in the wild (2019). https://doi.org/10.48550/ARXIV.1905.00641, https://arxiv.org/abs/1905.00641
5. Duong, C., Quach, K., Luu, K., Le, T., Savvides, M.: Temporal non-volume preserving approach to facial age-progression and age-invariant face recognition. In: 2017 IEEE International Conference on Computer Vision (ICCV), pp. 3755–3763. IEEE Computer Society, Los Alamitos, CA, USA (2017). https://doi.org/10.1109/ICCV.2017.403, https://doi.ieeecomputersociety.org/10.1109/ICCV.2017.403
6. Geng, X., Zhou, Z., Smith-Miles, K.: Automatic age estimation based on facial aging patterns. IEEE Trans. Pattern Anal. Mach. Intell. **29**(12), 2234–2240 (2007). https://doi.org/10.1109/TPAMI.2007.70733
7. Gong, D., Li, Z., Lin, D., Liu, J., Tang, X.: Hidden factor analysis for age invariant face recognition. In: 2013 IEEE International Conference on Computer Vision, pp. 2872–2879 (2013). https://doi.org/10.1109/ICCV.2013.357
8. Gong, D., Li, Z., Tao, D., Liu, J., Li, X.: A maximum entropy feature descriptor for age invariant face recognition. In: 2015 IEEE Conference on Computer Vision and Pattern Recognition (CVPR), pp. 5289–5297 (2015). https://doi.org/10.1109/CVPR.2015.7299166
9. Hast, A.: Magnitude of semicircle tiles in Fourier-space : a handcrafted feature descriptor for word recognition using embedded prototype subspace classifiers. J. WSCG **30**(1–2), 82–90 (2022). https://doi.org/10.24132/JWSCG.2022.10
10. Hast, A., Lind, M.: Ensembles and cascading of embedded prototype subspace classifiers. J. WSCG **28**(1/2), 89–95 (2020). https://doi.org/10.24132/JWSCG.2020.28.11
11. Hast, A., Lind, M., Vats, E.: Embedded prototype subspace classification: a subspace learning framework. In: Vento, M., Percannella, G. (eds.) CAIP 2019. LNCS, vol. 11679, pp. 581–592. Springer, Cham (2019). https://doi.org/10.1007/978-3-030-29891-3_51
12. Hast, A., Vats, E.: Word recognition using embedded prototype subspace classifiers on a new imbalanced dataset. J. WSCG **29**(1–2), 39–47 (2021). https://wscg.zcu.cz/WSCG2021/2021-J-WSCG-1-2.pdf
13. Huang, Z., Zhang, J., Shan, H.: When age-invariant face recognition meets face age synthesis: a multi-task learning framework. In: Proceedings of the IEEE/CVF Conference on Computer Vision and Pattern Recognition, pp. 7282–7291 (2021)
14. Huang, Z., Zhang, J., Shan, H.: When age-invariant face recognition meets face age synthesis: a multi-task learning framework and a new benchmark. IEEE Trans. Pattern Anal. Mach. Intell. **45**, 7917–7932 (2022)
15. InsightFace: Insightface (2023). https://insightface.ai. Accessed 30 Feb 2023

16. Kohonen, T., Lehtiö, P., Rovamo, J., Hyvärinen, J., Bry, K., Vainio, L.: A principle of neural associative memory. Neuroscience **2**(6), 1065–1076 (1977)
17. Kohonen, T., Oja, E.: Fast adaptive formation of orthogonalizing filters and associative memory in recurrent networks of neuron-like elements. Biol. Cybern. **21**(2), 85–95 (1976)
18. Kohonen, T., Reuhkala, E., Mäkisara, K., Vainio, L.: Associative recall of images. Biol. Cybern. **22**(3), 159–168 (1976)
19. Kohonen, T.: Self-organized formation of topologically correct feature maps. Biol. Cybern. **43**(1), 59–69 (1982)
20. Laaksonen, J.: Subspace classifiers in recognition of handwritten digits. G4 monografiaväitöskirja, Helsinki University of Technology (1997). https://urn.fi/urn:nbn:fi:tkk-001249
21. Lanitis, A., Taylor, C., Cootes, T.: Toward automatic simulation of aging effects on face images. IEEE Trans. Pattern Anal. Mach. Intell. **24**(4), 442–455 (2002). https://doi.org/10.1109/34.993553
22. Li, Z., Gong, D., Li, X., Tao, D.: Aging face recognition: a hierarchical learning model based on local patterns selection. IEEE Trans. Image Process. **25**(5), 2146–2154 (2016). https://doi.org/10.1109/TIP.2016.2535284
23. Li, Z., Park, U., Jain, A.K.: A discriminative model for age invariant face recognition. IEEE Trans. Inf. Forensics Secur. **6**(3), 1028–1037 (2011). https://doi.org/10.1109/TIFS.2011.2156787
24. Ling, H., Soatto, S., Ramanathan, N., Jacobs, D.W.: A study of face recognition as people age. In: 2007 IEEE 11th International Conference on Computer Vision, pp. 1–8 (2007). https://doi.org/10.1109/ICCV.2007.4409069
25. Ling, H., Soatto, S., Ramanathan, N., Jacobs, D.W.: Face verification across age progression using discriminative methods. IEEE Trans. Inf. Forensics Secur. **5**(1), 82–91 (2010). https://doi.org/10.1109/TIFS.2009.2038751
26. Maaten, L.V.D., Hinton, G.: Visualizing data using t-SNE. J. Mach. Learn. Res. **9**(11), 2579–2605 (2008)
27. Mahalingam, G., Kambhamettu, C.: Age invariant face recognition using graph matching. In: 2010 Fourth IEEE International Conference on Biometrics: Theory, Applications and Systems (BTAS), pp. 1–7 (2010). https://doi.org/10.1109/BTAS.2010.5634496
28. McInnes, L., Healy, J.: UMAP: Uniform Manifold Approximation and Projection for Dimension Reduction. ArXiv e-prints (2018)
29. Moschoglou, S., Papaioannou, A., Sagonas, C., Deng, J., Kotsia, I., Zafeiriou, S.: Agedb: The first manually collected, in-the-wild age database. In: 2017 IEEE Conference on Computer Vision and Pattern Recognition Workshops (CVPRW), pp. 1997–2005 (2017). https://doi.org/10.1109/CVPRW.2017.250
30. Oja, E., Kohonen, T.: The subspace learning algorithm as a formalism for pattern recognition and neural networks. In: IEEE 1988 International Conference on Neural Networks, vol. 1, pp. 277–284 (1988). https://doi.org/10.1109/ICNN.1988.23858
31. Park, U., Tong, Y., Jain, A.K.: Age-invariant face recognition. IEEE Trans. Pattern Anal. Mach. Intell. **32**(5), 947–954 (2010). https://doi.org/10.1109/TPAMI.2010.14
32. Ramanathan, N., Chellappa, R.: Modeling age progression in young faces. In: 2006 IEEE Computer Society Conference on Computer Vision and Pattern Recognition (CVPR 2006), vol. 1, pp. 387–394 (2006). https://doi.org/10.1109/CVPR.2006.187

33. Sawant, M.M., Bhurchandi, K.M.: Age invariant face recognition: a survey on facial aging databases, techniques and effect of aging. Artif. Intell. Rev. **52**, 981–1008 (2019). https://doi.org/10.1007/s10462-018-9661-z
34. Serengil, S.I., Ozpinar, A.: Lightface: a hybrid deep face recognition framework. In: 2020 Innovations in Intelligent Systems and Applications Conference (ASYU), pp. 23–27. IEEE (2020). https://doi.org/10.1109/ASYU50717.2020.9259802
35. Serengil, S.I., Ozpinar, A.: Hyperextended lightface: a facial attribute analysis framework. In: 2021 International Conference on Engineering and Emerging Technologies (ICEET), pp. 1–4. IEEE (2021). https://doi.org/10.1109/ICEET53442.2021.9659697
36. Serengil, S.I., Ozpinar, A.: An evaluation of SQL and NOSQL databases for facial recognition pipelines. https://www.cambridge.org/engage/coe/article-details/63f3e5541d2d184063d4f569 (2023). 10.33774/coe-2023-18rcn, https://doi.org/10.33774/coe-2023-18rcn. preprint
37. Wang, Y., et al.: Orthogonal deep features decomposition for age-invariant face recognition. In: Ferrari, V., Hebert, M., Sminchisescu, C., Weiss, Y. (eds.) Computer Vision - ECCV 2018, pp. 764–779. Springer, Cham (2018)
38. Watanabe, S., Pakvasa, N.: Subspace method in pattern recognition. In: 1st International Joint Conference on Pattern Recognition, Washington DC. pp. 25–32 (1973)
39. Watanabe, W., Lambert, P.F., Kulikowski, C.A., Buxto, J.L., Walker, R.: Evaluation and selection of variables in pattern recognition. In: Tou, J. (ed.) Computer and Information Sciences, vol. 2, pp. 91–122. Academic Press, New York (1967)
40. Xu, C., Liu, Q., Ye, M.: Age invariant face recognition and retrieval by coupled auto-encoder networks. Neurocomputing **222**, 62–71 (2017)
41. Yan, C., et al.: Age-invariant face recognition by multi-feature fusionand decomposition with self-attention. ACM Trans. Multimedia Comput. Commun. Appl. **18**(1s) (2022). https://doi.org/10.1145/3472810
42. Yi, D., Lei, Z., Liao, S., Li, S.Z.: Learning face representation from scratch (2014). https://doi.org/10.48550/ARXIV.1411.7923. https://arxiv.org/abs/1411.7923
43. Zheng, T., Deng, W., Hu, J.: Age estimation guided convolutional neural network for age-invariant face recognition. In: 2017 IEEE Conference on Computer Vision and Pattern Recognition Workshops (CVPRW), pp. 503–511 (2017). https://doi.org/10.1109/CVPRW.2017.77

BENet: A Lightweight Bottom-Up Framework for Context-Aware Emotion Recognition

Tristan Cladière(✉), Olivier Alata, Christophe Ducottet, Hubert Konik, and Anne-Claire Legrand

Laboratoire Hubert Curien UMR CNRS 5516, Institut d'Optique Graduate School, Université Jean Monnet Saint-Etienne, 42023 Saint-Etienne, France
{tristan.cladiere,olivier.alata}@univ-st-etienne.fr

Abstract. Emotion recognition from images is a challenging task. The latest and most common approach to solve this problem is to fuse information from different contexts, such as person-centric features, scene features, object features, interactions features and so on. This requires specialized pre-trained models, and multiple pre-processing steps, resulting in long and complex frameworks, not always practicable in real time scenario with limited resources. Moreover, these methods do not deal with person detection, and treat each subject sequentially, which is even slower for scenes with many people. Therefore, we propose a new approach, based on a single end-to-end trainable architecture that can both detect and process all subjects simultaneously by creating emotion maps. We also introduce a new multitask training protocol which enhances the model predictions. Finally, we present a new baseline for emotion recognition on EMOTIC dataset, which considers the detection of the agents. Our code and more illustrations are available at https://github.com/TristanCladiere/BENet.git.

Keywords: Emotion recognition · Detection · Bottom-Up · Multitask · Deep learning

1 Introduction

Understanding emotions is a difficult yet essential task in our daily life. They can be defined as discrete categories or as coordinates in a continuous space of affect dimensions [4]. For the discrete categories, Ekman and Friesen [5] defined six basic ones: anger, disgust, fear, happiness, sadness, and surprise. Later, contempt was added to the list [12]. Concerning the continuous space, valence, arousal, and dominance form the commonly used three-dimensional frame [13].

Regarding non-verbal cues, facial expression is one of the most important signals to convey emotional states and intentions [11]. However, the context is also essential in some cases, because it can be misleading to infer emotions using only the face [1]. For images, the context can include many cues, and the recent authors have built different deep learning architectures to process it. Lee

J. Blanc-Talon et al. (Eds.): ACIVS 2023, LNCS 14124, pp. 100–111, 2023.
https://doi.org/10.1007/978-3-031-45382-3_9

et al. [10] proposed an attention mechanism to extract features from everything else than the face. Zhang *et al.* [20] inferred emotions with a Graph Convolutional Network, using the features generated by a Region Proposal Network as nodes. Kosti *et al.* [9] created a two-stream Convolutional Neural Network which extracts body and scene features, and fuses them. Similarly, Bendjoudi *et al.* [2] used a two-stream network, and studied the synergy between continuous and categorical loss functions. Mittal *et al.* [14] combined agent, scene, and depth features with multiplicative fusion. Here, the agent features are computed from facial landmarks and body pose, both obtained with off-the-shelves models as pre-processing steps. Instead of depth features, Hoang *et al.* [6] designed a reasoning stream that explores relationships between the main subject and the adjacent objects in the scene, using an existent and pre-trained objects detector. Wang *et al.* [17] introduced the tubal transformer, a shared features representation space that facilitates the interactions among the face, body, and context features. Yang *et al.* [19] developed an adaptive relevance fusion module for learning the shared representations among multiple contexts, some of whom depend on external models.

Although the above approaches provide good results, they are all composed of multiple streams that use different kinds of inputs and are processed sequentially. Therefore, both training and inference become slower and more complicated, especially when memory and computational power are restricted. Moreover, none of them consider the detection task, which not only makes their method not directly usable for real world applications, but also gives emotion scores that are not representative of the whole process. Indeed, many pre-processing steps depend on the bounding boxes provided by the annotations, but in real scenario they would be obtained with a person detector, which may be inaccurate or even miss subjects. Based on these observations, we propose a totally different approach, designed to be later embedded in a robot for real-time uses, and leading to the three main contributions presented in this work. Firstly, we built a model that simultaneously assesses the emotions of all subjects present in an image. It is end-to-end trainable, relies on a single shared backbone, takes directly the raw image as input, and does not require specialized pre-trained modules. Secondly, we made the model multitask capable, which means that the same architecture can also predict the bounding boxes of the subjects by itself, estimate the emotions of a particular agent using only its person-centric features, and give all the emotions in the image using only background features. Thirdly, we share a new baseline that evaluates simultaneously the detection and the recognition parts of our model.

2 Proposed Method

In this section, the components of our multitask approach will be detailed. Each head of the architecture is dedicated to a specific task. First, the bottom-up head is introduced. It allows to estimate the emotions of multiple people simultaneously, unlike the usual methods which treat them sequentially. Next, the

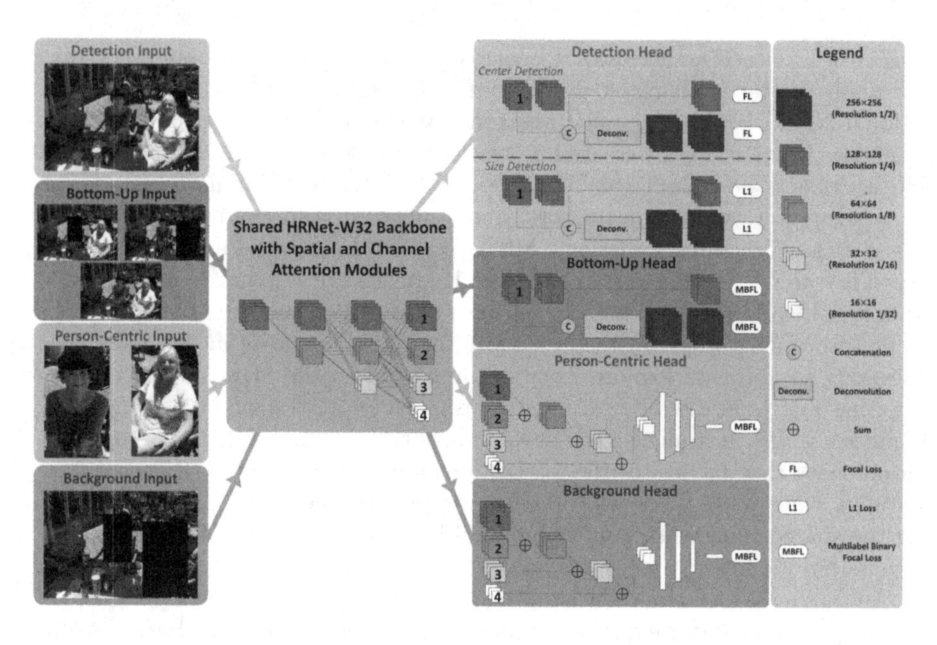

Fig. 1. Architecture overview

detection head is presented. Combined with the bottom-up one, these blocks make the model fully autonomous, since it becomes independent of the annotated bounding boxes. After, the person-centric head is described. It is specifically used to predict the emotions of a single subject given as input. Here we only rely on the subject's features, without processing neither other people nor the background information. On the contrary, the background head is finally shown. It makes a global prediction using only the scene features, i.e. everything except the annotated subjects. An overview of our architecture is given in Fig. 1.

2.1 Bottom-Up Approach

The authors cited in Sect. 1 use methods considered as top-down approaches: they first have to detect the subject (or use the ground truth) before inferring his emotions. With these approaches, each subject is treated sequentially and independently, which is slow and redundant for images with multiple people. Therefore, we proposed a new way to handle this problem, that can be considered as a bottom-up solution. Inspired by [3], a bottom-up head is used to produce E discrete emotion maps directly from the raw image, as illustrated in Fig. 2. The value of E depends on the number of discrete emotion categories used in the considered database. On these maps, only the value of the pixel at the center of each subject's bounding box is imposed: 1 if the emotion is present, 0 otherwise. For all the other pixels, the model is free to output anything that helps it in making its predictions. However, at test time, the bounding boxes coordinates are necessary to extract and attribute the predictions. They can be given either by the ground truth or by a person detector.

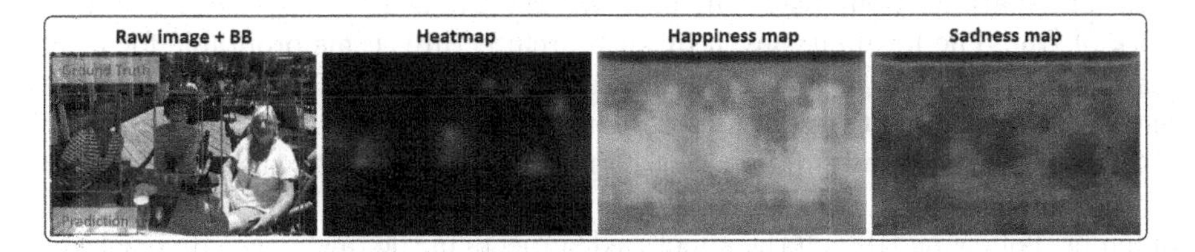

Fig. 2. Example of the heatmap produced by the detection head, and two emotion maps given by the bottom-up head. Normalisation is between 0 (black, emotion is absent) and 1 (white, emotion is present). The predicted and annotated bounding boxes are also added to the raw image.

2.2 Detection Head

To become fully autonomous, our model needs to be able to automatically extract its predictions from the emotion maps. Thus, similar to [21], a bottom-up detection head is added and trained to predict the center of the bounding boxes by creating a heatmap (see Fig. 2), and to regress their dimensions. During the inference, these centers coordinates can be extracted and used to retrieve the emotions of the subjects, and also the predicted dimensions of the bounding boxes. Therefore, the model both detects all the subjects in the raw image and estimates their emotions in a single forward pass.

At this point, the whole architecture is still end-to-end trainable, and the two tasks can be jointly trained. Moreover, the framework is composed of a unique backbone, which means that the heads share common features. It benefits the bottom-up approach since the model becomes better at predicting emotions at the correct coordinates of the emotion maps.

2.3 Person-Centric and Background Features

The bottom-up approach introduced in Sect. 2.1 uses as input the raw image to produce its emotions maps. The architecture has a global view of the scene, that contains both person-centric and background features. Depending on the situation, one of these features may be prevailing over the other one, and the model should still be able to perform well with this single source of information. This is why two heads were added to the framework, one that will be specialised in person-centric features, the other in background ones.

To extract features from the main subject, some authors used pre-trained deep-learning architectures to detect his face, his facial landmarks, and to estimate his posture from the portion of the input image corresponding to his bounding box [6,10,14,17,19]. These outputs served as intermediate information that further helps to infer the emotions of the subject, but they are also dependent on external resources. In our case, a simpler method is used: a classification block inspired by [16] is added to the model, and is referred as the person-centric head. The combination of our backbone and this specific head is very similar to the work of [16]. The flexibility given by such architecture seems profitable for processing in-the-wild images, including close range faces and far range silhouettes, mainly due to its multi-resolutions design.

For background information, the corresponding head has the same design as the person-centric one, but both the input and the training objective are different. Following [19], all the annotated subjects in the raw image are masked, forcing the model to rely on other sources of features (see Fig. 1). However, rather than predicting the emotions of a single person, the architecture has to estimate all the emotions present in the image, i.e. each emotion that is labelled for at least one subject. Given N people annotated for E emotions on a single image, a one-hot-encoded matrix of dimension $N \times E$ is therefore created, and the maximum along the N axis is taken, resulting in a vector of shape $1 \times E$.

Here again the whole architecture is end-to-end trainable, and all the tasks are jointly trained. In this configuration, the shared backbone learns to extract rich common features, so that each task benefits from the others.

3 Framework Details

In this section, the multitask architecture is first presented (see Fig. 1). Then the used data and their processing to jointly train all heads are detailed. Finally the loss functions are explained.

3.1 Network Architecture

Our Bottom-up Emotions Network (BENet) uses HRNet-W32 [16] as backbone. It contains four stages with four parallel convolution streams. The resolutions are 1/4, 1/8, 1/16, and 1/32, while the widths (numbers of channels) of the convolutions are C, $2C$, $4C$, and $8C$ ($C = 32$). We also integrated spatial attention modules and channel attention modules, inspired by [18].

For the detection and the bottom-up heads, a structure very similar to [3] has been implemented. The main point is to use a deconvolution module on top of the highest resolution feature maps in HRNet, increasing the resolution from 1/4 to 1/2. Such process mainly helps to detect smaller people in the image, since bottom-up approaches must deal with subjects of very different scales. At the end, the model is trained to output its predictions at two resolutions, 1/4 and 1/2.

For the two classification heads, the design of [16] is used, but the bottleneck expansion is reduced by a factor 4. Indeed, we do not need to have too many channels, considering that there are far fewer emotions in emotion recognition than classes in image classification. It also helps to reduce the global number of parameters of the architecture, which is profitable considering that there is not a lot of available data.

3.2 Databases

EMOTIC. database [8] contains 23,571 images of 34,320 annotated people in unconstrained environments. Each image has at least one subject, which is annotated with a bounding box, 26 discrete categories, and 3 continuous dimensions of emotions. The subject can be assigned multiple labels. The standard partition of the dataset is 70% for the training set, 10% for the validation one, and the remaining 20% are used for testing.

HECO. database [19] regroups 9,385 images and 19,781 annotated people, with rich context information and various agent interaction behaviours. The annotations include 8 discrete categories and 3 continuous dimensions, but also the novel *Self-assurance* (Sa) and *Catharsis* (Ca) labels, which describe the degree of interaction between subjects and the degree of adaptation to the context. Unfortunately, the authors do not provide any partition of their dataset, which makes the evaluations difficult to compare. Thus, we only used HECO as extra data for training.

3.3 Data Processing

The data augmentation is divided into two parts. The first part is designed to randomly apply a specific pre-transformation on each image of the training batch. Depending on the pre-transformations drawn, the images will be dispatched at the end of the backbone, and fed to the corresponding heads. Thus, each image is designed to train a specific task. These pre-transformations are named *ExtractSubject*, *MaskAllSubjects*, and *RandomMaskSubjects*. They will be briefly explained, and examples of the inputs are shown in Fig. 1.

ExtractSubject uses the ground truth to crop the image around the bounding box of a given subject. It will be used to train the model to extract person-centric features. This pre-transformation can only be drawn if the bounding box of the selected subject does not contain other people.

MaskAllSubjects uses the ground truth to mask all the annotated subjects in the image. With such images, the background head will have to extract features from everything except the people. To ensure that there is still enough information left for the model to learn useful features, this pre-transformation can only be applied on images in which the sum of the areas of the bounding boxes do not exceed 40% of the total area of the input.

RandomMaskSubjects is the transformation used to train the bottom-up head. If there are multiple annotated subjects in the image, we will randomly mask them, but always make sure to keep at least one. The idea is to augment and diversify the combinations of emotions presented to the model.

The last option is to keep the raw image. In this case, it will be used to train the detection head. Given a batch size B, each image will be pre-processed by picking one of the above pre-transformations, with probabilities of (namely) 0.25, 0.25, 0.25, and 0.25. This equiprobability is chosen as the default experiment.

The second part of the data augmentation consists of adding random gaussian noise, random blur, random colour jittering, random horizontal flip, and random perspective transformations to the pre-transformed images.

3.4 Loss Functions

Since the architecture is multitask, and also multi-resolutions for the bottom-up and detection heads, a loss function must be defined for each task at each resolution.

For the detection, following [21], the focal loss is used to train the generation of heatmaps, and the L1 loss for the regression of the bounding boxes dimensions. The focal loss is defined as follows:

$$L_{center} = \frac{-1}{N} \sum_{xy} \begin{cases} \left(1 - \hat{Y}_{xy}\right)^{\alpha} \log\left(\hat{Y}_{xy}\right) & \text{if } Y_{xy} = 1 \\ \left(1 - Y_{xy}\right)^{\beta} \left(\hat{Y}_{xy}\right)^{\alpha} \log\left(1 - \hat{Y}_{xy}\right) & \text{otherwise} \end{cases} \tag{1}$$

where $\alpha = 2$ and $\beta = 4$ are hyper-parameters, N is the number of subjects, $\hat{Y}_{xy}$ and Y_{xy} are namely the prediction and the ground truth at pixel (x, y). The normalization by N is chosen in order to normalize all positive focal loss instances to 1. For the size loss, given a subject k whose bounding box coordinates are $(x_1^k, y_1^k, x_2^k, y_2^k)$, his center point lies at $p_k = \left(\frac{x_1^k + x_2^k}{2}, \frac{y_1^k + y_2^k}{2}\right)$, and his dimensions are $s_k = \left(x_2^k - x_1^k, y_2^k - y_1^k\right)$. Therefore, the size loss is defined as follows:

$$L_{size} = \frac{1}{N} \sum_{k=1}^{N} \left| \hat{S}\left(p_k\right) - s_k \right| \tag{2}$$

where $\hat{S} \in \mathbb{R}^{w \times h \times 2}$ are the width and height prediction maps of size $w \times h$ for a given resolution. Hence, the detection loss at this resolution is:

$$L_{det} = \lambda_{center} L_{center} + \lambda_{size} L_{size} \tag{3}$$

λ_{center} is set to 1, and λ_{size} to 0.1. Since the model gives predictions at resolutions 1/4 and 1/2, the overall detection loss is therefore:

$$L_{DET} = L_{det-1/4} + L_{det-1/2} \tag{4}$$

For the emotion recognition task, which concerns person-centric, background, and bottom-up heads, a loss similar to [2] is used. It is a multi-label and binary focal loss, which gives better results while dealing with unbalanced data. It is defined as follows:

$$L_{cat-emo} = \frac{-1}{N} \sum_{N} \sum_{i=1}^{E} Y_i \left(1 - \hat{Y}_i\right)^{\alpha} \log\left(\hat{Y}_i\right) + \left(1 - Y_i\right) \left(\hat{Y}_i\right)^{\alpha} \log\left(1 - \hat{Y}_i\right) \tag{5}$$

where N is the number of subjects in the image, E is the number of emotions, $\hat{Y}_i$ and Y_i are namely the prediction and the ground truth for the $i-th$ emotion, and $\alpha = 2$ is a hyper-parameter. For the person-centric and background heads, we have $N = 1$ and directly a predicted array of size $1 \times E$. However, the bottom-up head outputs emotions maps of shape $E \times w \times h$, where w and h depend on the resolution considered (either 1/4 or 1/2). To extract the matrix of $N \times E$ predictions, the N bounding boxes centers given by the ground truth are used. Therefore, the global categorical emotions loss is:

$$L_{CAT} = L_{cat-emo}^{person-centric} + L_{cat-emo}^{background} + L_{cat-emo-1/4}^{bottom-up} + L_{cat-emo-1/2}^{bottom-up} \quad (6)$$

Finally, the total loss is defined as:

$$L_{TOT} = L_{DET} + L_{CAT} \quad (7)$$

4 Experiments and Results

4.1 Training Details

The method is built with the Pytorch toolbox [15]. The models are trained during 250 epochs, using the EMOTIC database. We kept the standard train, validation and test sets provided. When extra data are used, it means that HECO has been merged with the training set of EMOTIC. We use the Adam optimizer [7] with an initial learning rate of $1e^{-3}$. The *best model* is defined as the one with the lowest total validation loss. The *final model* is the one obtained at the end of the 250 epochs.

4.2 Evaluation Metrics

Since our method includes the detection of the subjects, we propose two evaluation metrics. The first one is the standard Average Precision score (AP) for all emotion categories, that can also be averaged (mAP). The predictions are extracted using the ground truth bounding boxes. Considering that this case is independent of the model's detection head, and assuming that all other methods in the literature also use the annotations, this metric can be considered the most appropriate for comparison with the state-of-the-art.

Nevertheless, in real world applications, such annotations are not provided, so we must rely on a detector whose performance can have a significant impact on emotions scores. To know if a detection is successful or not, the commonly used method is to compute the Intersection over Union (IoU) between the detected bounding box and the ground truth, which indicates how much they are superimposed, and count as True Positive the values superior to a given threshold. The IoU values are between 0 and 1, 1 being a perfect detection. In the COCO API, the final detection score is the mean of the AP values obtained with 10 thresholds from 0.5 to 0.95. Therefore, we use this API to evaluate not only our person detector on EMOTIC, but also the performances of the whole framework during autonomous inferences. To do so, the bounding boxes successfully predicted for a given IoU threshold are used to extract the emotion predictions from the emotions maps. Otherwise, when the detection fails for an annotated subject, his predicted emotions are treated as False Negative, i.e. a vector of zeros of shape $1 \times E$ is created. With this new evaluation protocol, the scores obtained are more representative to what could be achieve during real inferences.

4.3 Analysis of the Results

To evaluate the proposed method, both *best model* and *final model* were tested. It appears that 250 epochs are enough to witness over-fitting through the validation loss. However, in some cases the *final model* still give better results. Since only the best results are reported in this paper, we specify if they come from the *final model* by underlining the value in the Tables 1, 2, 3, and 4.

In Table 1, the results for emotion recognition following different training strategies are summarized. These mAP are obtained without considering the detection part, because the annotations were used to extract the emotion predictions, instead of the integrated person detector. As we can see, when the bottom-up head (BU) is trained alone, it does not perform well. However, when the detection head (Det) is added and jointly trained, the score are improved by a good margin. There is also a little improvement by adding person-centric (PC) and background (BG) heads. The use of additional data from HECO (ED) helps to improve even more the performance of the model.

Table 1. Ablation Experiments on EMOTIC Dataset for emotion recognition. Underlined values come from the *final model* instead of the *best model.*

Heads	BU	BU+Det	BU+Det+PC	BU+Det+PC+BG	BU+Det+PC+BG+ED
mAP	23.10	27.22	27.49	27.73	**28.75**

Regarding the detection task, the best score is obtained when all the heads are trained together with extra data, as it is illustrated in Table 2. Yet, the main drawback with EMOTIC database is that it is not fully annotated. Indeed, there are many images with several people but where only a few of them are labeled. Thus, the detector tends to produce many False Positive (as illustrated in Fig. 2), that are penalized during the training and may confuse the model, and also reduce the precision during the evaluations.

The scores presented in Table 3 correspond to the new evaluation protocol which considers the tasks of detection and emotion recognition together, introduced in Sect. 4.2. As expected, the results are worse than those obtained with the ground truth, but surprisingly the model using all heads and giving the best results in both detection and emotion recognition is no longer the best with this new metric. This can be explained by the fact that the latter detects more subjects, even people whose emotions are particularly difficult to assess, for example

Table 2. Ablation Experiments on EMOTIC Dataset for person detection, using the COCO API. Underlined values come from the *final model* instead of the *best model.*

Heads	BU+Det	BU+Det+PC	BU+Det+PC+BG	BU+Det+PC+BG+ED
AP	49.66	49.16	51.49	**51.71**

those who are partially occluded or quite distant in the background. In these situations, the model is more likely to be wrong and produce more False Positives and less True Positives, which decreases its precision. However, using external data still leads to better results.

Table 3. mAP scores for emotion recognition on EMOTIC Dataset, depending on the detector predictions for thresholds from 0.50 to 0.95. (1): BU+Det; (2): BU+Det+PC; (3): BU+Det+PC+BG; (4): BU+Det+PC+BG+ED. Underlined values in the "Average" column indicate that the scores in the whole row come from the *final model* instead of the *best model.*

Det. thr.	*0.50*	*0.55*	*0.60*	*0.65*	*0.70*	*0.75*	*0.80*	*0.85*	*0.90*	*0.95*	**Avg.**
(1)	25.73	25.38	25.04	24.68	24.37	24.00	23.62	22.83	21.71	19.37	23.67
(2)	26.19	25.85	25.51	25.32	25.13	24.81	24.20	23.24	21.91	19.37	24.15
(3)	26.01	25.86	25.57	25.24	24.97	24.58	23.99	23.29	22.00	19.48	24.10
(4)	**26.96**	**26.66**	**26.38**	**26.07**	**25.82**	**25.28**	**24.61**	**23.71**	**22.21**	**19.57**	**24.73**

Even if our framework and our objectives are quite different from the other authors, we finally compared our scores with those of the state-of-the-art in Table 4. The baseline on EMOTIC, provided by [9], is outperformed. Our model is multitask, but possible ways to fuse the predictions of the different heads have not been explored yet. Indeed, the person-centric and background heads are only used to help the model during its training, but not while inferring. Nevertheless, we still tried to average all the outputs, which requires to pre-process the raw image for the person-centric and background heads. It finally appears that the mean between the bottom-up and the person-centric outputs gives the best refined prediction, which means that 2 streams are used here. However, the most recent methods are still quite ahead, due to their rich and complex framework, and a well-made fusion.

Table 4. State-of-the-art on EMOTIC Dataset. NERR: Number of External Resources Required (off-the-shelves models).

Authors	[10]	[9]	[2]	[20]	[17]	[6]	[14]	[19]	**Ours**
nb. of streams	2	2	2	2	3	6	4	7	2
fusion module	✓	✓	✓	✓	✓	✓	✓	✓	✗
NERR	1	0	0	1	1	2	3	3	0
mAP	20.84	27.38	28.33	28.42	30.17	35.16	35.48	**37.73**	29.30

5 Conclusion and Future Work

In this paper, we present a innovative method to simultaneously detect people on an image, and predict their categorical emotions. Since all subjects are treated simultaneously, our approach can be referred as a bottom-up method, and we are the first ones to explore this path. We also introduce a multitask training strategy to improve the performance of the model. Finally, we propose a new evaluation protocol that consider both detection and emotion recognition tasks, in order to better represent the true capabilities of the method during real life inferences. As part of future work, we would also treat continuous emotions (valence, arousal, and dominance), and explore fusion methods to combine the bottom-up predictions with the person-centric and background ones, already available in our multitask model.

Acknowledgements. This work was sponsored by a public grant overseen by the French National Research Agency as part of project muDialBot (ANR-20-CE33-0008-01).

References

1. Barrett, L.F., Mesquita, B., Gendron, M.: Context in emotion perception. Curr. Dir. Psychol. Sci. **20**(5), 286–290 (2011)
2. Bendjoudi, I., Vanderhaegen, F., Hamad, D., Dornaika, F.: Multi-label, multi-task CNN approach for context-based emotion recognition. Inf. Fusion **76**, 422–428 (2021). https://doi.org/10.1016/j.inffus.2020.11.007
3. Cheng, B., Xiao, B., Wang, J., Shi, H., Huang, T.S., Zhang, L.: HigherHRNet: scale-aware representation learning for bottom-up human pose estimation. In: Proceedings of the IEEE/CVF Conference on Computer Vision and Pattern Recognition, pp. 5386–5395 (2020)
4. Ekman, P., Friesen, W.V.: Head and body cues in the judgement of emotion: a reformulation. Percept. Mot. Skills **24**(3), 711–724 (1967)
5. Ekman, P., Friesen, W.V.: Constants across cultures in the face and emotion. J. Pers. Soc. Psychol. **17**, 124–129 (1971). https://doi.org/10.1037/h0030377. Place: US Publisher: American Psychological Association

6. Hoang, M.H., Kim, S.H., Yang, H.J., Lee, G.S.: Context-aware emotion recognition based on visual relationship detection. IEEE Access **9**, 90465–90474 (2021). https://doi.org/10.1109/ACCESS.2021.3091169. Conference Name: IEEE Access
7. Kingma, D.P., Ba, J.: ADAM: a method for stochastic optimization. arXiv preprint arXiv:1412.6980 (2014)
8. Kosti, R., Alvarez, J.M., Recasens, A., Lapedriza, A.: EMOTIC: emotions in context dataset. In: Proceedings of the IEEE Conference on Computer Vision and Pattern Recognition Workshops, pp. 61–69 (2017)
9. Kosti, R., Alvarez, J.M., Recasens, A., Lapedriza, A.: Context based emotion recognition using EMOTIC dataset. IEEE Trans. Pattern Anal. Mach. Intell. **42**(11), 2755–2766 (2020). https://doi.org/10.1109/TPAMI.2019.2916866. conference Name: IEEE Transactions on Pattern Analysis and Machine Intelligence
10. Lee, J., Kim, S., Kim, S., Park, J., Sohn, K.: Context-aware emotion recognition networks. In: Proceedings of the IEEE/CVF International Conference on Computer Vision, pp. 10143–10152 (2019)
11. Li, S., Deng, W.: Deep facial expression recognition: a survey. IEEE Trans. Affect. Comput. **13**(3), 1195–1215 (2022). https://doi.org/10.1109/TAFFC.2020.2981446. Conference Name: IEEE Transactions on Affective Computing
12. Matsumoto, D.: More evidence for the universality of a contempt expression. Motiv. Emot. **16**(4), 363–368 (1992). https://doi.org/10.1007/BF00992972
13. Mehrabian, A.: Framework for a comprehensive description and measurement of emotional states. Genet. Soc. Gen. Psychol. Monogr. **121**, 339–361 (1995). Place: US Publisher: Heldref Publications
14. Mittal, T., Guhan, P., Bhattacharya, U., Chandra, R., Bera, A., Manocha, D.: EmotiCon: context-aware multimodal emotion recognition using Frege's principle. In: Proceedings of the IEEE/CVF Conference on Computer Vision and Pattern Recognition, pp. 14234–14243 (2020)
15. Paszke, A., et al.: PyTorch: an imperative style, high-performance deep learning library. In: Advances in Neural Information Processing Systems, vol. 32 (2019)
16. Wang, J., et al.: Deep high-resolution representation learning for visual recognition. IEEE Trans. Pattern Anal. Mach. Intell. **43**(10), 3349–3364 (2021). https://doi.org/10.1109/TPAMI.2020.2983686. Conference Name: IEEE Transactions on Pattern Analysis and Machine Intelligence
17. Wang, Z., Lao, L., Zhang, X., Li, Y., Zhang, T., Cui, Z.: Context-dependent emotion recognition. SSRN Electron. J. (2022). https://doi.org/10.2139/ssrn.4118383
18. Woo, S., Park, J., Lee, J.-Y., Kweon, I.S.: CBAM: convolutional block attention module. In: Ferrari, V., Hebert, M., Sminchisescu, C., Weiss, Y. (eds.) ECCV 2018. LNCS, vol. 11211, pp. 3–19. Springer, Cham (2018). https://doi.org/10.1007/978-3-030-01234-2_1
19. Yang, D., et al.: Emotion recognition for multiple context awareness. In: Avidan, S., Brostow, G., Cissé, M., Farinella, G.M., Hassner, T. (eds.) ECCV 2022. LNCS, vol. 13697, pp. 144–162. Springer, Cham (2022). https://doi.org/10.1007/978-3-031-19836-6_9
20. Zhang, M., Liang, Y., Ma, H.: Context-aware affective graph reasoning for emotion recognition. In: 2019 IEEE International Conference on Multimedia and Expo (ICME), pp. 151–156 (2019). https://doi.org/10.1109/ICME.2019.00034. ISSN: 1945-788X
21. Zhou, X., Wang, D., Krähenbühl, P.: Objects as points. arXiv preprint arXiv:1904.07850 (2019)

YOLOPoint: Joint Keypoint and Object Detection

Anton Backhaus(✉), Thorsten Luettel, and Hans-Joachim Wuensche

Institute of Autonomous Systems Technology, University of the Bundeswehr Munich, Neubiberg, Germany
anton.backhaus@unibw.de

Abstract. Intelligent vehicles of the future must be capable of understanding and navigating safely through their surroundings. Camera-based vehicle systems can use keypoints as well as objects as low- and high-level landmarks for GNSS-independent SLAM and visual odometry. To this end we propose YOLOPoint, a convolutional neural network model that simultaneously detects keypoints and objects in an image by combining YOLOv5 and SuperPoint to create a single forward-pass network that is both real-time capable and accurate. By using a shared backbone and a light-weight network structure, YOLOPoint is able to perform competitively on both the HPatches and KITTI benchmarks.

Keywords: Deep learning · Keypoint detection · Autonomous driving

1 Introduction

Keypoints are low-level landmarks, typically points, corners, or edges that can easily be retrieved from different viewpoints. This makes it possible for moving vehicles to estimate their position and orientation relative to their surroundings and even perform loop closure (i.e., SLAM) with one or more cameras. Historically, this task was performed with hand-crafted keypoint feature descriptors such as ORB [17], SURF [2], HOG [5], SIFT [14]. However, these are either not real-time capable or perform poorly under disturbances such as illumination changes, motion blur, or they detect keypoints in clusters rather than spread out over the image, making pose estimation less accurate. Learned feature descriptors aim to tackle these problems, often by applying data augmentation in the form of random brightness, blurring and contrast. Furthermore, learned keypoint descriptors have shown to outperform classical descriptors. One such keypoint descriptor is SuperPoint [6], a convolutional neural network (CNN) which we use as a base network to improve upon.

SuperPoint is a multi-task network that jointly predicts keypoints and their respective descriptors in a single forward pass. It does this by sharing the feature outputs of one backbone between a keypoint detector and descriptor head. This makes it computationally efficient and hence ideal for real-time applications.

This research is funded by dtec.bw – Digitalization and Technology Research Center of the Bundeswehr. dtec.bw is funded by the European Union - NextGenerationEU.

J. Blanc-Talon et al. (Eds.): ACIVS 2023, LNCS 14124, pp. 112–123, 2023.
https://doi.org/10.1007/978-3-031-45382-3_10

Fig. 1. Example output of YOLOPointM on a KITTI scene with keypoint tracks from 3 frames and object bounding boxes.

Furthermore, after making various adjustments to the SuperPoint architecture, we fuse it with YOLOv5 [11], a real-time object detection network. The full network thus uses one shared backbone for all three tasks. Moreover, using the YOLOv5 framework we train four models of various sizes: *nano*, *small*, *middle* and *large*. We call the combined network *YOLOPoint*.

Our motivation for fusing the keypoint detector with the object detector is as follows: Firstly, for a more accurate SLAM and a better overview of the surrounding scene, using a suite of multi-directional cameras is beneficial. But processing multiple video streams in parallel means relying on efficient CNNs. In terms of computational efficiency, the feature extraction part does most of the heavy lifting. This is why it has become common practice for multiple different tasks to share a backbone. SuperPoint's architecture is already such that is uses a shared backbone for the keypoint detection and description, hence it was an adequate choice for fusion with another model. Secondly, visual SLAM works on the assumption of a static environment. If keypoints are detected on moving objects, it would lead to localization errors and mapping of unwanted landmarks. However, with classified object bounding boxes we simply filter out all keypoints within a box of a dynamic class (e.g., pedestrian, car, cyclist). Finally, although it seems that keypoint and object detection are too different to jointly learn, they have been used in conjunction in classical methods (e.g., object detection based on keypoint descriptors with a support vector machine classifier [27]) (Fig. 1).

Our contributions can be summarized as follows:

- We propose YOLOPoint, a fast and efficient keypoint detector and descriptor network which is particularly well suited for performing visual odometry.
- We show that object and keypoint detection are not mutually exclusive and propose a network that can do both in a single forward pass.
- We demonstrate the efficacy of using efficient cross stage partial network (CSP) blocks [25] for point description and detection tasks.

Code will be made available at https://github.com/unibwtas/yolopoint.

2 Related Work

Classical keypoint descriptors use hand-crafted algorithms designed to overcome challenges such as scale and illumination variation [2,5,14,17,20,22] and have been thoroughly evaluated [16,21]. Although their main utility was keypoint description, they have also been used in combination with support vector machines to detect objects [27].

Since then, deep learning-based methods have dominated benchmarks for object detection (i.e., object localization and classification) and other computer vision tasks [26–28]. Therefore, research has increasingly been dedicated to finding ways in which they can also be used for point description. Both using CNN-based transformer models COTR [10] and LoFTR [24] achieve state-of-the-art results on multiple image matching benchmarks. COTR finds dense correspondences between two images by concatenating their respective feature maps and processing them together with positional encodings in a transformer module. Their method, however, focuses on matching accuracy rather than speed and performs inference on multiple zooms of the same image. LoFTR has a similar approach, the main difference being their "coarse-to-fine" module that first predicts coarse correspondences, then refines them using a crop from a higher level feature map. Both methods are detector-free, and while achieving excellent results in terms of matching accuracy, neither method is suitable for real-time applications. Methods that yield both keypoint detections and descriptors are generally faster due to matching only a sparse set of points and include R2D2 [19], D2-Net [7] and SuperPoint [6]. R2D2 tackles the matching problems of repetitive textures by introducing a reliability prediction head that indicates the discriminativeness of a descriptor at a given pixel. D2-Net has the unique approach of using the output feature map both for point detection and description, hence sharing all the weights of the network between both tasks. In contrast, SuperPoint has a shared backbone but seperate heads for the detection and description task. What sets it apart from all other projects is its self-supervised training framework. While other authors obtain ground truth point correspondences from depth images gained from structure from motion, i.e., using video, the SuperPoint framework can create labels for single images. It does this by first generating a series of labeled grayscale images depicting random shapes, then training an intermediate model on this synthetic data. The intermediate model subsequently makes point-predictions on a large data set (here: MS COCO [13]) that are refined using their "homographic adaptation" method. The final model is trained on the refined point labels.

While there exist several models that jointly predict keypoints and descriptors, there are to our knowledge none that also detect objects in the same network. Maji et al's work [15] comes closest to ours. They use YOLOv5 to jointly predict keypoints for human pose estimation as well as bounding boxes in a single forward pass. The main differences are that the keypoint detection training uses hand-labelled ground truth points, the object detector is trained on a single class (human), and both tasks rely on similar features.

3 Model Architecture

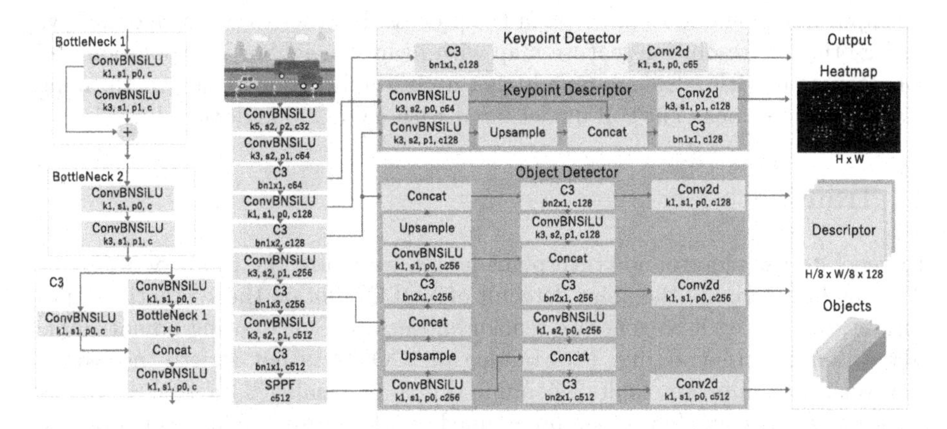

Fig. 2. Full model architecture exemplary for YOLOPointS. The two types of bottlenecks, C3 block (left) and a sequence of convolution, batch normalization and SiLU activation form the main parts of YOLO and by extension YOLOPoint. *k*: kernel size, *s*: stride, *p*: pad, *c*: output channels, *bn*: bottleneck, *SPPF*: fast spacial pyramid pooling [11].

Our keypoint detection model is an adaptation of SuperPoint with CSPDarknet [25] elements. SuperPoint uses a VGG-like encoder [23] that breaks down a grayscale image of size $H \times W \times 1$ into a feature vector of size $H/8 \times W/8 \times 128$. In each decoder head the feature vector is further processed and reshaped to the original image size. The keypoint decoder branch thus produces a heat map representing the pixel-wise probability of "point-ness", which is passed through a non-maximum suppression method that produces the final keypoints. The output of the descriptor branch is a normed descriptor vector linearly up-scaled to $H \times W \times 256$. Both outputs are combined into a list of keypoints and corresponding descriptor vectors, with which keypoints can be matched from frame to frame.

Our version of SuperPoint substitutes the VGG-like backbone with part of CSPDarknet and has a CSP bottleneck with additional convolutions in each head. CSP bottlenecks are the cornerstone of YOLO and provide a good speed to accuracy trade-off for various different computer vision tasks but have so far not been used in keypoint detection for visual landmarks. We keep the same final layers as SuperPoint, i.e., softmax and reshape in the detector head and 2D convolution in the descriptor head. Using YOLOv5's scaling method we create four networks, each varying in width, depth and descriptor length. The descriptors $\mathcal{D}$ are sized at 256, 196, 128 and 64 for models YOLOPointL, -M, -S and -N respectively. A shorter descriptor reduces the computational cost of matching at the cost of accuracy. For comparison: SIFT's descriptor vector has 128 elements and ORB's has only 32. Since reducing the vector size comes at an accuracy trade-off, the smallest vector is left at 64. Furthermore, in order for the descriptor to be able to match and distinguish between other keypoints, it requires a large receptive field [24]. This, however, comes with down-sampling the input image and loosing detail in the process. In order to preserve detail, we enlarge the low-resolution feature map with nearest-neighbor interpolation and concatenate it with a feature map higher up in the backbone before performing further convolutions. The full model fused with YOLOv5 is depicted in Fig. 2.

4 Training

To generate pseudo ground truth point labels, we follow the protocol of SuperPoint by first training the point detector of YOLOPoint on the synthetic shapes dataset, then using it to generate refined outputs on COCO using homographic adaptation for pre-training. Pre-training on COCO is not strictly necessary, however it can improve results when fine-tuning on smaller data sets, as well as reduce training time. Thus, the pre-trained weights are later fine-tuned on the KITTI dataset [8]

For training the full model, pairs of RGB images that are warped by a known homography are each run through a separate forward pass. The model subsequently predicts "point-ness" heat maps, descriptor vectors and object bounding boxes. When training on data sets of variable image size (e.g., MS COCO) the images must be fit to a fixed size in order to be processed as a batch. A common solution is to pad the sides of the image such that $W = H$, also known as letterboxing. However, we found that this causes false positive keypoints to be predicted close to the padding due to the strong contrast between the black padding and the image, negatively impacting training. Therefore, when pre-training on COCO, we use mosaic augmentation, that concatenates four images side-by-side to fill out the entire image canvas, eliminating the need for image padding [4]. All training is done using a batch size of 64 and the Adam optimizer [12] with a learning rate of 10^{-3} for pre-training and 10^{-4} for fine-tuning.

For fine-tuning on KITTI we split the data into 6481 training and 1000 validation images resized to 288×960. To accommodate the new object classes we replace the final object detection layer and train for 20 epochs with all weights frozen except those of the detection layer. Finally, we unfreeze all layers and train the entire network for another 50 epochs.

4.1 Loss Functions

The model outputs for the warped and unwarped image are used to calculate the keypoint detector and descriptor losses. However, only the output of the unwarped images is used for the object detector loss, as to not teach strongly distorted object representations.

The keypoint detector loss $\mathcal{L}_{\text{det}}$ is the mean of the binary cross-entropy losses over all pixels of the heatmaps of the warped and unwarped image of size $H \times W$ and corresponding ground truth labels and can be expressed as follows:

$$\mathcal{L}_{\text{det}} = -\frac{1}{HW}\sum^{H,W}(y_{ij}\cdot\log x_{ij} + (1-y_{ij})\cdot\log(1-x_{ij})) \tag{1}$$

where $y_{ij} \in \{0,1\}$ and $x_{ij} \in [0,1]$ respectively denote the target and prediction at pixel ij.

The original descriptor loss is a contrastive hinge loss applied to all correspondences and non-correspondences of a low-resolution descriptor $\mathcal{D}$ of size $H_c \times W_c$ [6] creating a total of $(H_c \times W_c)^2$ matches. This, however, becomes computationally unfeasible for higher resolution images. Instead, we opt for a sparse version adapted from DeepFEPE [9] that samples N matching pairs of feature vectors d_{ijk} and $d_{i'j'k}$ and M non-matching pairs of a batch of sampled descriptors $\tilde{\mathcal{D}} \subset \mathcal{D}$ and their warped counterpart $\tilde{\mathcal{D}}' \subset \mathcal{D}'$. Using the known homography, each descriptor $\mathbf{d}$ of pixel ij of the kth image of a mini-batch can be mapped to its corresponding warped pair at $i'j'$. m_p furthermore denotes the positive margin of the sampled hinge loss.

$$\mathcal{L}_{\text{desc}} = \mathcal{L}_{\text{corr}} + \mathcal{L}_{\text{n.corr}} \tag{2}$$

where

$$\mathcal{L}_{corr} = \frac{1}{N}\sum_{i,j,k,i',j'}^{\tilde{\mathcal{D}},\tilde{\mathcal{D}}'} \max(0, m_p - \mathbf{d}_{ijk}^T\mathbf{d}'_{i'j'k}) \tag{3}$$

$$\mathcal{L}_{n.corr} = \frac{1}{M}\sum_{i,j,k}^{\tilde{\mathcal{D}}}\sum_{o,p,q}^{\tilde{\mathcal{D}}'}(\mathbf{d}_{ijk}^T\mathbf{d}'_{opq}), \quad (o,p,q) \neq (i',j',k) \tag{4}$$

The object detector loss $\mathcal{L}_{\text{obj}}$ is a linear combination of intermittent losses based on the objectness, class probability and bounding box regression score and is identical to the loss function used in YOLOv5.

The final loss is calculated as the weighted sum of the keypoint detector, descriptor and object loss.

$$\mathcal{L} = \mathcal{L}_{\text{det}} + \mathcal{L}_{\text{det,warp}} + w_{\text{desc}}\cdot\mathcal{L}_{\text{desc}} + w_{\text{obj}}\cdot\mathcal{L}_{\text{obj}} \tag{5}$$

5 Evaluation

In the following sections we present our evaluation results for keypoint detection and description on HPatches [1] and using all three task heads for visual odometry (VO) estimation on the KITTI benchmark. For evaluation on HPatches the models trained for 100 epochs on MS COCO are used, for VO the models are fine-tuned on KITTI data.

Fig. 3. HPatches matches between two images with viewpoint change estimated with YOLOPointS. Matched keypoints are used to estimate the homography matrix describing the viewpoint change.

5.1 Repeatability and Matching on HPatches

The HPatches dataset comprises a total of 116 scenes, each containing 6 images. 57 scenes exhibit large illumination changes and 59 scenes large viewpoint changes. The two main metrics used for evaluating keypoint tasks are repeatability which quantifies the keypoint detector's ability to consistently locate keypoints at the same locations despite illumination and/or viewpoint changes and homography estimation which tests both repeatability and discrimination ability of the detector *and* descriptor. Our evaluation protocols are kept consistent with SuperPoint's where possible.

Repeatability. Repeatability is computed at 256×320 resolution with up to 300 points detected per image using a non-maximum suppression of 8 pixels. A keypoint counts as repeated if it is detected within $\epsilon = 3$ pixels in both frames. The repeatability score determines the ratio of repeated keypoints to overall detected keypoints [6]. Table 1 summarizes the repeatability scores under viewpoint and illumination changes. SuperPoint outperforms YOLOPoint in both illumination and viewpoint changes by a small margin. Surprisingly, all four versions of YOLOPoint perform virtually the same, despite varying strongly in number of parameters. A likely explanation for this is that the keypoint detector branch has more parameters than it actually needs with regard to the simplicity of the task.

Homography Estimation. Homography estimation is computed at 320×480 resolution with up to 1000 points detected per image using a non-maximum suppression of 8 pixels. By using matched points between two frames (cf. Fig. 3) a homography matrix that describes the transformation of points between both frames is estimated. The estimated homography is then used to transform the

Table 1. Repeatability score on HPatches in scenes of strong illumination and viewpoint changes. SuperPoint outperforms YOLOPoint by a small margin in both scenarios. Best results are boldfaced.

	Repeatability	
	Illumination	Viewpoint
YOLOPointL	.590	.540
YOLOPointM	.590	.540
YOLOPointS	.590	.540
YOLOPointN	.589	.529
SuperPoint	**.611**	**.555**

corners of one image to another. A homography counts as correct if the l2 norm of the distance of the corners lies within a margin ϵ [6].

Table 2 shows the results of the homography estimation on the viewpoint variation scenes. Overall, both models are roughly on par, although SuperPoint yields slightly better results within larger error margins ϵ. SuperPoint's descriptors have a significantly higher nearest neighbor mean average precision (NN mAP), indicating good discriminatory ability, however a worse matching score than YOLOPoint. The matching score is a measure for both detector and descriptor, as it measures the ratio of ground truth correspondences over the number of proposed features within a shared viewpoint region [6].

Table 2. Homography estimation on HPatches. YOLOPoint and SuperPoint perform comparably to each other. Additional descriptor metrics are included. Best results are boldfaced.

	Homography estimation			Descriptor metrics	
	$\epsilon = 1$	$\epsilon = 3$	$\epsilon = 5$	NN mAP	Matching Score
YOLOPointL	.390	.739	.841	.650	**.459**
YOLOPointM	.410	.712	.817	.625	.449
YOLOPointS	**.420**	.729	.817	.592	.436
YOLOPointN	.292	.668	.803	.547	.389
SuperPoint	.312	**.742**	**.851**	**.756**	.409

5.2 Visual Odometry Estimation on KITTI

The KITTI odometry benchmark contains 11 image sequences of traffic scenes with publicly available ground truth camera trajectories captured from a moving vehicle from the ego perspective.

Odometry is estimated using only frame-to-frame keypoint tracking (i.e., no loop closure, bundle adjustment, etc.), in order to get an undistorted evaluation of the keypoint detectors/descriptors. In our tests we evaluate different

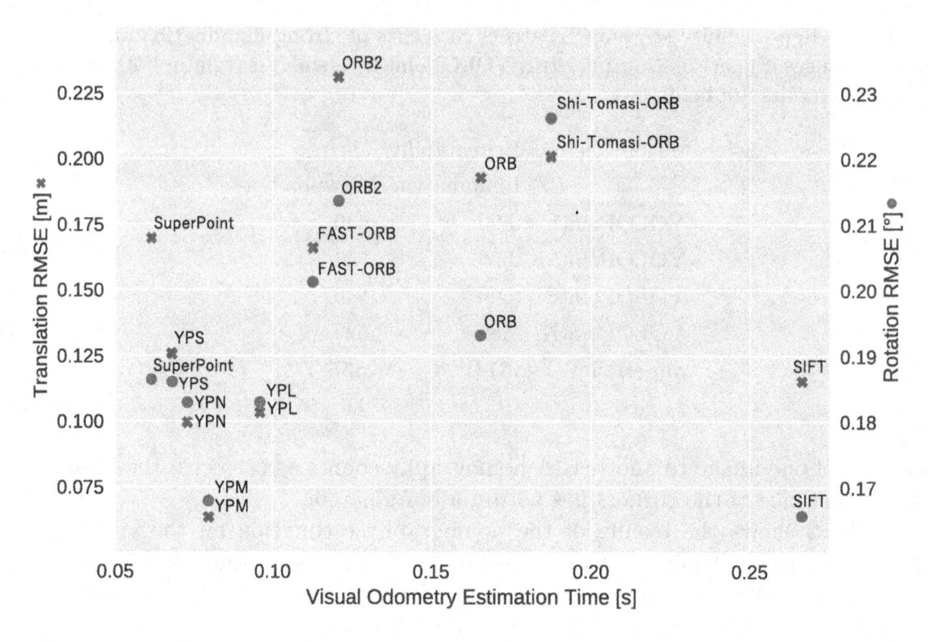

Fig. 4. Translation and rotation RMSE over all KITTI sequences plotted against mean VO estimation time for YOLOPointL (YPL), M, S and N with filtered points as well as SuperPoint and classical methods for comparison (lower left is better). VO estimation was done with 376×1241 images, NVIDIA RTX A4000 and Intel Core i7-11700K.

versions of YOLOPoint with filtering out keypoints on dynamic objects using object bounding boxes and compare them to SuperPoint and other real-time classical methods. Figure 4 plots the RMSE over all sequences against the mean VO estimation time (i.e., keypoint detection, matching and pose estimation) for several models. Despite the overhead of detecting objects, YOLOPoint provides a good speed-to-accuracy trade-off with a full iteration taking less than 100 ms. The pure inference times of YOLOPoint (i.e. without pose estimation) are 49, 36, 27 and 25 ms for L, M, S and N, respectively. Note that the VO time using YOLOPointN is longer despite having a faster inference since its keypoints have more outliers and noisier correspondences. Consequently, the pose estimation algorithm needs more iteration steps to converge, resulting in a slower VO.

The efficacy of filtering out dynamic keypoints can best be demonstrated on sequence 01 as it features a drive on a highway strip where feature points are sparse and repetitive, i.e., hard to match. Consequently, points detected on passing vehicles are no longer outweighed by static points (Fig. 5), causing significant pose estimation error. Figure 6 shows some trajectories including YOLOPointM with and without filtered points. Without filtering, YOLOPoint performs comparatively poorly on this sequence due to it detecting many points on passing vehicles. Finally, although all points on vehicles are removed, regardless of whether or not they are in motion, we do not find that this has a significant

Fig. 5. Sequence 01: Driving next to a car on a highway. Top: All keypoints. Bottom: Keypoints on car removed via its bounding box.

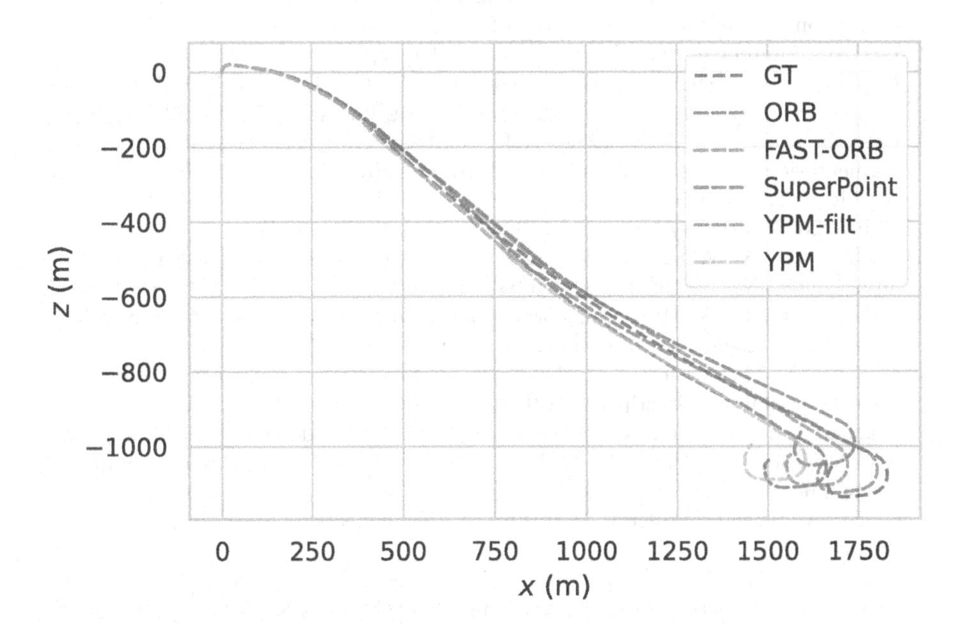

Fig. 6. Sequence 01 trajectories of YOLOPointM (YPM) using filtered and unfiltered points and other keypoint detectors for comparison.

negative impact on pose estimation accuracy in scenes featuring many parked cars.

6 Conclusion and Future Work

In this work, we propose YOLOPoint, a convolutional neural network that jointly predicts keypoints, their respective descriptors and object bounding boxes fast and efficiently in a single forward pass, making it particularly suitable for real-time applications such as autonomous driving. Our tests show that the key-point detector heads that use a Darknet-like architecture perform similarly to SuperPoint in matching and repeatability tasks featuring strong view-point and illumination changes. On the KITTI visual odometry dataset we compare our model's pose estimation performance to SuperPoint and some classical methods. Using predicted bounding boxes to filter out non-static keypoints, YOLOPoint shows the best accuracy to speed trade-off of all tested methods.

Future work will concentrate on incorporating YOLOPoint into our SLAM framework [3] by using keypoints and static objects as landmarks and increasing the robustness of object tracking [18] by matching keypoints.

References

1. Balntas, V., Lenc, K., Vedaldi, A., Mikolajczyk, K.: HPatches: a benchmark and evaluation of handcrafted and learned local descriptors. In: CVPR (2017)
2. Bay, H., Tuytelaars, T., Van Gool, L.: SURF: speeded up robust features. In: Leonardis, A., Bischof, H., Pinz, A. (eds.) ECCV 2006. LNCS, vol. 3951, pp. 404–417. Springer, Heidelberg (2006). https://doi.org/10.1007/11744023_32
3. Beer, L., Luettel, T., Wuensche, H.J.: GenPa-SLAM: using a general panoptic segmentation for a real-time semantic landmark SLAM. In: Proceedings of IEEE Intelligent Transportation Systems Conference (ITSC), pp. 873–879. IEEE, Macau, China (2022). https://doi.org/10.1109/ITSC55140.2022.9921983
4. Bochkovskiy, A., Wang, C., Liao, H.M.: YOLOv4: optimal speed and accuracy of object detection. CoRR abs/2004.10934 (2020)
5. Dalal, N., Triggs, B.: Histograms of oriented gradients for human detection, vol. 1, pp. 886–893 (2005). https://doi.org/10.1109/CVPR.2005.177
6. DeTone, D., Malisiewicz, T., Rabinovich, A.: SuperPoint: self-supervised interest point detection and description (2018). http://arxiv.org/abs/1712.07629
7. Dusmanu, M., et al.: D2-Net: a trainable CNN for joint detection and description of local features. In: Proceedings of the 2019 IEEE/CVF Conference on Computer Vision and Pattern Recognition (2019)
8. Geiger, A., Lenz, P., Stiller, C., Urtasun, R.: Vision meets robotics: the kitti dataset (2013)
9. Jau, Y.Y., Zhu, R., Su, H., Chandraker, M.: Deep keypoint-based camera pose estimation with geometric constraints, pp. 4950–4957 (2020). https://doi.org/10.1109/IROS45743.2020.9341229
10. Jiang, W., Trulls, E., Hosang, J., Tagliasacchi, A., Yi, K.M.: COTR: correspondence transformer for matching across images. CoRR abs/2103.14167 (2021)

11. Jocher, G.: YOLOv5 by Ultralytics: v7.0 (2020). https://doi.org/10.5281/zenodo.7347926. http://github.com/ultralytics/yolov5
12. Kingma, D.P., Ba, J.: ADAM: a method for stochastic optimization. arXiv preprint arXiv:1412.6980 (2014)
13. Lin, T.Y., et al.: Microsoft coco: common objects in context (2014). http://arxiv.org/abs/1405.0312
14. Lowe, D.: Object recognition from local scale-invariant features. In: Proceedings of the Seventh IEEE International Conference on Computer Vision, vol. 2, pp. 1150–1157 (1999). https://doi.org/10.1109/ICCV.1999.790410
15. Maji, D., Nagori, S., Mathew, M., Poddar, D.: Yolo-pose: enhancing yolo for multi person pose estimation using object keypoint similarity loss (2022). https://doi.org/10.48550/ARXIV.2204.06806
16. Mikolajczyk, K., Schmid, C.: A performance evaluation of local descriptors. IEEE Trans. Pattern Anal. Mach. Intell. **27**(10), 1615–1630 (2005)
17. Mur-Artal, R., Montiel, J.M.M., Tardós, J.D.: ORB-SLAM: a versatile and accurate monocular SLAM system. CoRR abs/1502.00956 (2015). http://arxiv.org/abs/1502.00956
18. Reich, A., Wuensche, H.J.: Fast detection of moving traffic participants in LiDAR point clouds by using particles augmented with free space information. In: Proceedings of IEEE/RSJ International Conference on Intelligent Robots and Systems (IROS). IEEE, Kyoto, Japan (2022)
19. Revaud, J., et al.: R2D2: repeatable and reliable detector and descriptor (2019). https://doi.org/10.48550/ARXIV.1906.06195
20. Rosten, E., Drummond, T.: Machine learning for high-speed corner detection. In: Leonardis, A., Bischof, H., Pinz, A. (eds.) ECCV 2006. LNCS, vol. 3951, pp. 430–443. Springer, Heidelberg (2006). https://doi.org/10.1007/11744023_34
21. Schmid, C., Mohr, R., Bauckhage, C.: Evaluation of interest point detectors. Int. J. Comput. Vision **37**(2), 151–172 (2000)
22. Schweitzer, M., Wuensche, H.J.: Efficient keypoint matching for robot vision using GPUs. In: Proceedings of the 12th IEEE International Conference on Computer Vision, 5th IEEE Workshop on Embedded Computer Vision (2009). https://doi.org/10.1109/ICCVW.2009.5457621
23. Simonyan, K., Zisserman, A.: Very deep convolutional networks for large-scale image recognition (2014). https://doi.org/10.48550/ARXIV.1409.1556
24. Sun, J., Shen, Z., Wang, Y., Bao, H., Zhou, X.: LoFTR: detector-free local feature matching with transformers (2021). https://doi.org/10.48550/ARXIV.2104.00680
25. Wang, C.Y., et al.: CSPnet: a new backbone that can enhance learning capability of CNN. In: Proceedings of the IEEE/CVF Conference on Computer Vision and Pattern Recognition Workshops, pp. 390–391 (2020)
26. Wang, M., Leelapatra, W.: A review of object detection based on convolutional neural networks and deep learning. Int. Sci. J. Eng. Technol. (ISJET) **6**(1), 1–7 (2022)
27. Xiao, Y., et al.: A review of object detection based on deep learning. Multimedia Tools Appl. **79**(33), 23729–23791 (2020)
28. Zhao, Z.Q., Zheng, P., Xu, S.T., Wu, X.: Object detection with deep learning: a review. IEEE Trans. Neural Netw. Learn. Syst. **30**(11), 3212–3232 (2019)

Less-than-One Shot 3D Segmentation Hijacking a Pre-trained Space-Time Memory Network

Cyril Li[1(✉)], Christophe Ducottet[1], Sylvain Desroziers[2], and Maxime Moreaud[3]

[1] Université Jean Monnet Saint-Etienne, CNRS, Institut d Optique Graduate School, Laboratoire Hubert Curien UMR 5516, 42023 Saint-Etienne, France
{cyril.li,ducottet}@univ-st-etienne.fr

[2] Manufacture Française des Pneumatiques Michelin, 23 Place des Carmes Déchaux, 63000 Clermont-Ferrand, France
sylvain.desroziers@michelin.com

[3] IFP Energies nouvelles, Rond-point de l'échangeur de Solaize BP 3, 69360 Solaize, France
maxime.moreaud@ifpen.fr

Abstract. In this paper, we propose a semi-supervised setting for semantic segmentation of a whole volume from only a tiny portion of one slice annotated using a memory-aware network pre-trained on video object segmentation without additional fine-tuning. The network is modified to transfer annotations of one partially annotated slice to the whole slice, then to the whole volume. This method discards the need for training the model. Applied to Electron Tomography, where manual annotations are time-consuming, it achieves good segmentation results considering a labeled area of only a few percent of a single slice. The source code is available at https://github.com/licyril1403/hijacked-STM.

Keywords: Deep Neural Network · Electron Tomography · Weakly Annotated Data · Memory Network · Semi-Supervised Segmentation

1 Introduction

Electron Tomography (ET) [7] is a powerful technique to reconstruct 3D nanoscale material microstructure. A Transmission Electron Microscope (TEM) acquires sets of projections from several angles, allowing the reconstruction of 3D volumes. However, the resulting data contain noisy reconstruction artifacts because the number of projections is limited, and their alignment remains a challenging task [25] (Fig. 1). Thus, standard segmentation methods often fail [8], requiring the input of an expert to achieve a good segmentation [9,13,26].

Deep learning (DL) based approaches have achieved excellent results in this area [1,11,14], as advances are made in semantic segmentation of 2D and 3D images [2,5,19,22,24]. Standard approaches rely on training a neural network on fully labeled datasets, which requires many annotated 3D samples. To address

J. Blanc Talon et al. (Eds.): ACIVS 2023, LNCS 14124, pp. 124–135, 2023.
https://doi.org/10.1007/978-3-031-45382-3_11

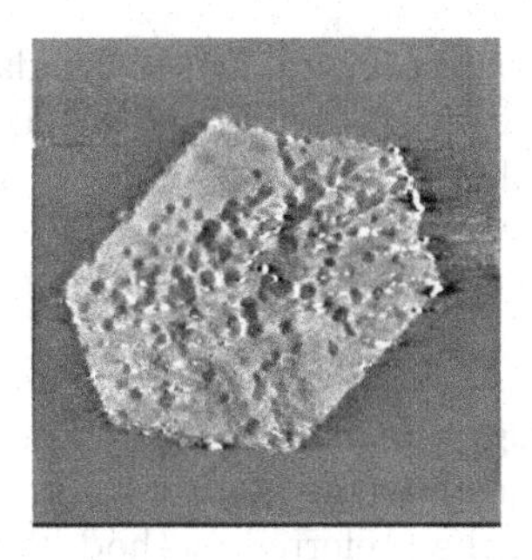

Fig. 1. Reconstruction of a zeolite slice (resolution: 1 nm/pixel) showing many artifacts and noise, making automated segmentation difficult.

the problem of low availability of annotated data, transfer learning [21,28] or semi-supervised learning methods have been proposed with various learning strategies [6,12,15,29]. These methods still require a specific training step for each new image type or object class to segment. Specifically for ET, a new training step will be necessary for every new material and acquisition condition. Moreover, the training process requires some expertise in machine learning to be done properly.

In video object segmentation, a problem close to volume segmentation, significant progress has been made using memory networks [4,20,27]. The system is fed with an annotated query frame and provides a complete segmentation of the corresponding video. The query frame is usually the first frame of the video, manually segmented by an annotator. This setup, similar to one-shot segmentation, has the benefit of being class independent and does not require any learning for new query images. Memory networks encode the annotated frame in the memory and segment the remaining frames using that memory. However, they have two main drawbacks. First, they need various tedious training steps and large datasets for training. Second, they require a whole segmented frame as input.

In this paper, we propose to hijack a Space-Time Memory (STM) network pre-trained for video object segmentation and use it to segment ET volumes with only a few annotated pixels from one slice. The structure of the hijacked network is slightly modified to take only a few pixels of a slice as a query at the inference step and does not require any training. To the best of our knowledge, this is the first time this type of general-purpose pre-trained video object segmentation network has been used to segment ET images.

Our main contributions are:

- A new semi-supervised volume segmentation method reusing a pre-trained object segmentation STM network without additional training.
- A mask-oriented memory readout module to provide a partially segmented query image at the inference step.

- A detailed experimentation on several actual ET data showing that an accurate segmentation is possible with only one slice and a very small portion of annotated pixels in this slice.

2 Related Works

Electron Tomography Segmentation. Segmentation of tomograms remains challenging because of reconstruction artifacts and low signal-to-noise ratio. Manual segmentation is still the preferred method [9] with the support of visualization tools [13] and various image processing methods such as watershed transform [26]. DL-based methods have been applied in electron tomography in more recent work [1,14], with DL models performing better in general semantic segmentation tasks [2,5,19,22,24]. The main bottleneck for ET segmentation tasks is the low availability of labeled training data. Recent works addressed the issue with either a semi-supervised setup with contrastive learning [15] or a scalable DL model, which only requires small- and medium-sized ground-truth datasets [14]. Our method goes further by repurposing a VOS model without any training phase.

Memory Network. Memory networks have an external module that can access past experiences [23]. Usually, an object in the memory can be referred to as a key feature vector and encoded by a value feature vector. Segmentation memory networks such as STM [20], SwiftNet [27], or STCN [4] encode the first video frame, annotated by the user, into the memory component. The next frame (query) is encoded into key and value feature vectors. The query keys and the memory are then compared, resulting in a query value feature vector used to segment the object on that frame. The memory component is then completed with the new key and value. This technique is often used in video segmentation as the object to segment, whose shapes change as time passes, is constantly added to the memory, providing several examples to help segmentation [20]. Unlike our approach, where only a fraction of the frame is needed, these methods require the annotation of the first entire frame.

Video and Volumic Interactive Segmentation. Memory networks are effective but require the segmentation of the entire first frame. An interaction loop can be added where the user is asked to segment the first frame with clicks or scribble to annotate the object of interest to the network [3]. The user corrects the result until they are satisfied. Networks for interactive segmentation are standard semantic segmentation models trained with an image channel, a mask channel, and an interaction channel [17]. By combining the interactive network for the first frame and the memory network to propagate the mask, recent works produce a segmentation mask for the whole video with minimal input from the user [3]. Similar methods have been applied in volumic segmentation [16,30,31]. However, for complex porous networks imaged with ET, standard interactive methods struggle to segment correctly. Adapting an interactive model requires training data composed of many segmented volumes not available for ET. We propose an approach similar to interactive methods using a partially segmented slice. Our method does not require any prior training.

3 Proposed Method

Our method uses the same model to reconstruct the partially annotated frame and the entire volume. The images and masks in the memory are stored as key and value feature vectors. The key encodes a visual representation of the object so that objects with similar keys have similar shapes and textures. The value contains information for the decoder on the segmentation. Our intuition is that if we disable areas containing unlabelled data in the memory, we can encode useful information to segment whole slices even with a small amount of labeled data.

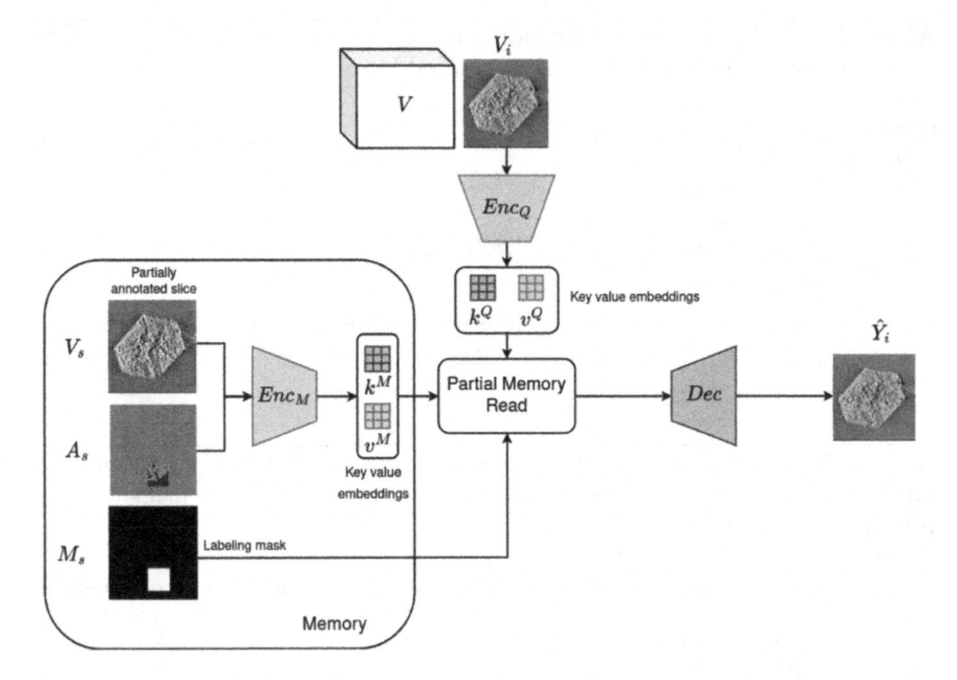

Fig. 2. The partially annotated slice is encoded by the memory encoder into the memory. At inference time, the query slice is encoded by the query encoder into a key and a value and compared with the memory keys and values. During the memory reading, the labeling mask selects only keys from labeled pixels, indicating whether a pixel is annotated. The result is given to the decoder, which reconstructs the whole segmentation.

In our framework, we ask an expert to annotate a small part of a slice A_s of the volume V to get the segmentation $\hat{Y}$ of the entire volume. From the annotations given by the expert, a labeling mask M_s is built where labeled and unlabeled pixels are denoted. The image and the annotations are encoded to one key and one value $\{k^M, v^M\}$ stored in the memory. There are two ways to propagate the annotation into the entire volume. In the first one, the memory is directly used to segment other slices (Algorithm 1 and Fig. 2). On the other hand, we pass the same image V_s as a query into the network to get pseudo-labels of the entire slice $\hat{Y}_s$. The entire slice and the newly acquired pseudo-label mask are then encoded into a key and a value $\{k^S, v^S\}$ to segment other slices $V_i, i \in [1, N]$ with N the number of slices in the volume (Algorithm 2).

During the memory read, the key in memory is modified to mask unknown zones to get the value to segment the whole slice. We use an STM network [20] as the backbone of our method, as it was the first network to use memory networks for 2D semantic segmentation. Moreover, as other methods in the field are based on the STM network, our approach is generalizable to other networks.

Algorithm 1. Procedure for segmenting the whole volume with a portion of a single slice annotated, with only the annotated bit in the memory.

Input: V, s, A_s, M_s, N
Output: $\hat{Y}$
  $\{k^M, v^M\} \leftarrow Enc_M(V_s, A_s)$
  while $i \in \{1, ..., N\}$ **do**
    $\{k^Q, v^Q\} \leftarrow Enc_Q(V_i)$
    $f \leftarrow PartialMemoryRead(M_s, k^M, k^Q, v^M, v^Q)$
    $\hat{Y}_i \leftarrow Decoder(V_i, f)$
  end while

Algorithm 2. Procedure for segmenting the whole volume with a portion of a single slice annotated, with pseudo-labels of the entire slice in the memory.

Input: V, s, A_s, M_s, N
Output: $\hat{Y}$
  $\{k^M, v^M\} \leftarrow Enc_M(V_s, A_s)$ ▷ First slice reconstruction
  $\{k^Q, v^Q\} \leftarrow Enc_Q(V_s)$
  $f \leftarrow PartialMemoryRead(M_s, k^M, k^Q, v^M, v^Q)$
  $\hat{Y}_s \leftarrow Decoder(V_s, f)$
  $\{k^S, v^S\} \leftarrow Enc_M(V_s, \hat{Y}_s)$
  while $i \in \{1, ..., s-1\} \cup \{s+1, ..., N\}$ **do** ▷ Whole volume propation
    $\{k^Q, v^Q\} \leftarrow Enc_Q(V_i)$
    $f \leftarrow MemoryRead(k^S, k^Q, v^S, v^Q)$
    $\hat{Y}_i \leftarrow Decoder(V_i, f)$
  end while

3.1 Key Value Embedding

The memory and the query encodings are slightly different. We consider, for a set of image $I \in \mathbb{R}^{H\times W}$ and its annotation $A \in \mathbb{R}^{H\times W}$, the memory encoder Enc_M composed of a backbone network and followed by two parallel convolutional layers, outputting a memory key $k^M \in \mathbb{R}^{H/8\times W/8\times C/8}$ and a memory value $v^M \in \mathbb{R}^{H/8\times W/8\times C/2}$ such as:

$$Enc_M(I, A) = \{k^M, v^M\} \tag{1}$$

where W and H are the image size and C is the number of dimensions of the feature vectors at the output of the backbone network.

The query encoder shares the same architecture with different weights, but since the mask of the image is not available, only the slice passes through the query encoder Enc_Q:

$$Enc_Q(I) = \{k^Q, v^Q\} \tag{2}$$

with $k^Q \in \mathbb{R}^{H/8\times W/8\times C/8}$ the query key and $v^Q \in \mathbb{R}^{H/8\times W/8\times C/2}$ the query value.

3.2 Partial Memory Read

The standard method for memory reading from the STM network is modified to account for partially annotated slices. Let $M \in \mathbb{R}^{H\times W}$ be the annotation mask:

$$M_{i,j} = \begin{cases} 0 & \text{if } A_{i,j} \text{ is unannotated} \\ 1 & \text{if } A_{i,j} \text{ is annotated} \end{cases} \tag{3}$$

The mask M is then downsampled to be applied directly to the memory key:

$$M^D = Downsample(M, 8) \in \mathbb{R}^{H/8\times W/8} \tag{4}$$

A bilinear interpolation is used, which smoothes the boundaries. Next, the similarity map $S \in \mathbb{R}^{HW/16\times HW/16}$ is computed between a reshaped memory $k_r^M \in \mathbb{R}^{HW/16\times C/8}$ and a query $k_r^Q \in \mathbb{R}^{HW/16\times C/8}$ keys:

$$S = k_r^Q \times k_r^{M^T} \tag{5}$$

where $\times$ is the matrix product. A softmax is then applied to S. However, we mask S with M^D to cancel the unannotated areas' memory key's contribution k^M. Since S is the matrix product of k^Q and k^M, to properly mask k^M, M^D is reshaped into one dimension $M^L \in \mathbb{R}^{HW/16}$. We then multiply each row of S with M^L:

$$R_{i,j} = \frac{\exp(S_{i,j}) M_i^L}{\sum_{k,l} \exp(S_{k,l}) M_k^L} \tag{6}$$

The resulting matrix $R \in \mathbb{R}^{HW/16 \times HW/16}$ is the similarity between each zone of k^M and k^Q without the contribution of unannotated zones. The segmentation key $f \in \mathbb{R}^{H/8 \times W/8 \times C}$ is obtained by concatenating the query value and the memory value, weighted by R.

$$f = [v^Q, R \times v^M] \tag{7}$$

where $\times$ denotes the matrix product.

4 Experiments and Results

4.1 Implementation Details

We use the STM network architecture proposed in [20] as well as the weights proposed by the authors. The network's backbone is a ResNet, trained for a video segmentation task with Youtube-VOS and DAVIS as training datasets. The decoder outputs a probability map that is 1/4 of the initial input size, which degrades the results for ET where fine details on porous areas are needed. An upsampling operation is performed on the input slice before entering the network. The input slice is upscaled two times as a compromise between memory consumption and finer details.

All the results are computed with Intersection Over Union (IOU) on the entire volume V:

$$IOU(V) = \frac{\sum_{i=1}^{N} \hat{Y}_i \cap Y_i}{\sum_{i=1}^{N} \hat{Y}_i \cup Y_i} \tag{8}$$

with $\hat{Y}$ the segmented volume and Y the ground truth. The closer the IOU is to 1, the better the segmentation is.

4.2 Data

Chemical processes in the energy field often require using zeolites [10]. However, the numerous nanometric scale cavities make zeolites complex to segment. We evaluated our method on three volumes of hierarchical zeolites, NaX Siliporite G5 from Ceca-Arkema [18]. Volumes' sizes are $592 \times 600 \times 623$, $512 \times 512 \times 52$, and $520 \times 512 \times 24$.

The slices are automatically partially annotated to simulate real-world data. A rectangle window of the area A_w is considered labeled. The remaining slice is unlabelled. The center of the window is drawn randomly on the border between the object and the background (Fig. 3). The window is adjusted to fit entirely on the screen while maintaining its area. We define the labeling rate as $r = \frac{A_w}{H \times W}$.

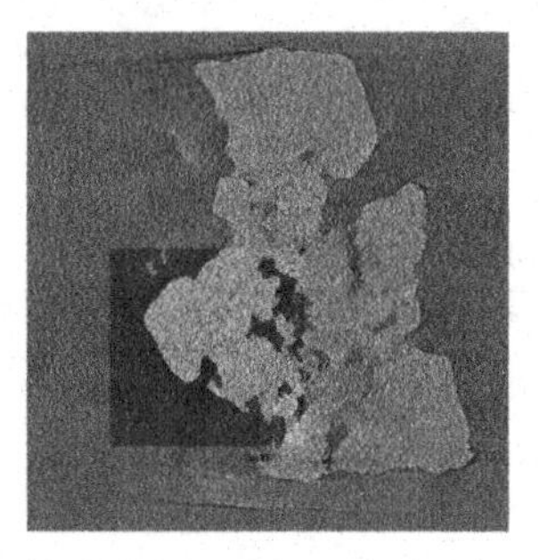

Fig. 3. A partially annotated slice of a zeolite. A window of an area A_w is considered to be annotated, while the pixel label on the outside of the window is unknown. The center of the window is randomly selected near the border between the object and the background to include pixels from both classes.

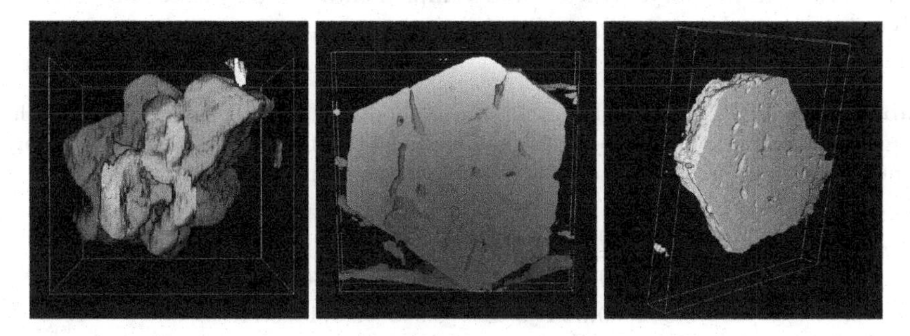

Fig. 4. 3D visualization of segmented volumes of hierarchical zeolites NaX Siliporite G5. A random window of 6% of one slice has been annotated. Segmentations are provided using our approach (Algorithm 1).

4.3 Results

For each volume, we run each experiment on the same five randomly selected slices. The mean IOU of the three volumes is reported.

Comparaison with the STM Network. We first compare our method with an unmodified STM network for several labeling rates r. We give the same partially annotated slices for the STM network and our method. The results are shown in Table 1. The STM network can not process the partially labeled slice because it was not intended to deal with such data. As a result, there is no way for the STM network to differentiate labeled and unlabeled pixels, which leads to poor segmentation. Our key masking allows the STM network to achieve significantly better results with accurate segmentation (Fig. 4).

First Slice Propagation. We then study the different approaches for the first slice. We tested our method with only the annotated parts in the memory shown

Table 1. Mean IOU on our volumes for our method and an unmodified STM network. The modification from our approach allows an STM-like model to produce good segmentation.

r	0.02	0.06	0.12	0.18	0.25
Ours	**0.551**	**0.625**	**0.693**	**0.725**	**0.758**
STM	0.013	0.024	0.070	0.150	0.285

(Ours) in the Algorithm 1 against our method with the pseudo-labels of the entire first slice in the memory (Ours+F) described by the Algorithm 2. The results in Table 2 demonstrate that better results are obtained with only the annotated bit in the memory. The STM network performs better with accurate data instead of more variety in memory. Our method's implementation only uses the partially labeled parts in the memory.

Table 2. Mean IOU on our volumes for our method with only the labeled parts in the memory (Ours) and our method with the pseudo-labels of the first slice in the memory (Ours+F). Our approach uses only the annotated zones in the memory.

r	0.02	0.06	0.12	0.18	0.25
Ours	0.550	**0.625**	**0.693**	**0.725**	**0.758**
Ours+F	**0.564**	0.588	0.650	0.671	0.710

Comparaison with Other Methods. Finally, we compare our methods with other approaches that can handle partially labeled slices [15]. We study the performance of our method against a UNET with a weighted cross entropy to train only on labeled zones and a contrastive UNET that uses contrastive learning to exploit unlabeled areas. Both methods require a training phase. The results are shown in Table 3. Our method performs better than a standard UNET but still lags behind the contrastive UNET. Nevertheless, our method shows promising results with scores close to methods with a training procedure. Figure 5 shows all the results from the experiments previously mentioned.

Table 3. Mean IOU on our volumes for our method, a UNET adapted for partially segmented areas, and a UNET using a contrastive loss to exploit both labeled and unlabeled zones. Our proposed method achieves results close to these methods despite no training phase.

r	0.02	0.06	0.12	0.18	0.25
Ours	0.551	0.625	0.693	0.725	0.758
UNET	0.544	0.600	0.671	0.695	0.839
Contrastive UNET	0.737	0.768	0.793	0.813	0.815

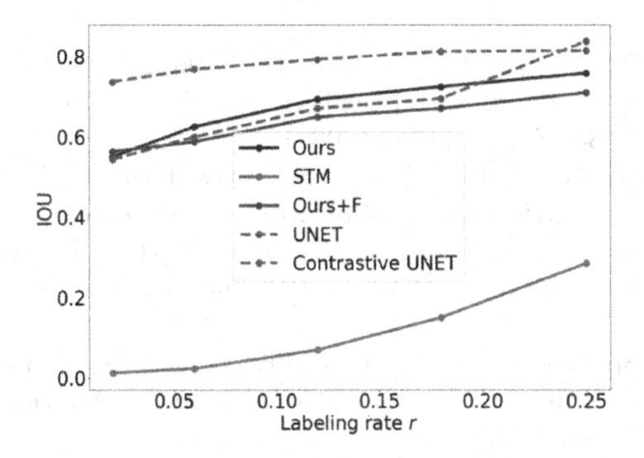

Fig. 5. Mean IOU for several labeling rates r. All methods do not need an additional training procedure except UNET and contrastive UNET [15].

5 Conclusion

In this paper, we illustrate that a slightly modified STM network handles accurate volumetric segmentation of 3D scans from ET with only a tiny portion of one slice labeled needed without any further fine-tuning. This approach achieves results close to methods that require a training procedure. The masking of the memory shows that semi-labeled slices can be used to propagate accurate segmentation in fields where annotated data are not widely available. A more detailed segmentation mask can be obtained with further investigations, as the original STM network output size is 1/4 of the original size.

Acknowledgments. This work was supported by the LABEX MILYON (ANR-10-LABX-0070) of Université de Lyon, within the program "Investissements d'Avenir" (ANR-11-IDEX- 0007) operated by the French National Research Agency (ANR).

References

1. Akers, S., et al.: Rapid and flexible segmentation of electron microscopy data using few-shot machine learning. NPJ Computat. Mater. **7**(1), 1–9 (2021)
2. Chen, L.C., Papandreou, G., Kokkinos, I., Murphy, K., Yuille, A.L.: DeepLab: semantic image segmentation with deep convolutional nets, Atrous convolution, and fully connected CRFs. IEEE Trans. Pattern Anal. Mach. Intell. **40**(4), 834–848 (2018)
3. Cheng, H.K., Tai, Y.W., Tang, C.K.: Modular interactive video object segmentation: interaction-to-mask, propagation and difference-aware fusion. In: Proceedings of the IEEE/CVF Conference on Computer Vision and Pattern Recognition, pp. 5559–5568 (2021)
4. Cheng, H.K., Tai, Y.W., Tang, C.K.: Rethinking space-time networks with improved memory coverage for efficient video object segmentation. Adv. Neural. Inf. Process. Syst. **34**, 11781–11794 (2021)
5. Çiçek, Ö., Abdulkadir, A., Lienkamp, S.S., Brox, T., Ronneberger, O.: 3D U-net: learning dense volumetric segmentation from sparse annotation. In: Ourselin, S., Joskowicz, L., Sabuncu, M.R., Unal, G., Wells, W. (eds.) MICCAI 2016. LNCS, vol. 9901, pp. 424–432. Springer, Cham (2016). https://doi.org/10.1007/978-3-319-46723-8_49
6. Dosovitskiy, A., Springenberg, J.T., Riedmiller, M., Brox, T.: Discriminative unsupervised feature learning with convolutional neural networks. Adv. Neural Inf. Process. Syst. **27** (2014)
7. Ersen, O., et al.: 3D-TEM characterization of nanometric objects. Solid State Sci. **9**(12), 1088–1098 (2007)
8. Evin, B., et al.: 3D analysis of helium-3 nanobubbles in palladium aged under tritium by electron tomography. J. Phys. Chem. C **125**(46), 25404–25409 (2021)
9. Fernandez, J.J.: Computational methods for electron tomography. Micron **43**(10), 1010–1030 (2012)
10. Flores, C., et al.: Versatile roles of metal species in carbon nanotube templates for the synthesis of metal-zeolite nanocomposite catalysts. ACS Appl. Nano Mater. **2**(7), 4507–4517 (2019)
11. Genc, A., Kovarik, L., Fraser, H.L.: A deep learning approach for semantic segmentation of unbalanced data in electron tomography of catalytic materials. arXiv preprint arXiv:2201.07342 (2022)
12. Hadsell, R., Chopra, S., LeCun, Y.: Dimensionality reduction by learning an invariant mapping. In: 2006 IEEE Conference on Computer Vision and Pattern Recognition (CVPR 2006), vol. 2, pp. 1735–1742 (2006)
13. He, W., Ladinsky, M.S., Huey-Tubman, K.E., Jensen, G.J., McIntosh, J.R., Björkman, P.J.: FcRn-mediated antibody transport across epithelial cells revealed by electron tomography. Nature **455**(7212), 542–546 (2008)
14. Khadangi, A., Boudier, T., Rajagopal, V.: EM-net: deep learning for electron microscopy image segmentation. In: 2020 25th International Conference on Pattern Recognition (ICPR), pp. 31–38 (2021)
15. Li, C., Ducottet, C., Desroziers, S., Moreaud, M.: Toward few pixel annotations for 3D segmentation of material from electron tomography. In: International Conference on Computer Vision Theory and Applications, VISAPP 2023 (2023)
16. Liu, Q., Xu, Z., Jiao, Y., Niethammer, M.: iSegFormer: interactive segmentation via transformers with application to 3D knee MR images. In: Wang, L., Dou, Q., Fletcher, P.T., Speidel, S., Li, S. (eds.) MICCAI 2022. LNCS, vol. 13435, pp. 464–474. Springer, Cham (2022). https://doi.org/10.1007/978-3-031-16443-9_45

17. Mahadevan, S., Voigtlaender, P., Leibe, B.: Iteratively trained interactive segmentation. In: British Machine Vision Conference (BMVC) (2018)
18. Medeiros-Costa, I.C., Laroche, C., Pérez-Pellitero, J., Coasne, B.: Characterization of hierarchical zeolites: combining adsorption/intrusion, electron microscopy, diffraction and spectroscopic techniques. Microporous Mesoporous Mater. **287**, 167–176 (2019)
19. Milletari, F., Navab, N., Ahmadi, S.A.: V-net: fully convolutional neural networks for volumetric medical image segmentation. In: 2016 Fourth International Conference on 3D Vision (3DV), pp. 565–571 (2016)
20. Oh, S.W., Lee, J.Y., Xu, N., Kim, S.J.: Video object segmentation using space-time memory networks. In: Proceedings of the IEEE/CVF International Conference on Computer Vision, pp. 9226–9235 (2019)
21. Pan, S.J., Yang, Q.: A survey on transfer learning. IEEE Trans. Knowl. Data Eng. **22**(10), 1345–1359 (2009)
22. Ronneberger, O., Fischer, P., Brox, T.: U-net: convolutional networks for biomedical image segmentation. In: Navab, N., Hornegger, J., Wells, W.M., Frangi, A.F. (eds.) MICCAI 2015. LNCS, vol. 9351, pp. 234–241. Springer, Cham (2015). https://doi.org/10.1007/978-3-319-24574-4_28
23. Sukhbaatar, S., Weston, J., Fergus, R., et al.: End-to-end memory networks. Adv. Neural Inf. Process. Syst. **28** (2015)
24. Sun, K., Xiao, B., Liu, D., Wang, J.: Deep high-resolution representation learning for human pose estimation. In: Proceedings of the IEEE/CVF Conference on Computer Vision and Pattern Recognition, pp. 5693–5703 (2019)
25. Tran, V.D., Moreaud, M., Thiébaut, É., Denis, L., Becker, J.M.: Inverse problem approach for the alignment of electron tomographic series. Oil Gas Sci. Technol.-Rev. d'IFP Energies Nouvelles **69**(2), 279–291 (2014)
26. Volkmann, N.: Methods for segmentation and interpretation of electron tomographic reconstructions. Methods Enzymol. **483**, 31–46 (2010)
27. Wang, H., Jiang, X., Ren, H., Hu, Y., Bai, S.: SwiftNet: real-time video object segmentation. In: Proceedings of the IEEE/CVF Conference on Computer Vision and Pattern Recognition, pp. 1296–1305 (2021)
28. Wurm, M., Stark, T., Zhu, X.X., Weigand, M., Taubenböck, H.: Semantic segmentation of slums in satellite images using transfer learning on fully convolutional neural networks. ISPRS J. Photogramm. Remote. Sens. **150**, 59–69 (2019)
29. Zhao, X., et al.: Contrastive learning for label efficient semantic segmentation. In: Proceedings of the IEEE/CVF International Conference on Computer Vision, pp. 10623–10633 (2021)
30. Zhou, T., Li, L., Bredell, G., Li, J., Konukoglu, E.: Quality-aware memory network for interactive volumetric image segmentation. In: de Bruijne, M., et al. (eds.) MICCAI 2021. LNCS, vol. 12902, pp. 560–570. Springer, Cham (2021). https://doi.org/10.1007/978-3-030-87196-3_52
31. Zhou, T., Li, L., Bredell, G., Li, J., Unkelbach, J., Konukoglu, E.: Volumetric memory network for interactive medical image segmentation. Med. Image Anal. **83**, 102599 (2023)

Segmentation of Range-Azimuth Maps of FMCW Radars with a Deep Convolutional Neural Network

Pieter Meiresone(✉), David Van Hamme, and Wilfried Philips

Ghent University, IPI-imec, 9000 Gent, Belgium
{pieter.meiresone,david.vanhamme,wilfried.philips}@ugent.be

Abstract. In this paper, we propose a novel deep convolutional neural network for the segmentation of Range-Azimuth (RA) maps produced by a low-power Frequency Modulated Continuous Wave (FMCW) radar to facilitate autonomous operation on a moving platform in a dense, feature-rich environment such as warehouses, storage and industrial sites. Our key contribution is a compact neural network architecture that estimates the extent of free space with high distance and azimuth resolution, yielding high-quality navigation cues at a much lower computational cost than full 2D space segmentation techniques. We demonstrate our method on a unmanned aerial vehicle (UAV) in an industrial warehouse environment.

Keywords: deep learning · FMCW radar · segmentation · U-Net

1 Introduction

Recent developments in low cost frequency modulated continuous wave (FMCW) mmWave radar sensors have opened up a new field of applications thanks to its small packaging and low power requirements. Thanks to those new properties, this type of sensor can now be mounted on a small sized unmanned aerial vehicle (UAV), commonly known as a drone, to be used for navigation and detection purposes. Classical signal processing techniques can be applied for navigation and detection applications, however those quickly come short in challenging environments [1]. In particular in indoor environments multipath propagation of radar signals creates a lot of radar clutter. Constant False Alarm Rate (CFAR) detectors often fail to correctly interpret the structure of such environments and point detections further make the path planning process more difficult as the precise spatial extent of an obstacle is lost in the processing.

In order to keep radar sensing at the forefront of technology, the radar community has made a push into deep learning methods [3]. A deep Convolutional Neural Network (CNN) is more suitable to more generally interpret and extract features from the environment. A CNN acts as an approximation component

This research received funding from the Flemish Government under the "Onderzoeksprogramma Artificiele Intelligentie (AI) Vlaanderen" program.

J. Blanc-Talon et al. (Eds.): ACIVS 2023, LNCS 14124, pp. 136–147, 2023.
https://doi.org/10.1007/978-3-031-45382-3_12

which is capable of learning a complex function given enough training data. To the authors knowledge, no prior research has been done for the segmentation of range-azimuth maps from a UAV in indoor environment for a single-chip FMCW radar. In this paper, we propose to create a neural network to segment range-azimuth maps by optimizing the existing neural network segmentation approach and adapting it to be more suitable to process range-azimuth maps in their natural sensor domain. We empirically show that our method achieves better segmentation performance while being more compact in parameters and memory requirements.

Our paper is further structured as follows: in Sect. 2, we discuss currently related deep learning methods for radar processing. In Sect. 3, we present our approach. We discuss the evaluation methodology and results in Sect. 4. Finally, in Sect. 5, we conclude our work and discuss interesting paths for further research.

2 Related Work

As mentioned in the previous section, classical methods such as CFAR quickly fall short in indoor environments to provide a complete picture of the surrounding obstacles, notably suffering from ghost detections caused by multipath reflection. The authors in [10] developed a low complexity obstacle detector that outperforms OS-CFAR specifically for indoor obstacle avoidance. However, the output of the method is insufficient for navigation, which requires either a dense spatial map or at the very least estimated object sizes in order to segment the scene into occupied and free space zones.

When looking at classical signal processing techniques to segment radar images we find that typically high-resolution radars are used. In [11], the authors propose a novel radar segmentation method using the watershed transform and statistical parameters of each radar image region. First, the radar image is pre-segmented using the watershed transform. Afterwards, each segmented region is classified by its statistical characteristics. Due to the usage of detailed radar features, this kind of segmentation only works on high resolution radar images.

Looking further to already applied deep learning methods for radar sensing, we find that the usage of convolutional neural networks is not new. In [7], the authors use a Convolutional Autoencoder to denoise range-doppler maps. The authors use parallel networks to do majority voting, as well as processing the current and multiple previous range-doppler maps simultaneously. In [5], the authors present a weakly-supervised multi-class semantic segmentation neural network for FMCW radar training on the Oxford radar dataset. They present a weakly-supervised training procedure based on a RGB and lidar sensor. The main drawback of this method is that a lidar sensor is required for labelling and that no evaluation is provided on indoor environments which are typically more cluttered. In [8], a novel architecture is used to directly process complex range-doppler signals instead of fully computed radar cubes. The architecture can be used for different tasks such as object detection and free space segmentation. These studies highlight the effectiveness of encoding-decoding networks in interpreting complex environments, although they use high-resolution radar data.

In [9], a convolutional network for biomedical image segmentation, U-Net, is proposed. The authors propose to use a contracting path and a expanding path with a bottleneck and skip-connections in between to capture features efficiently from biomedical images. The similarity between noise characteristics of biomedical and radar images allows to use similar techniques to process radar images.

In [4] authors use a integrated single-chip FMCW radar in a multi-resolution segmentation CNN to produce a probability of occupancy grid for the detection of Vulnerable Road Users (VRU). The network is based on a U-Net architecture, but modified by using 3D convolutions to process entire radar cubes. This allows the network to learn the micro-doppler characteristics of a VRU. While this network achieves high spatial resolution, its reliance on the doppler dimension makes the approach unsuitable for resolving static objects essential for navigation. The authors in [12] use the same single-chip FMCW radar and an encoder network to extract a latent feature space from an indoor warehouse environment, which is then used for recognizing its position relative to a prior experience map. These works show that encoding-decoding neural networks can deliver significant performance benefits for detection tasks using single-chip FMCW radar, successfully extracting useful information from cluttered radar scenes and overcoming the lack of resolution for objects of interest. However, these methods fall short of producing a complete segmentation of the scene into free and occupied regions, and therefore do not enable safe navigation in dense environments.

3 Proposed Method

The goal is to train a radar neural network for free space detection in the ground plane using manually annotated free space maps. Note that reconstructing a complete map of the environment is generally neither possible nor required because of scene self-occlusion: mmwave radar has only limited penetration depth for most commonly encountered materials and hence cannot look far beyond the first obstacle encountered by the wavefront, but these areas behind an obstacle are also unreachable, so irrelevant for navigation. Therefore, we will train our free space estimation network to predict per azimuth bin the distance to the nearest obstacle. More formally, our radar CNN $f_{CNN}()$ computes a one-dimensional vector $\hat{\mathbf{Y}}$ with $\hat{y}_i \in [0, 1]$ representing the estimated normalized distance to the closest obstacle in the ith azimuth bin. A value of 0 represents no free space at all, and a value of 1 represents no obstacles within the maximum range of the radar. This estimation vector is then compared to a ground truth vector $\mathbf{Y}$ obtained from manually annotated free space maps, normalized similarly on the maximum radar range.

3.1 Sensorial System Setup

Recording and processing of the radar data on the UAV happens in the local coordinate system. In Fig. 1 a side view of our UAV is displayed. While only the pitch angle θ is shown, roll and yaw are also present during recording. Input and output of the neural network is always relative to the local coordinate system $x'y'z'$. Since we are not building a map of the environment, we don't have to take this variation into account and simplifies our processing significantly.

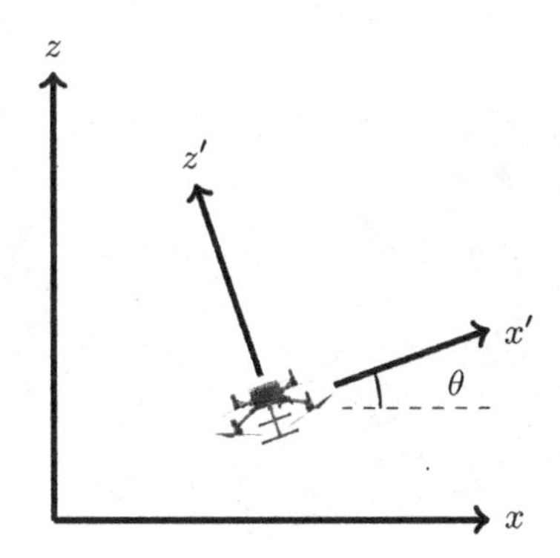

Fig. 1. Side view of sensorial system setup and its coordinate system. The UAV is displayed with pitch angle θ. Our segmentation, including labelling, is done in local coordinate system $x'y'z'$.

3.2 Pre-processing

Our capture platform saves the ADC samples from the radar. The 3D radar cube is subsequently constructed in software. This gives us increased flexibility in the exploration of our algorithms. To improve the poor azimuth resolution of our radar, we opted to use the CAPON [2] beamforming algorithm. After beamforming, we obtain a range-azimuth map with increased angular resolution that will be processed by our neural network. While CAPON represents a not-negligible computation cost in our pipeline, we expect natural evolutions in commercial FMCW radars (i.e., larger antenna arrays) will realize similar resolution at no cost in the near future.

3.3 Radar CNN Architecture

Our range-azimuth maps are fed into a contracting head that consists of multiple layers of convolutions and max-pooling operators. The contracting head maps the range azimuth images to a more dense feature space. Each contracting block consists of a `Conv2D`, `BatchNorm2d` and `LeakyReLu` layer, repeated twice. A contracting blocks reduces the spatial dimension of the feature map by a factor of 2 through max-pooling.

While a typical U-Net would have a similar expanding path to predict a segmentation result at the original 2D resolution, our problem formulation removes the need to scale up to the input resolution. Instead we feed the feature vector into two fully connected (FC) layers that allow for prediction of the nearest obstacle per azimuth bin. The final layer is a sigmoid activation which outputs a normalized distance between 0 and 1. A detailed overview of the different layers of the neural network is shown in Fig. 2.

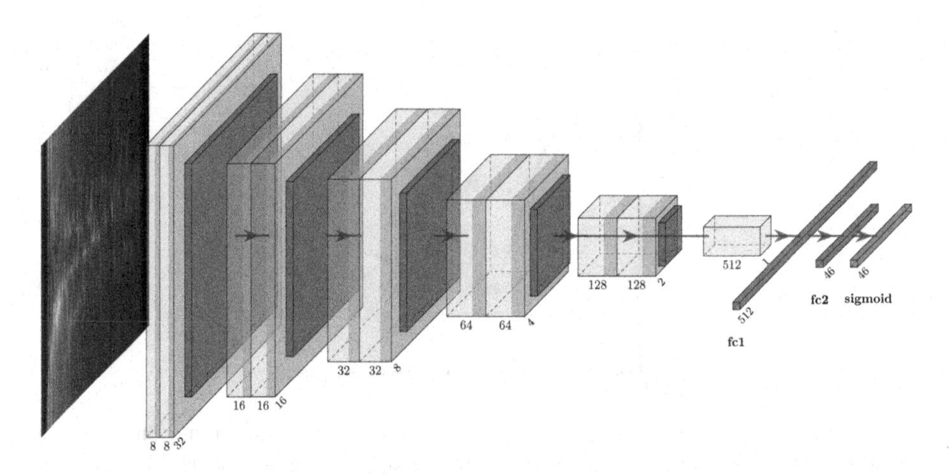

Fig. 2. Structure of the proposed segmentation network. Range-azimuth maps are passed into multiple contracting blocks that are then passed into two fully connected layers. Finally, per azimuth bin the distance to the nearest obstacle is predicted. The output is compared with the manually annotated labels and backpropagated through the network.

3.4 Loss Function and Regularization

Our segmentation problem is defined as a prediction problem between 0 and 1 for each azimuth bin. The loss function is the mean squared error function. Thus, the loss function between the network output $\hat{\mathbf{Y}}$ and ground truth label $\mathbf{Y}$ is defined as:

$$loss(\mathbf{Y}, \hat{\mathbf{Y}}) = \sum_i l(y_i, \hat{y}_i) = \sum_i (y_i - \hat{y}_i)^2$$

To minimize overfitting, we use a combination of regularization techniques. Every contraction consists of a drop-out layer and batch-normalization layer. Secondly, we use the (Adaptive Moment Estimation) ADAM optimizer [6]. The global learning rate starts at $10e-3$ and is reduced by a factor of $10e-1$ every 10 epochs.

Finally, we also augment every training sample by flipping and rotating the sample around its azimuth axis, $\theta = 0$. Rotating the sample is done over $i\Delta\theta$ where $\Delta\theta = 10°$ and $i \in \{-8, -7, ..., 8\}$. This operation is achieved in practice by rolling the array over the azimuth axis. It should be noted that this is not strictly equivalent to rotating the scene, as the radar cross section (RCS) of an object is typically angle dependent. However, for our purposes, the inaccuracy of the resulting RCS does not negatively impact the results. After augmentation, we have an abundance of training samples where the typical characteristics of radar clutter is retained which is exactly what we want our network to suppress.

4 Experiments and Implementation Details

4.1 Dataset and Evaluation Protocol

In order to evaluate our proposed method, we operated our UAV through our industrial IoT lab environment, which consists of a 10 m × 30 m space divided in different aisles. Our lab environment, shown in Fig. 3a, strongly resembles a typical industrial warehouse. In a flight of around 5 min, all areas of the warehouse were covered. This dataset contains scenes where the UAV is flying through the aisles, as well as flying through the open space in front of the aisles. At the time of recording, no people or other dynamic moving objects where present. The UAV is equipped with a RGB camera (Intel Realsense), a 77 GHz FMCW radar (Texas Instruments IWR1443) and a processing backbone to process and store the captured data and control the drone (Jetson Nano). Data was captured at 5 Hz due to the decision to store the raw ADC radar response. Our drone platform and warehouse is shown in Fig. 3.

(a) Warehouse setting.

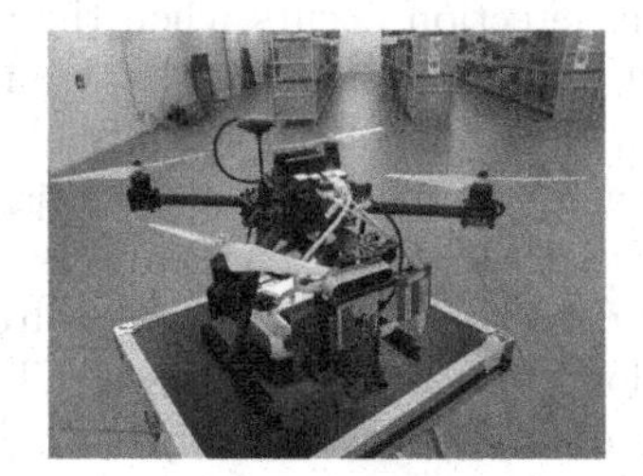

(b) Platform.

Fig. 3. IoT lab acting as industrial warehouse and drone platform with Jetson Nano (top), Intel RealSense depth camera (middle), TI IWR1443 mmWave radar (bottom left) and Infineon 24 Ghz radar (not used, bottom right).

The warehouse is conceptually divided in two areas, which results in a training and test set which are content-independent. An overview of the trajectory and split is shown in Fig. 4. A human annotator was tasked to label both the training and test set based on the free space visible via the RGB camera. In total, 1430 labels were annotated which took 16 h. The labelling was done in Cartesian space to achieve the highest possible accuracy. Afterwards, polar and nearest obstacle per aximuth bin can be automatically calculated for all our experiments. It should be noted that the obtained labels are imperfect due to the poor radar resolution especially at high azimuths. In order to make a qualitative and quantitative comparison of our method, we compared its performance to segmentation based on CFAR and by using a U-Net convolutional neural network.

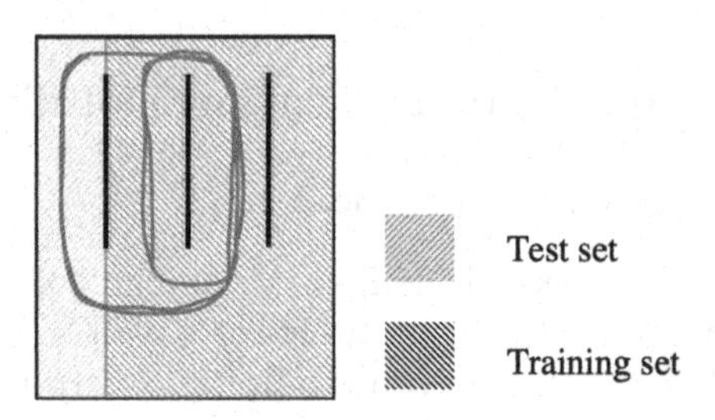

Fig. 4. A floor plan of our warehouse with the trajectory of the UAV that is used for data generation. The three black lines represent the shelves between the different aisles. Data captured in the green area is used for the test set, the blue area is used for the training set. This results in a content-independent training and test dataset. (Color figure online)

Comparison with Segmentation Based on CFAR. In a standard CFAR algorithm detection occurs when the Cell Under Test (CUT) exceeds a certain threshold value. This threshold value is calculated by estimating the noise floor on the surrounding cells. Typically also a few guard cells are employed to avoid leaks of energy of the CUT in the adjacent cells. This approach is known as Cell Averaging CFAR (CA-CFAR) and is used to compare our method with.

After detections are obtained, a free space map still has to be constructed. This is done by taking the closest CFAR detection per azimuth. This method can lead to detections in multipath reflections which illustrates one of the problems why CFAR detections are suboptimal to provide free space maps useful for navigation purposes.

Comparison with U-Net Segmentation. To make a fair comparison, we constructed a U-Net with a contracting head equal to our neural network, the expanding layers and skip through connections are identical as in [9]. This neural network is trained on the Cartesian labels. This segmentation is defined as a binary classification problem where 1 encodes the free space and 0 otherwise. As loss, we use a binary cross-entropy function combined with a dice function to handle the data imbalance between our two classes. After training, this network is also evaluated on the test set. A binary mask is obtained after thresholding. This threshold is determined empirically.

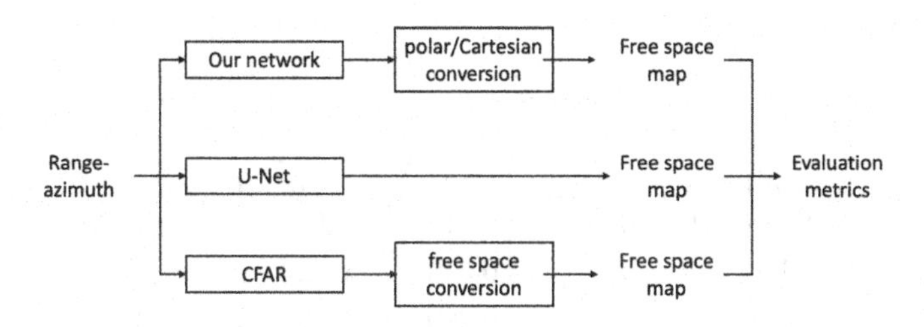

Fig. 5. Schematic overview of the evaluation procedure for our neural network compared to CFAR and U-Net based segmentation.

To compare between both outputs, an occupancy map is also generated from our proposed neural network. From the network's prediction, a segmented image in polar space is created. This occupancy map is finally compared with the segmentation output of the general U-Net by comparing the F1-scores with respect to the labels. We evaluated our network both in polar and Cartesian space. The U-Net was trained immediately on polar and Cartesian labels respectively. Unfortunately because our neural network doesn't output probabilities we can't compare precision-recall curves between the two architectures. An overview of our evaluation procedure in shown in Fig. 5.

4.2 Experimental Results

A summary of the results is presented in Table 1. Firstly we observe that CFAR thresholding performs very poorly. This is mainly due to the CFAR detections in multipath reflections. Our method outperforms the U-Net segmentation approach when evaluating the segmentation result in the polar system while also being significantly more compact. The increase is best visible through the F1-score due to the data imbalance between class labels. When evaluated in Cartesian space, our method underperforms compared to the classical approach due to only predicting the distance to the nearest obstacle per azimuth bin.

On Fig. 6 we have shown some results when our method is evaluated on the test set. We compare our method with the classical U-Net approach in polar coordinates. In the two examples, the U-Net outputs at first sight a slightly more detailed segmented image. Although appearing more detailed, this is not conformed in the evaluation metrics already discussed in the Table 1. In Fig. 7, a representation of the estimated free space is displayed as an overlay on the RGB camera.

Table 1. Evaluation results on our content-independent test dataset containing 463 frames.

Method	F1-score	Precision	Recall	Network information
Our method (polar)	0.760	0.809	0.710	Params: 321K
Our method (cart)	0.587	0.713	0.500	Size: 2,36 MB
CFAR (polar)	0.578	0.518	0.652	N/A
CFAR (cart)	0.435	0.390	0.491	
U-Net (polar)	0.713	0.936	0.528	Params: 1,94M
U-Net (cart)	0.639	0.940	0.486	Size: 63,4 MB

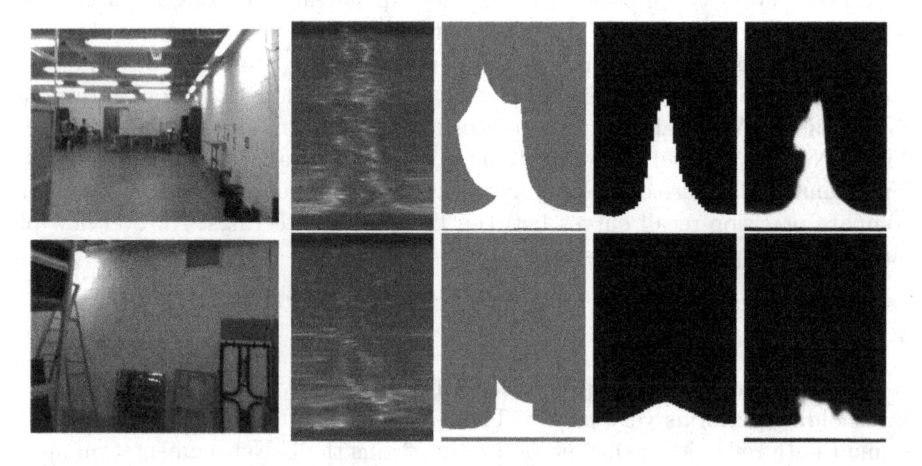

Fig. 6. Output comparison between our neural network and the U-Net neural network for two frames from the test dataset in polar coordinates. From left to right: RGB camera, range-azimuth image from radar, label, output of our network converted to an occupancy map, segmentation output of U-Net.

Fig. 7. Overlay of free space estimation on the RGB camera. The overlay is constructed by mapping $\hat{\mathbf{Y}}$ in to a color map of values between red and green. Red represents no free space at all, green represents no obstacles within the maximum range of the radar. The left frame corresponds with the top row in Fig. 6, the right frame corresponds with the bottom row. (Color figure online)

4.3 Implementation Details

FMCW radar frames are recorded at 4 Hz and our represented in our pipeline by the ADC samples after which it is converted to the range-azimuth-doppler space. The range is 20m spaced equally over 256 bins. After applying CAPON beamforming, the 180° field of view of the radar is divided over 181 equally spaced azimuth bins. The 181×256 range-azimuth image is zero-padded up to 256×256 and then downsampled to 32×32 before being passed through the network.

The network architecture consists first of 5 contraction blocks. Each consist of 2 groups of a convolutional, batch-normalization and ReLu layer with additionaly a dropout layer. Each contraction block ends with a max pooling layer. At the bottleneck, we have another convolutional layer which reduces the resolution to 2×2 and 128 dimensional feature space. This is passed through two fully connected layers which finally outputs a vector of 46 elements representing azimuth bins. These are compared with our labelled data which also contains 46 azimuth bins.

The network was trained during 50 epochs which is just below 25K backpropagations. No early stopping was applied. We used a constant training batch size (BS) of $BS = 64$ for all epochs.

5 Conclusion

In our work, we show that a low power millimeter wave radar can be used as an navigation sensor on a UAV in an indoor environment by using a segmentation approach. We presented a novel CNN for free space estimation in range-azimuth images of a single-chip FMCW radar in an indoor environment. Our method improves the classical U-Net segmentation approach while also having fewer parameters to train. Furthermore, our network is also smaller in size.

The main downside of our method is that it requires training on pre-labelled images which are dependent on the environment of operation. Furthermore, the manual segmentation of radar images is time consuming and error prone due to the amount of radar clutter.

The aim of future research is to investigate weakly supervised learning approaches by incorporating automatic labelling techniques by using a depth or lidar camera. By creating an abundance of imperfect training samples, the neural network should be able to better detect features in the complex environment. Furthermore, we will also investigate the performance when dynamic targets are included in the scene. Finally, we will also investigate new CNN architectures that predict immediately the evasive action that needs to be undertaken by the UAV.

Acknowledgments. This research received funding from the Flemish Government under the "Onderzoeksprogramma Artificiele Intelligentie (AI) Vlaanderen" program.

References

1. Bilik, I., Longman, O., Villeval, S., Tabrikian, J.: The rise of radar for autonomous vehicles: signal processing solutions and future research directions. IEEE Signal Process. Mag. **36**(5), 20–31 (2019). https://doi.org/10.1109/MSP.2019.2926573
2. Capon, J.: High-resolution frequency-wavenumber spectrum analysis. Proc. IEEE **57**(8), 1408–1418 (1969). https://doi.org/10.1109/PROC.1969.7278
3. Dickmann, J., et al.: Automotive radar the key technology for autonomous driving: from detection and ranging to environmental understanding. In: 2016 IEEE Radar Conference (RadarConf), pp. 1–6 (2016). https://doi.org/10.1109/RADAR.2016.7485214,ISSN: 2375-5318
4. Dimitrievski, M., Shopovska, I., Hamme, D.V., Veelaert, P., Philips, W.: Weakly supervised deep learning method for vulnerable road user detection in FMCW radar. In: 2020 IEEE 23rd International Conference on Intelligent Transportation Systems (ITSC), pp. 1–8 (2020). https://doi.org/10.1109/ITSC45102.2020.9294399
5. Kaul, P., de Martini, D., Gadd, M., Newman, P.: RSS-net: weakly-supervised multi-class semantic segmentation with FMCW radar. In: 2020 IEEE Intelligent Vehicles Symposium (IV), pp. 431–436 (2020). https://doi.org/10.1109/IV47402.2020.9304674. ISSN 2642-7214
6. Kingma, D.P., Ba, J.: Adam: a method for stochastic optimization (2014). https://doi.org/10.48550/arXiv.1412.6980, http://arxiv.org/abs/1412.6980
7. de Oliveira, M.L.L., Bekooij, M.J.G.: Deep convolutional autoencoder applied for noise reduction in range-doppler maps of FMCW radars. In: 2020 IEEE International Radar Conference (RADAR), pp. 630–635 (2020). https://doi.org/10.1109/RADAR42522.2020.9114719. ISSN 2640-7736
8. Rebut, J., Ouaknine, A., Malik, W., Pérez, P.: Raw high-definition radar for multi-task learning. In: 2022 IEEE/CVF Conference on Computer Vision and Pattern Recognition (CVPR), pp. 17000–17009 (2022). https://doi.org/10.1109/CVPR52688.2022.01651. ISSN 2575-7075

9. Ronneberger, O., Fischer, P., Brox, T.: U-net: convolutional networks for biomedical image segmentation. In: Navab, N., Hornegger, J., Wells, W.M., Frangi, A.F. (eds.) MICCAI 2015. LNCS, vol. 9351, pp. 234–241. Springer, Cham (2015). https://doi.org/10.1007/978-3-319-24574-4_28
10. Safa, A., et al.: A low-complexity radar detector outperforming OS-CFAR for indoor drone obstacle avoidance. IEEE J. Sel. Top. Appl. Earth Observ. Remote Sens. **14**, 9162–9175 (2021). https://doi.org/10.1109/JSTARS.2021.3107686
11. Xiao, Y., Daniel, L., Gashinova, M.: Image segmentation and region classification in automotive high-resolution radar imagery. IEEE Sens. J. **21**(5), 6698–6711 (2020). https://doi.org/10.1109/JSEN.2020.3043586
12. Çatal, O., Jansen, W., Verbelen, T., Dhoedt, B., Steckel, J.: LatentSLAM: unsupervised multi-sensor representation learning for localization and mapping. In: 2021 IEEE International Conference on Robotics and Automation (ICRA), pp. 6739–6745 (2021). https://doi.org/10.1109/ICRA48506.2021.9560768. ISSN 2577-087X

Upsampling Data Challenge: Object-Aware Approach for 3D Object Detection in Rain

Richard Capraru[1,2](✉), Jian-Gang Wang[1], and Boon Hee Soong[2]

[1] Institute for Infocomm Research, Agency for Science, Technology and Research, Singapore 138632, Singapore
{richard_capraru,jgwang}@i2r.a-star.edu.sg
[2] Department of Electrical and Electronic Engineering, Nanyang Technological University, Singapore 639798, Singapore
ebhsoong@ntu.edu.sg

Abstract. Lidar-based 3D object detection has been widely adopted for autonomous vehicles. However, adverse weather conditions, such as rain, pose significant challenges by reducing both detection distance and accuracy. Intuitively, one could adopt upsampling to improve detection accuracy. Nevertheless, the task of increasing the number of target points, especially the key detection points crucial for object detection, remains an open issue. In this paper, we explore how an additional data upsampling pre-processing stage to increase the density of the point cloud can potentially benefit deep-learning object detection. Unlike the state-of-the-art upsampling approaches which aim to improve point cloud appearance and uniformity, we are interested in object detection. The object of interest, rather than full scenarios or small patches, is used to train the network - we call it object-aware learning. Additionally, data collection and labelling are time-consuming and expensive, especially for rain scenarios. To tackle this challenge, we propose a semi-supervised upsampling network that can be trained using a relatively small number of labelled simulated objects. Lastly, we verify a well-established sensor/rain simulator, using a publicly available database. The experimental results on a database generated by this simulator are promising and have shown that our object-aware networks can extend the detection range in rainy scenarios and can achieve improvements in Bird's-eye-View Intersection-over-Union (BEV IoU) detection accuracy.

Keywords: object detection · point cloud upsampling · generative adversarial network · object-aware learning · semi-supervised learning

1 Introduction

LiDAR object detection in adverse weather conditions poses a significant challenge in autonomous vehicle research and remains an open issue. Given an

Supported by A*STAR.

J. Blanc-Talon et al. (Eds.): ACIVS 2023, LNCS 14124, pp. 148–159, 2023.
https://doi.org/10.1007/978-3-031-45382-3_13

unordered sparse point cloud received by LiDAR in the rain, one could explore point cloud upsampling to achieve a denser point cloud in order to improve the detection of different targets. Traditionally, upsampling methods have primarily been employed to support tasks like object classification [1] and surface smoothness reconstruction [2]. For upsampling experiments, the different patches of the samples present in the training dataset are selected for training and downsampled using methods such as Poisson Disk Sampling. Subsequently, during the training phase, the network is tasked with upsampling the point cloud and comparing it with the ground truth to evaluate its performance. Traditional approaches for LiDAR point clouds typically employ the Farthest Point Sampling (FPS) method to select seed points, followed by the application of the K-Nearest Neighbours (KNN) algorithm to acquire input patches. These patches are then merged after the upsampling process. While this approach presents good results for many applications, it has lower performance in scenarios involving adverse weather driving due to the extreme sparsity of data and the lack of emphasis on the downsampling pattern observed in natural settings.

Reconstructing complex geometry or topology from a sparse point cloud is still an open problem. Recently there has been some work done on optimizing object detection improving the resolution from low-resolution LiDAR (32 Ch) to high-resolution LiDAR (64 Ch), using 2D interpolation methods [3]. Their work demonstrated improved mAP (mean Average Precision) for different objects provided by the publicly available Kitti dataset [4]. However, their method is not adequate for rainy scenarios, where LiDARs can only receive point clouds with high sparsity. We aim to enhance and detect objects rather than obtain high-density point clouds. Building upon this motivation, we propose a novel object-aware upsampling approach, trained using the object of interest instead of small patches, to extend the detection range. The main contributions include:

- We propose a few object-aware upsampling strategies (an angle-invariant approach, a semi-supervised approach and an object-aware-traditional-patch-based combined approach) to increase the LiDAR detection range and to overcome the difficulty of collecting labelled data.
- We verify a well-established simulator and generate a rain database, which we can adopt as a benchmark for object detection in rain.

The paper is organized as follows. In Sect. 2, we present a rain model for our experiments. Section 3 provides an overview of existing upsampling technologies. In Sect. 4, we introduce our novel object-aware approach. The experimental results and concluding remarks are presented in Sect. 5.

2 Sensor and Rain Models, Verification and Simulated Database

Prior to conducting experiments on actual rain data, simulations can be employed to efficiently verify our methods. For this purpose, we will utilize a state-of-the-art simulator [5] in our experiments. This simulator offers both the physical sensor model and the rain model. However, it should be noted that the simulator does not account for the noise introduced by rain.

The theoretical model used in this simulator regarding the impact of rain on LiDAR measurements can be found in [6]. The power received by a LIDAR sensor [7] (reflected intensity) is,

$$P_r(z) = E_l * \frac{c * \rho(z) * A_r}{2 * z^2} * \tau_T * \tau_R * exp((-2) * \int_0^z \alpha(z^{'})\, dz^{'}\ ) \tag{1}$$

where E_l is the laser pulse energy, c is the speed of light, z is the detection range, $\rho(z)$ is the back-scattering coefficient of a target, $\alpha(z^{'})$ is the scattering coefficient of rain along the path to a target, A_r is the effective receiver area, τ_T and τ_R are the transmitter and the receiver efficiencies. Without loss of generality, assuming a homogeneous environment, the constant coefficient $C_s = c * E_l * A_r * \tau_T * \tau_R$ represents the particular characteristics of the sensor, and can be ignored when calculating the relative sensor power, which is,

$$P_n(z) = \frac{\rho}{z^2} * e^{(-2)*\alpha*z} \tag{2}$$

Under clear weather conditions, corresponding to α=0.0, at the maximum detection range of a LiDAR (z_{max}), the maximum detectable power for a high reflectivity object (ρ=0.9/π) will be,

$$P_n^{min} = \frac{0.9}{\pi * z_{max}^2} \tag{3}$$

To calculate the relative sensor power measured under rainy conditions, the scattering coefficient, α, can be defined according to the power law [8],

$$\alpha = a * R^b \tag{4}$$

where R is the rainfall rate (mm/h), and a and b are empirical coefficients. The authors obtain the values a=0.01 and b=0.6 using the measurements of another paper [9]. Therefore the final model for the relative intensity returned by the LiDAR as a function of rainfall rate is:

$$P_n(z) = \frac{\rho}{z^2} * e^{(-0.02)*R^{0.6}*z} \tag{5}$$

The last equation is used to simulate rain in terms of rain rate. The points with a power/intensity less than the value defined by (3) will be eliminated.

3 Point Cloud Upsampling for Object Detection

Several upsampling approaches have been proposed in the literature, such as those mentioned in [1] and [11]. However, these methods have not been specifically evaluated or tested in the context of enhancing point clouds for rainy scenes to improve object detection performance. To the best of our knowledge, we are the first to explore upsampling methods to improve detection in rain.

The first algorithm explored in our experiments is the Multi-step Point cloud Upsampling network (MPU) [11]. The MPU network offers a notable advantage through its feature extraction unit, which employs K-nearest neighbours (KNN) search based on feature similarity to transform inputs into a fixed number of features. These features are further refined through a densely connected multilayer perceptron (MLP) chain. Furthermore, MPU's multi-step progressive upsampling and dynamic graph convolution capabilities make it particularly suitable for our research objectives, leading us to adopt it in our work.

Given the increasing interest in generative adversarial networks (GANs) and their remarkable success in image generation, we further investigate the PU-GAN algorithm [1]. PU-GAN offers a distinct advantage over MPU in terms of its ability to generate uniform point clouds. This improved performance is attributed to the utilization of uniform loss and adversarial loss mechanisms within its framework. Considering these strengths, we include PU-GAN as the second algorithm explored in this paper.

4 Object-Aware Upsampling

Building upon the insights provided in the previous section, it has been established that traditional upsampling approaches primarily emphasize patches and treat individual points equally, overlooking the specific objects of interest. To enhance the accuracy of object detection, this study proposes a novel rain object-aware upsampling technique that extends the capabilities of patch-based approaches. This approach leverages the objects of interest to guide the upsampling process, thereby improving the overall effectiveness of object detection.

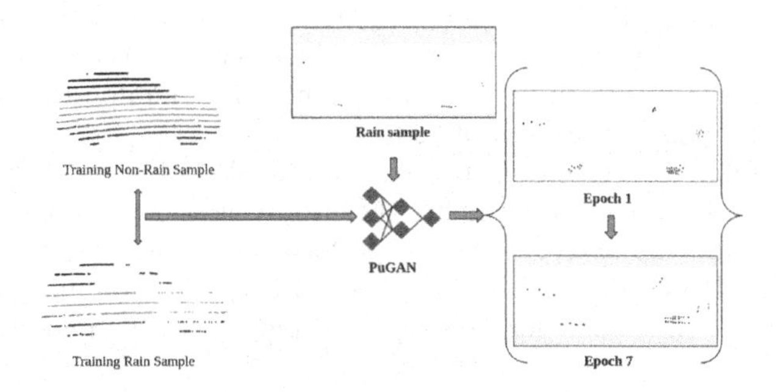

Fig. 1. Upsampling methodology

In this paper, we leverage the up-and-down sampling structure and adversarial training strategy of PuGAN to facilitate our research. Specifically, we implement PuGAN in Pytorch [16], which has been verified in [17]. As shown in Figure: 1, our upsampling network learns weights from some rain and non-rain data. The training data was generated as follows:

- Samples were recorded using the MSU Autonomous Vehicle Simulator (MAVS) [5] simulator on a clear day by placing the target at some distances and angles, e.g. 7 m and 90° in Fig. 1, which served as ground truth. Each sample in the ground truth training data contains the same number of points, e.g. 4096 points in this work.
- An angle-invariant upsampling approach was used: The network was trained using measurements of the car positioned at different angles, e.g. 400 samples are used in our experiments, which are generated with angle values from 0 to 90° in intervals of 10°.

Without loss of generality, an upsampling ratio of 4 is applied in our experiments. The angle-invariant approach has been trained for 19 epochs (with two different settings: 1. unsupervised trained using only non-rain object samples, 2. semi-supervised using 90% non-rain samples and 10% rain samples). In the final pre-processing experiment, we merged the points obtained from the semi-supervised approach with the pre-trained MPU upsampled points.

4.1 Rain Object-Aware Framework

The first stage of our rain object-aware framework involves enhancing the point cloud by upsampling it to restore and augment 3D object features. Taking inspiration from Wu et al. [13], we employ a density-based spatial clustering with noise (DBSCAN) method for vehicle clustering. As presented in their approach, a recommended value for ϵ is 1.1 m, as prior research has demonstrated that the average headway of vehicles in saturated flow conditions at signalized intersections is 2.18 m [14].

After identifying the car cluster, the target point cloud is passed through the pre-trained MPU to assess its initial performance. For performance comparison with traditional methods, we employ the MPU (implemented in Tensorflow as described in [15]) pre-trained using the Sketchfab dataset. Without loss of generality, we can set the minimum number of points to 33, an up ratio of 16, a step ratio of 3, and a patch num ratio of 400.

In the second stage of our framework, we employ object detection on the enhanced point cloud data. To evaluate the effectiveness of our upsampling approach, we measure the object detection performance using the state-of-the-art network, CenterPoint [12] (utilizing the Second backbone [18] and voxel size of 0.1). We choose CenterPoint for its simplicity, high speed, and good accuracy.

5 Experimental Results

5.1 Simulation Setting and Analysis

To validate the effectiveness of the simulator adopted in this work, we conducted experiments using a publicly available database. Firstly, we focused on verifying the rain model using NuScenes [10]. In their paper [6], Goodin et al. validated the rain model by measuring the maximum detection range based on the rain rate and object reflectivity. To assess the capabilities of our rain model, we reproduced these experiments and compared our results with the real-world findings presented by the BMW Research group in [19]. The experimental results, shown in Fig. 2, illustrate the distance to the farthest point in a random NuScenes scene after applying the model (Eq. (5)). We assumed uniform reflectivity for all points, employing the same values as [19]: 0.2 for the red line and 0.07 for the blue line.

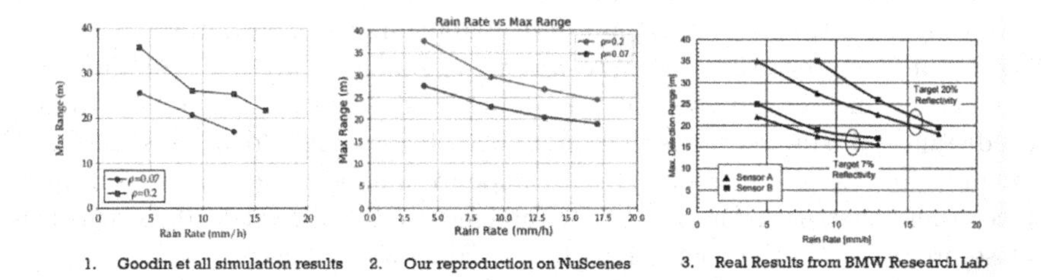

Fig. 2. Max Range vs Rain Rate for 1. Goodin et al. simulation [6], 2. Our reproduction on NuScenes, 3. Real Results from BMW Research Lab [19]

As Fig. 2, the validity of the model matches with the real experiments, motivating us to use the MAVS [5] simulator for this paper.

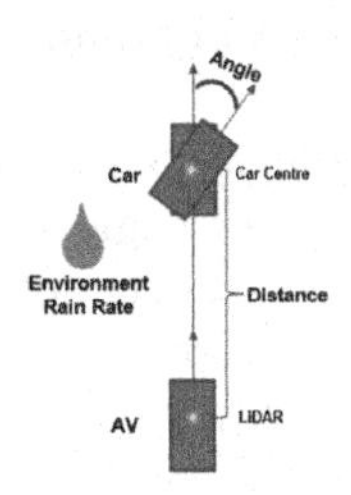

Fig. 3. Simulation setting

To assess the effectiveness of our upsampling method in enhancing LiDAR object detection performance, we generated a new database using the MAVS simulator. The MAVS simulator is developed with an MPI-based framework that enables parallel process coupling, along with a physics-based sensor simulator for LIDAR, GPS, cameras, and radars [5]. In this study, we utilized a Velodyne HDL-32E sensor positioned 1.84 m above the host vehicle (AV) to generate simulation data. This LiDAR position was selected to match the recording configuration of the NuScenes dataset. The data recording process is depicted in Fig. 3.

- A simulated car is placed in front of the AV's LiDAR at 10 m with an orientation of 0° (parallel to the direction of the host vehicle)

- We incrementally placed the car 1 m away from the AV until it became undetectable. This iterative process was repeated for various angles of the car, ranging from 0 to 90° (perpendicular to the AV), with an interval of 10°. Additionally, we varied the rain rate from 30 mm/s to 70 mm/s during these experiments.

A total of 5,170 scenarios have been recorded (with nine samples for each). The hierarchy of this dataset is a series of subtrees which represent the data from different rain rates, angles and distances.

We evaluated the impact of rain on the above-mentioned dataset using the CenterPoint detector [12]. The detector was applied to predict bounding box (bbox) scores and coordinates. For each sample scenario, we considered a minimum distance of 5 m below the maximum detectable range, focusing specifically on challenging cases where detection confidence and accuracy are significantly reduced. The threshold for the bbox confidence score was set to 0.28, which is in close proximity to the commonly chosen value of 0.3 [20].

5.2 Semi-supervised and unsupervised learning for angle-invariant upsampling analysis

In the subsequent stage of our experiments, we aimed to determine the impact of different training approaches on detection performance. Specifically, we investigated whether employing a semi-supervised approach with a combination of rain non-rain pair training samples and non-rain training samples, or an unsupervised approach using only non-rain training samples, would influence the detection performance.

Through our experimental findings, we have observed that non-rain training samples provide advantages for smaller angles, while rain-non-rain pair training samples provide the greatest benefits for larger angles. As illustrated in Fig. 4, the semi-supervised object-aware approach offers substantial benefits for both small angles (e.g., 0°) and large angles (e.g., 80°).

5.3 Extension of Maximum Detection Range

Moreover, we visually depicted the enhancement of the detection range for three different methods: (1) PuGAN semi-supervised object-aware method combined with MPU points, (2) Traditional state-of-the-art patch-based approach (MPU), and (3) PuGAN semi-supervised object-aware method. The visualization, presented in Fig. 5, showcases the improvements achieved for two angles, namely 0° and 80°. The results illustrate the promising performance of combining points from different upsampling methods (object-aware and patch-based) in improving detection confidence in certain cases, while the semi-supervised method displays better improvements in other cases. Notably, both methods outperform traditional approaches in terms of performance.

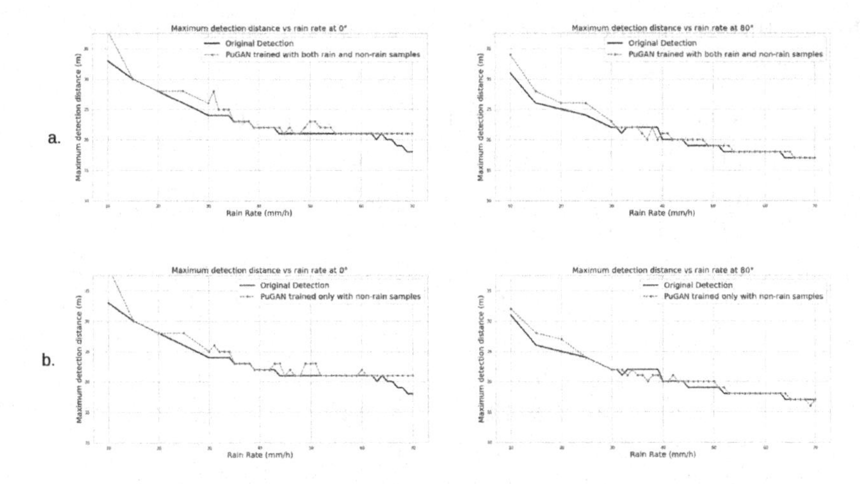

Fig. 4. Maximum detection distance vs rain rate for the target placed at an orientation of 0 and 80° for a. Unsupervised PuGAN object-aware method trained only with non-rain samples, b. Semisupervised PuGAN object-aware method trained with both rain non-rain pairs and non-rain samples

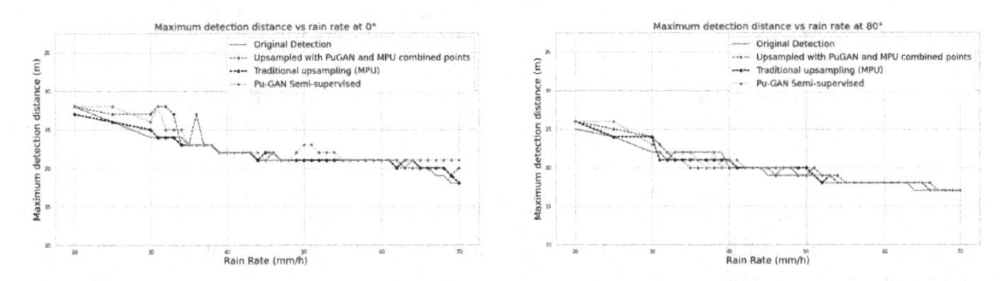

Fig. 5. Maximum detection distance vs rain rate for the target placed at an orientation of 0 and 80° for: PuGAN semi-supervised object-aware method combined with MPU points (green line); Traditional state-of-the-art patch-based approach (MPU) (blue line) and PuGAN semi-supervised object-aware method (red line) (Color figure online)

Lastly, it is also relevant to note that the PuGAN Angle-invariant method achieved a speed of 18 FPS using NVIDIA Corporation GM200 [GeForce GTX TITAN X], further bolstering the relevance of these methods. The speed improvement can be attributed to the exclusion of multiple patches, which would otherwise result in a substantial increase in computational power requirements.

5.4 Bird-Eye-View IoU

The evaluation of our method's improvements includes the use of the BEV IoU metric. Our methods can achieve good improvements in various scenarios, particularly in situations with extreme sparsity. Table 1 highlights the most significant improvements in IoU concerning rain rate, achieved through the combination of the PuGAN semi-supervised angle invariant object-aware method and the MPU

Table 1. Highest Improvement of BEV IoU for different angles (mm/h) rain rates and distances (m)

Rain	Dist.	Ang.	Max 2D IoU	Up Max 2D IoU	Impr.
10	32	90°	72.1%	88.3%	16.2%
15	31	70°	14.5%	65.3%	50.7%
20	24	90°	26.5%	89.2%	62.7%
25	25	80°	0%	85.6%	85.6%
34	25	70°	3.3%	69.0%	65.7%
35	24	70°	7.6%	60.4%	52.8%
37	25	60°	0%	66.7%	66.7%
39	21	70°	28.5%	81.2%	52.7%
41	20	80°	27.4%	88.1%	60.7%
42	21	80°	26.7%	87.6%	61.0%
45	23	0°	0%	60.8%	60.8%
50	23	20°	0%	89.2%	**89.2%**
53	22	0°	0%	69.8%	69.8%
62	19	60°	29.6%	83.6%	54.0%
70	19	0°	23.9%	78.6%	54.7%
Average Improvement					**60.2%**

points method. These improvements can be attributed to the inherent capability of our approach to enhance crucial object features that are essential for existing object detection methods.

Table 1 illustrates the upsampling improvements across various rain rates. In several instances, the CenterPoint model failed to detect the car before upsampling, resulting in an IoU value of 0. However, following the application of upsampling techniques, the network successfully detected the car with an accuracy ranging from 60% to 89%.

5.5 Experiment on Real Data

Finally, we conducted experiments on the Kitti dataset [4] to further evaluate the benefits of upsampling for object detection. In Fig. 6, we present an example of the detection results obtained using the 3D-SSD detector [21] on a randomly selected scene. Initially, the original detection (with a bbox threshold score of 0.6) failed to detect certain cars. However, by individually selecting and upsampling the non-detected cars using the semi-supervised method, successful detection was achieved. This result emphasizes the significance of upsampling in enhancing detection confidence, particularly for objects situated at long distances. In our experiments, the cars were positioned at distances of 38, 43, and 45 m, respectively.

Despite the promising nature of the findings, we acknowledge that there are anticipated limitations in our method. Expected and current limitations encompass the following aspects: potential instability in performance, leading to a degradation in object detection confidence or accuracy when employed with various object detectors across diverse scenarios, necessitating further fine-tuning (e.g. For experiments conducted on Kitti, there is a minimal-to-none improvement in the average precision metric using the semi-supervised learning method); lack of generalizability, for example, domain knowledge or a segmentation algorithm are required to select the targets of interest. These limitations primarily stem from the ongoing research challenge of point cloud upsampling, as well as the fact that existing approaches have not been specifically designed to address object detection in adverse weather conditions. Specifically, the reconstruction of a target's shape and the utilization of LiDAR measurements for learning purposes prove challenging due to the inherent sparsity of the data. Moreover, introducing an excessive number of points or misplacing points during the upsampling process can adversely affect the detection performance. To overcome these challenges, it is imperative to devise a more robust upsampling algorithm that takes into account the unique difficulties associated with upsampling sparse LiDAR point clouds.

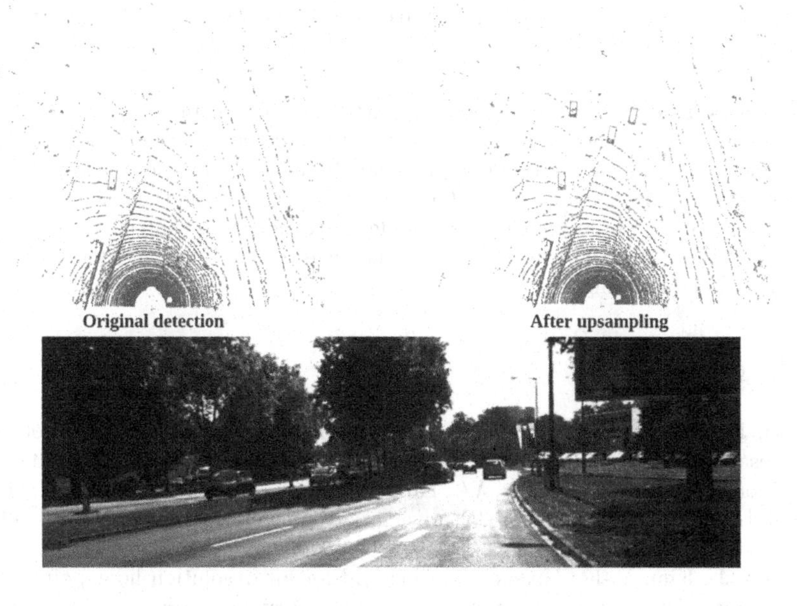

Fig. 6. Experiment on Real Kitti Data

6 Conclusion

Lidar object detection range as well as accuracy will be reduced significantly in rain. In this paper, we explore an object-aware upsampling method to increase the LiDAR object detection range and accuracy in the rain. Different from the existing upsampling approach, which increases the density of different patches equally, we aim to detect an object. We verified a well-established simulator and the experiments on a database generated by this simulator have shown that our object-aware networks can extend the detection range from traditional patch-based upsampling approaches by several meters in rain conditions. In addition, it can improve object detection accuracy in terms of BEV IoU.

Although some preliminary results obtained by applying our novel approach presented in this study are very encouraging, more experiments and optimization are needed to make it a stable solution to perception in rain. Currently, limited experimentation has been conducted to ascertain the characteristics including the conditions under which the approach is effective or unsuccessful, as well as the accuracy limitations associated with it.

A benchmark database could be built based on more simulation data in the near future. Furthermore, experiments on real rain data, collected using our autonomous vehicle, will allow us to evaluate our methods under challenging adverse weather conditions. Finally, We could extend our work to improve robustness and adaptability across various targets, angles, and sparsity characteristics in future.

Acknowledgment. This research is supported in part by grant no. I2001E0063 from the Singapore government's Research, Innovation and Enterprise 2020 plan (Advanced Manufacturing and Engineering domain) and administered by the Agency for Science, Technology and Research. The authors would like to extend their sincere gratitude to Dr. Teoh Eam Khwang for his invaluable suggestions and contributions during the writing of this paper. The authors would also like to thank the anonymous reviewers for their valuable comments.

References

1. Li, R., Li, X., Fu, C.-W., Cohen-Or, D., Heng, P.-A.: PU-GAN: a point cloud upsampling adversarial network (2019). https://arxiv.org/abs/1907.10844
2. Zhou, H., Chen, K., Zhang, W., Fang, H., Zhou, W., Yu, N.: DUP-net: denoiser and upsampler network for 3D adversarial point clouds defense (2019). https://arxiv.org/abs/1812.11017
3. You, J., Kim, Y.-K.: Up-sampling method for low-resolution lidar point cloud to enhance 3D object detection in an autonomous driving environment. Sensors **23**(1) (2023). https://www.mdpi.com/1424-8220/23/1/322
4. Geiger, A., Lenz, P., Stiller, C., Urtasun, R.: Vision meets robotics: the KITTI dataset. Int. J. Robot. Res. **32**, 1231–1237 (2013)
5. MSU autonomous vehicle simulator. https://www.cavs.msstate.edu/capabilities/mavs.php. Accessed 29 Jan 2023

6. Goodin, C., Carruth, D., Doude, M., Hudson, C.: Predicting the influence of rain on LiDAR in ADAS. Electronics **8**, 89 (2019)
7. Dannheim, C., Icking, C., Mader, M., Sallis, P.: Weather detection in vehicles by means of camera and LiDAR systems. In: 2014 Sixth International Conference on Computational Intelligence, Communication Systems and Networks, pp. 186–191 (2014)
8. Lewandowski, P.A., Eichinger, W.E., Kruger, A., Krajewski, W.F.: LiDAR-based estimation of small-scale rainfall: Empirical evidence. J. Atmos. Oceanic Technol. **26**(3), 656–664 (2009). https://journals.ametsoc.org/view/journals/atot/26/3/2008jtecha11221.xml
9. Filgueira, A., Gonzalez-Jorge, H., Laguela, S., Diaz-Vilarino, L., Arias, P.: Quantifying the influence of rain in LiDAR performance. Measurement **95**, 143–148 (2017). https://www.sciencedirect.com/science/article/pii/S0263224116305577
10. Caesar, H., et al.: nuScenes: a multimodal dataset for autonomous driving (2019). https://arxiv.org/abs/1903.11027
11. Yifan, W., Wu, S., Huang, H., Cohen-Or, D., Sorkine-Hornung, O.: Patch-based progressive 3D point set upsampling (2018). https://arxiv.org/abs/1811.11286
12. Yin, T., Zhou, X., Krahenbuhl, P.: Center-based 3D object detection and tracking (2020). https://arxiv.org/abs/2006.11275
13. Wu, J., Xu, H., Zheng, J., Zhao, J.: Automatic vehicle detection with roadside lidar data under rainy and snowy conditions. IEEE Intell. Transp. Syst. Mag. **13**(1), 197–209 (2021)
14. Wu, J., Xu, H.: The influence of road familiarity on distracted driving activities and driving operation using naturalistic driving study data. Traffic Psychol. Behav. **52**, 75–85 (2018)
15. MPU tensorflow implementation. https://github.com/yifita/3PU. Accessed 29 Jan 2023
16. PyTorch unofficial implementation of PU-net and PUGAN. https://github.com/UncleMEDM/PUGAN-pytorch. Accessed 29 Jan 2023
17. Yang, Q., Zhang, Y., Chen, S., Xu, Y., Sun, J., Ma, Z.: MPED: quantifying point cloud distortion based on multiscale potential energy discrepancy. IEEE Trans. Pattern Anal. Mach. Intell. 1–18 (2022)
18. Yan, Y., Mao, Y., Li, B.: SECOND: sparsely embedded convolutional detection. Sensors **18**, 3337 (2018)
19. Rasshofer, R., Spies, M., Spies, H.: Influences of weather phenomena on automotive laser radar systems. Adv. Radio Sci. **9**, 07 (2011)
20. MMDetection3D: OpenMMlab next-generation platform for general 3D object detection. https://github.com/open-mmlab/mmdetection3d. Accessed 29 Jan 2023
21. Yang, Z., Sun, Y., Liu, S., Jia, J.: 3DSSD: point-based 3D single stage object detector. In: Proceedings of the IEEE/CVF Conference on Computer Vision and Pattern Recognition (2020)

A Single Image Neuro-Geometric Depth Estimation

George Dimas, Panagiota Gatoula, and Dimitris K. Iakovidis(✉)

Department of Computer Science and Biomedical Informatics, School of Science, University of Thesssaly, Volos, Greece
{gdimas,pgatoula,diakovids}@uth.gr

Abstract. This paper introduces a novel neuro-geometric methodology for single image object depth estimation, abbreviated as NGDE. The proposed methodology can be described as a hybrid methodology since it combines a geometrical and a deep learning component. NGDE leverages the geometric camera model to initially estimate a set of probable depth values between the camera and the object. Then, these probable depth values along with the pixel coordinates that define the boundaries of an object, are propagated to a deep learning component, appropriately trained to output the final object depth estimation. Unlike previous approaches, NGDE does not require any prior information about the scene, such as the horizon line or reference objects. Instead, NGDE uses a virtual 3D point cloud projected to the 2D image plane that is used to estimate probable depth values indicated by 3D-2D point correspondences. Then by leveraging a multi-layer perceptron (MLP) that considers both the probable depth values and the 2D bounding box of the object, NGDE is capable of accurately estimating the depth of an object. A major advantage of NGDE over the state-of-the-art deep learning-based methods is that it utilizes a simple MLP instead of computationally complex Convolutional Neural Networks (CNNs). The evaluation of NGDE on KITTI indicates its advantageous performance over relevant state-of-the-art approaches.

Keywords: Machine Learning · Visual Measurements · Computer Vision · Single Image Measurements

1 Introduction

Depth estimation is a fundamental task in computer vision that aims to estimate the distance of each pixel in an image from the camera [16]. It has many applications in areas such as 3D scene perception, autonomous driving, robotics, augmented reality *etc.* [5, 21, 22]. The computational estimation of depth can be performed using different methods [4, 6, 14]. Among these methods, monocular depth estimation is particularly challenging because it requires learning complex and ambiguous relationships between image features and depth values [19]. However, it has some advantages over other methods, such as lower cost and simplified system complexity [5]. Therefore, many researchers have been developing deep learning models to tackle this problem using large-scale datasets and powerful neural networks [16].

J. Blanc-Talon et al. (Eds.): ACIVS 2023, LNCS 14124, pp. 160–171, 2023.
https://doi.org/10.1007/978-3-031-45382-3_14

Various depth prediction methods have been proposed in the literature aiming at estimating the depth of object of interest. Liu *et al.* [15] proposed a methodology for monocular depth prediction based on semantic class knowledge learning under geometrical constrains. However, the accuracy of this methodology is highly constrained by the quality of the training data. In another work, the camera and the various sensors of a mobile device have been leveraged to perform depth estimation using a single image [3]. A limitation of this work is that its performance is dependent on the accuracy of mobile sensors and its requirement for user input. In [2], monocular dash-cam images are employed to estimate the relative depth between vehicles in real-time. Nevertheless, this methodology relies on certain geometric assumptions related to the scene geometry such as the detection of the horizon. Other studies [17, 18] measure the distance between the camera and objects of interest and perform size measurements, using RGB-D sensors. These methodologies can produce predictions of enhanced accuracy; however, they rely on specialized sensors which increase the hardware complexity.

Recently, research interest has been focused on deep learning (DL)-based methods for monocular depth estimation. Fu *et al.* [7] proposed a Deep Ordinal Regression Network (DORN) for performing monocular depth estimation given a single RGB image. DORN is based on a supervised CNN model following a multi-scale feature extraction scheme with an ordinal training regression loss. In addition, DORN employs a spacing-increasing discretization strategy for the depth values to predict high resolution depth maps. In [1], a methodology based on transfer-learning was presented for addressing the issue of monocular depth estimation. A pre-trained CNN, using the DenseNet architecture [12] as a backbone and trained via a loss considering edge consistency and structural similarity, has been proposed to predict high quality depth maps. In [13], a supervised deep learning framework that includes a synergy network architecture and an attention-driven loss, was presented for jointly learning semantic labelling and depth estimation from a single image. Godard *et al.* [9], proposed a CNN framework, trained in an unsupervised manner, for performing monocular depth estimation. In that study, a training scheme based on pairs of stereo images was leveraged for predicting a disparity map to infer depth indirectly. The methodology presented in [10] expanded the work of [9] by introducing a self-supervised training approach which incorporates an occlusion aware loss along with an auto-masking training loss. Furthermore, a full-resolution multi-scale estimation procedure was applied to reduce artifacts. The methodologies of [9, 10] facilitate depth estimation from a single image; however, they require stereoscopic datasets for their training.

Considering the importance of single depth estimation, this paper proposes a novel neuro-geometric method, in the sense that combines a geometrical and deep learning approach, for object depth estimation, abbreviated as NGDE. NGDE infers the depth information of an object, *i.e.*, the object-to-camera distance, by propagating a set of probable depth values and the 2D pixel coordinates of the bounding box of an object to an appropriately trained MLP. Then, the MLP model is tasked to approximate an accurate estimation of the object depth given the respective inputs. The set of probable depth values is estimated using a virtual, automatically generated 3D point cloud (PC) that is subsequently projected to the 2D image plane using the intrinsic parameters of the camera. In this way, NGDE establishes correspondences between 2D pixel coordinates

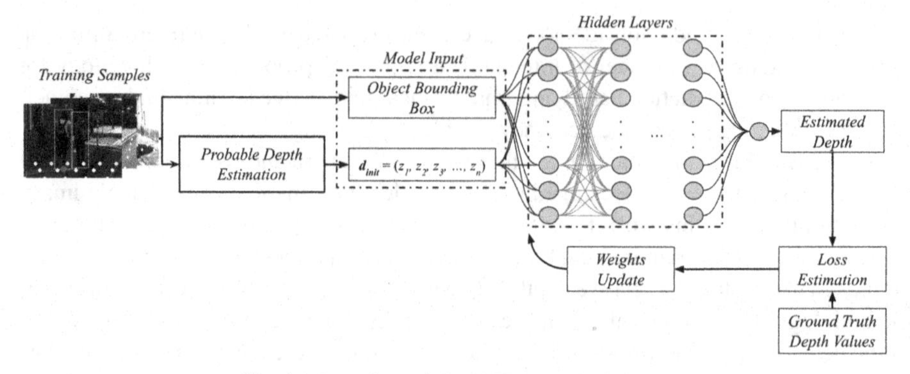

Fig. 1. Overview of the NGDE methodology.

and 3D world points. By leveraging these correspondences and the 2D bounding box of an object, a set of probable depth values between the object and the camera is estimated.

An advantage of NGDE over state-of-the-art DL methods is that the object depth estimation is solely based on the bounding box defining an object of interest, and the geometric properties of the camera model, *i.e.*, the depth estimation process does not consider pixel values of the image content. Hence, NGDE is not directly affected by changes in the environment, *e.g.*, illumination conditions, or by compression artifacts. Additionally, in contrast to NGDE, other single-image depth estimation approaches require prior knowledge regarding the horizon line [2]. Nevertheless, horizon detection may be challenging in various scenarios, such as in urban areas or indoors. Instead, NGDE leverages the parameters of the geometric camera model to establish 3D-2D correspondences between the real world and the image plane to predict the object depth accurately.

The rest of this paper is organized into 3 sections. Section 2 describes the proposed methodology; Sect. 3 presents the experimental setup and the evaluation results of NGDE against state-of-the-art models, and Sect. 4 that provides the conclusions of this study.

2 Methodology

NGDE can be regarded as a two-stage methodology aiming at estimating the object depth using a single image. An overview of the NGDE methodology is illustrated in Fig. 1. Initially, in the first stage, NGDE considers the geometrical camera model to approximate probable depth values between the camera and the target object. This is achieved by leveraging the intrinsic parameters of the camera to establish 3D-2D direct correspondences between the real world and image plane. This process is based on a recent approach we proposed for single image size-measurements [4]. The 3D-2D point correspondences are determined through a 2D point matching procedure using an automatically generated virtual 3D point cloud, $\boldsymbol{V}$, consisting of virtual 3D points. In detail, using the intrinsic parameters of the camera, each point of $\boldsymbol{V}$ is projected on the image plane. Having these initial 3D-2D correspondences and a bounding box defining a target object, a set of possible depth values between the camera and that object can be extracted. Then, in the second stage, NGDE uses both the estimated set of probable depth values and the bounding box pixel coordinates as inputs to an MLP trained to estimate an accurate approximation of the depth between the camera and the object.

2.1 Estimation of Probable Depth Values

Initially, NGDE requires the generation of a virtual point cloud (PC) $\boldsymbol{V}$ (Fig. 1) and the projection of each point $\boldsymbol{v}_i = (x_i, y_i, z_i)^{\mathrm{T}} \in \boldsymbol{V}$ to the image plane. This is feasible by using the intrinsic parameters of the camera, *i.e.*, focal length (f_x, f_y) and principal point (pp_x, pp_y) through the use of the pinhole camera model [11]:

$$\begin{pmatrix} u_i \\ v_i \end{pmatrix} = \begin{pmatrix} \frac{f_x x_i}{z} + pp_x \\ \frac{f_y y_i}{z} + pp_y \end{pmatrix} \tag{1}$$

Each point $\boldsymbol{v}_i$ is a coordinate in the *XYZ* Cartesian system with the *X* and *Y* axes spanning along the height and width of a 3D scene whereas the *Z*-axis spans along the depth of the scene. The 2D projection of $\boldsymbol{V}$, in the image plane following the pinhole camera model, is a set $\boldsymbol{P}$ comprising pixel coordinates $\boldsymbol{p}_i = (u_i, v_i)^{\mathrm{T}}$, where each $\boldsymbol{p}_i$ has a corresponding virtual 3D point $\boldsymbol{v}_i \in \boldsymbol{V}$ with known 3D world coordinates.

For the needs of this study, the virtual PC $\boldsymbol{V}$ consists of points defining a horizontal plane that approximates the ground, *i.e.*, coordinate y of $\boldsymbol{v}_i$ has value in the range of $(-\delta -h, -h + \delta)$, where h is the height of the camera, and δ denotes the offset from $-h$. Hence, a rough estimation of the height of the camera is required for the successful application of NGDE. In addition, the target objects are assumed to lie on the ground plane. By conditioning the PC generation to roughly approximate the ground plane the uncertainty deriving from the 3D-2D projection is minimized. In detail, by projecting the PC $\boldsymbol{V}$ at the height of the camera, due to the properties of the pinhole camera model that enables the 3D-2D projection, a multiple point mapping is possible, *i.e.*, different 3D points may share the same pixel coordinates on the image plane.

Once the virtual PC $\boldsymbol{V}$ and its 2D projection $\boldsymbol{P}$ are defined, a target object needs to be identified with a bounding box. This study assumes that the objects are on the ground plane (therefore, one side of the bounding box lies on the ground plane). To estimate the approximate depth between the camera and the target object a point matching process is performed between the point set $\boldsymbol{P}$ and a point of the linear segment of the bounding box of the object. Since PC $\boldsymbol{V}$ is an approximation of the ground plane, the middle point $\boldsymbol{p}_m$ of the linear segment that denotes the side of the bounding box lying on the ground plane is used (Fig. 2). The 2D point matching process uses the Euclidean distance to identify a subset $\boldsymbol{P}'$ comprising N 2D points $\boldsymbol{p}_i \in \boldsymbol{P}$ that share similar pixel.

coordinates with $\boldsymbol{p}_m$. Since, each point $\boldsymbol{p}_i \in \boldsymbol{P}'$ has a direct corresponding 3D point $\boldsymbol{v}_i$ with known z values, a set of probable depth values $\boldsymbol{d}_{init} = (z_1, z_2, z_3, \ldots, z_N)$ can be extracted. This set of probable depth values $\boldsymbol{d}_{init}$ along with the pixel coordinates of the bounding box are used as inputs to the MLP in the next stage of the proposed methodology (Fig. 3).

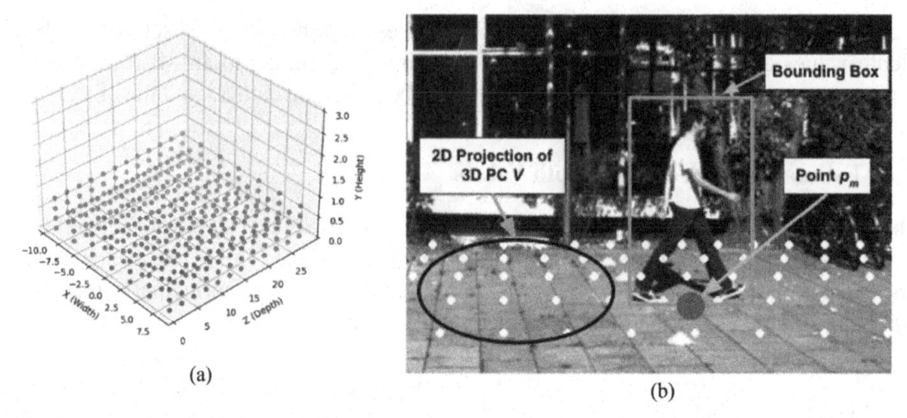

Fig. 2. Example of the 3D and 2D components used for the estimation of the probable depth values. (a) Illustration of the 3D PC $\boldsymbol{V}$; (b) Illustration of the 2D projection of $\boldsymbol{V}$, object bounding box, and $\boldsymbol{p_m}$.

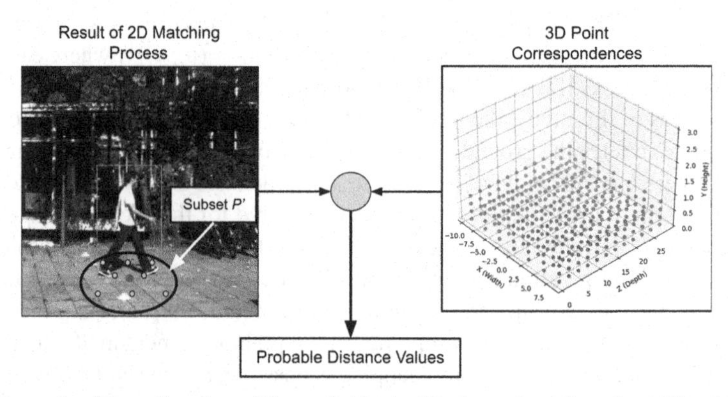

Fig. 3. Example of the estimation of the probable depth values given the subset $\boldsymbol{P'}$ and 3D PC $\boldsymbol{V}$.

2.2 Depth Estimation with Multilayer Perceptron

The probable depth estimation process produces a set of rough object depth approximations. The selection of a depth value estimated by this process can be subject to potential deviations in the depth estimation originating from the systematic bias due to the calibration and the position of the camera [4]. To cope with this problem an additional component for the final depth estimation is integrated in the proposed methodology. In detail, an MLP network is trained to estimate the final depth value given a set of rough approximations and the pixel coordinates of the bounding box.

Given the initial set of probable depth values $\boldsymbol{d_{init}}$, and the bounding box, $\boldsymbol{bb}$, defining the object of interest, NGDE presumes that there is a non-linear function $f(d_{init}, bb; \boldsymbol{w})$, parametrized by $\boldsymbol{w}$, capable of inferring the final object depth. That function $f(d_{init}, bb; \boldsymbol{w})$ is approximated using an MLP network composed of ↕ hidden layers and a single output neuron. Each hidden layer contains a fixed number of k neurons that use the hyperbolic tangent (*tanh*) as activation function. The prediction layer consists of a single neuron that employs the *log*-sigmoid function as its activation. The final architecture of the MLP has been experimentally determined (see Sect. 3.2).

To train the MLP of NGDE the Mean Squared Logarithmic Error (MSLE) was used as loss function formally expressed as:

$$\mathcal{L}(y, \hat{y}) = \frac{1}{M}\sum_{i=1}^{M}\left(\log(y_i + 1) - \log(\hat{y}_i + 1)\right)^2 \tag{2}$$

where the real and predicted depth values are denoted by y and $\hat{y}$, respectively, whereas M denotes the total amount of training data. MSLE was selected on the basis that is less sensitive to large errors compared to other loss functions, such as the Mean Squared Error (MSE) [20]. MSE tends to emphasize larger errors because it regards their squared differences resulting in increment of the total training error for a model. Additionally, the range in which the objects are detected is wide. As a result, MSE loss would emphasize the estimation errors occurring when objects are detected in larger distances compared to the errors in small distances. Furthermore, to cope with the difference regarding the range of the depth and the bounding box pixel coordinate values, the min-max normalization has been applied separately to the bounding box and probable depths.

3 Experiments and Results

3.1 Experimental Setup

The experimental evaluation of the proposed methodology was performed on the KITTI dataset [8]. KITTI dataset contains RGB images representing various outdoor urban and rural scenarios along with 3D point clouds. KITTI dataset is considered a benchmark for a variety of computer vision applications including stereo, optical flow, tracking, visual odometry, SLAM and 3D object detection. The dataset was recorded with RGB and grayscale stereo cameras and multiple sensors, *e.g.*, depth lidar sensors. It provides 3D bounding boxes for pedestrians, cars, cyclists, and other objects that are depicted at each image (Fig. 4) as well as ground truth poses, IMU measurements, GPS coordinates and calibration data. The MLP of NGDE has been trained using the training data partition of the KITTI dataset. For one-to-one comparison with other similar methodologies, only un-occluded objects were considered as well as the same test data partitioning with [2]. The comparison of the trained MLP models in the task of depth estimation performed with the metrics of Mean Absolute Error (MAE), Relative Error (RE), Relative Squared Error (RSER), Root Mean Squared Error (RMSE), logarithmic RMSE (RMSE-*log*) and Ratio threshold δ, defined as:

$$MAE(y, \hat{y}) = \frac{1}{M}\sum_{i=0}^{M}|y_i - \hat{y}_i| \tag{3}$$

$$RE(y, \hat{y}) = \frac{1}{M}\sum_{i=0}^{M}\frac{|y_i - \hat{y}_i|}{y} \tag{4}$$

Fig. 4. Images of the KITTI dataset. (a) Multiple objects of class person; (b) single object of class person; and (c) single object of class vehicle

$$RSER(y,\hat{y}) = \frac{1}{M}\sum_{i=0}^{M}\frac{(y_i - \hat{y}_i)^2}{y} \tag{5}$$

$$RMSE(y,\hat{y}) = \sqrt{\frac{1}{M}\sum_{i=0}^{M}(y_i - \hat{y}_i)^2} \tag{6}$$

$$RMSE_{log}(y,\hat{y}) = \sqrt{\frac{1}{M}\sum_{i=0}^{M}(log(y_i) - log(\hat{y}_i))^2} \tag{7}$$

$$\delta_\iota : \%\text{ of } y_i\ s.t.\ \max\left(\frac{y_i}{\hat{y}_i}, \frac{\hat{y}_i}{y_i}\right) = \delta < thr,\ \text{ for thr} = 1.25, 1.25^2, 1.25^3 \tag{8}$$

3.2 Ablation Study

An experimental evaluation was conducted for selecting the hyper-parameters of the MLP. For this reason, different combinations regarding the number of hidden layers $\updownarrow$ and number, k, of hidden neurons were employed to evaluate the performance of the MLP network. The ablation study for the selection of an optimized MLP, evaluated architectures with a number of hidden layers ranging from 1 to 4. For each trial, the MLPs were evaluated for a varying number of neurons, *i.e.*, 10, 100 and 256, comprising a hidden layer. Hence, a total of 12 (4×3) different MLPs have been trained and evaluated to investigate how different degrees of network complexity benefits the accuracy of NGDE. Each MLP architecture was trained for a maximum of 5000 epochs with a batch size of 128.

The Adam optimizer was used for the update of the weights of each MLP during the training process with an initial learning rate of 10^{-3} [22].

The results of this investigation are summarized in Table 1. As it can be observed, the best performance has been achieved with a model architecture composed of 4 hidden

Table 1. Comparative results for different MLP architectures used by NGDE. ↑ and ↓ means higher and lower is better, respectively.

Models		Metrics						
ℓ	k	$RE\downarrow$	$MAE\downarrow$	$RSER\downarrow$	$RMSE\downarrow$	$\delta < 1.25\uparrow$	$\delta < 1.25^2\uparrow$	$\delta < 1.25^3\uparrow$
1	10	0.16	3.18	0.84	5.25	0.85	0.96	0.98
	100	0.16	3.07	0.79	5.20	0.83	0.95	0.98
	256	0.16	3.16	0.80	5.20	0.83	0.96	0.98
2	10	0.12	2.32	0.56	4.24	0.89	0.96	0.98
	100	0.11	2.22	0.46	3.94	0.90	0.97	0.99
	256	0.11	2.13	0.42	3.77	0.89	0.97	0.99
3	10	0.11	2.07	0.57	3.89	0.90	0.97	0.98
	100	0.10	1.73	0.38	3.28	0.92	0.97	0.99
	256	0.11	1.92	0.45	3.49	0.92	0.97	0.99
4	10	0.11	2.04	0.45	3.88	0.90	0.97	0.99
	100	0.09	1.51	**0.30**	**2.95**	0.94	0.97	0.99
	256	**0.09**	**1.48**	0.34	3.21	**0.94**	**0.97**	**0.99**

Table 2. Comparisons of NGDE with state-of-the-art methods for depth estimation. ↑ and ↓ means higher and lower is better, respectively.

Methods	Metrics						
	$RE\downarrow$	$RSER\downarrow$	$RMSE\downarrow$	$RMSE_{log}\downarrow$	$\delta < 1.25\uparrow$	$\delta < 1.25^2\uparrow$	$\delta < 1.25^3\uparrow$
MonoDepth [9]	0.13	1.36	6.34	0.21	0.82	0.94	0.98
DORN [7]	0.11	0.44	**2.44**	0.18	0.92	0.96	0.98
Alhashim et al [1]	0.09	0.59	4.17	0.17	0.89	0.97	0.99
MonoDepth2 [10]	0.11	0.81	4.63	0.19	0.88	0.96	0.98
Ali et al [2]	0.29	2.02	6.24	0.30	0.53	0.89	0.98
DevNet [23]	0.10	0.70	4.41	0.17	0.89	0.97	0.99
Proposed	**0.09**	**0.30**	3.21	**0.16**	**0.94**	**0.97**	**0.99**

layers with 256 neurons each. However, considering the complexity of the network, the architecture with 4 hidden layers with 100 neurons each has been chosen as the optimal network configuration since it has comparative performance with the more complex model (Fig. 5).

Fig. 5. Examples of NGDE depth predictions

3.3 Comparisons with State-of-the-Art

The performance of the proposed NGDE methodology was assessed in comparison to state-of-the-art depth estimation models. For one-to-comparison with the state-of-the art, the metrics used in [1, 2, 7, 9, 10], have been adopted. Thus, object depth predictions have been evaluated in terms of Relative Error (RE), Relative Squared Error (RSER), Root Mean Squared Error (RMSE), logarithmic RMSE and ratio threshold δ. Ratio threshold accuracy indicates the percentage to which the ratio and inverse ratio between the predicted and ground truth depth values is under a pre-defined threshold. The results are summarized in Table 2. As it can observed NGDE model achieves lower RE and RSER scores among the models compared, which are equal to 0.09 and 0.30 respectively. Additionally, NGDE achieves the highest ratio threshold accuracy ranging from 0.94 to 0.99, for the different ratio threshold values. NGDE has the second lower RMSE score behind DORN. Overall, the CNN model of *Alhashim et al.* [1] achieves the second more accurate predictions with the scores of Relative Error and ratio threshold for threshold values of 1.25^2 and 1.25^3 to be equal with the scores of NGDE model. Furthermore, the CNN model of *Zhou et al.* [23] namely *DevNet,* achieves performance similar to the NGDE for logarithmic RMSE score and ratio threshold for threshold values of 1.25^2 and 1.25^3.

4 Conclusions

In this paper a novel hybrid, in the sense that incorporates both a geometrical and deep learning component, methodology is proposed for the estimation of object depths. Unlike other approaches, NGDE leverages the parameters of the geometric camera model to initially estimate a set of probable values denoting the depth between the object and the camera. These depth values, in combination with the bounding box defining the borders of an object, are propagated to an MLP model that makes an accurate estimation of the depth between the target object and the camera. The architecture of the MLP component of NGDE has been determined by an ablation study were various combinations of hyperparameters were tested.

Since NGDE does not consider any information regarding the image content, *i.e.*, the pixel values, it cannot be directly affected by various changes, *e.g.*, changes in illumination conditions or compression artifacts, and thus it can be easily applied in various settings including indoor environments. However, in the case that an object detector is used for the extraction of bounding boxes, inaccuracies in the bounding box estimation can affect the performance of NGDE.

The experimental study performed in the context of this paper shown that NGDE can achieve a relative error regarding the object depth estimation of 0.09, outperforming other state-of-the-art approaches that have been proposed for object depth estimation. A limitation of NGDE is that its MLP receives as input the bounding box of a target object. This is considered as a limitation because the dimensions of the bounding box of the same object captured under the same conditions by different cameras may vary; as a result, it can affect the generalization capacity of NGDE. NGDE assumes that the objects are on the ground plane and that the camera parameters and height are given. These can be acquired offline in many cases, such as in autonomous driving, using vehicle specification and camera calibration. However, the automatic estimation of these factors constitutes directions for future work. Additionally, future work the investigation of approaches that could make the performance of NGDE modality more computational efficient and independent of the camera model, *e.g.*, by normalizing the bounding box coordinates considering the intrinsic parameters of the camera, as well as comparing its performance in indoor environments.

Acknowledgements. We acknowledge support of this work by the project "Smart Tourist" (MIS 5047243) which is implemented under the Action "Reinforcement of the Research and Innovation Infrastructure", funded by the Operational Programme "Competitiveness, Entrepreneurship and Innovation" (NSRF 2014–2020) and co-financed by Greece and the European Union (European Regional Development Fund).

References

1. Alhashim, I., Wonka, P.: High quality monocular depth estimation via transfer learning. arXiv preprint arXiv:181211941 (2018)
2. Ali, A., Hassan, A., Ali, A.R., Khan, H.U., Kazmi, W., Zaheer, A.: Real-time vehicle distance estimation using single view geometry. In: Proceedings of the IEEE/CVF Winter Conference on Applications of Computer Vision (WACV) (2020)
3. Chen, S., Fang, X., Shen, J., Wang, L., Shao, L.: Single-image distance measurement by a smart mobile device. IEEE Trans. Cybernet. **47**, 4451–4462 (2016)
4. Dimas, G., Bianchi, F., Iakovidis, D.K., Karargyris, A., Ciuti, G., Koulaouzidis, A.: Endoscopic single-image size measurements. Meas. Sci. Technol. **31**, 074010 (2020)
5. Dimas, G., Gatoula, P., Iakovidis, D.K.: MonoSOD: monocular salient object detection based on predicted depth. In: 2021 IEEE International Conference on Robotics and Automation (ICRA), pp 4377–4383. IEEE (2021)

6. Falkenhagen, L.: Depth estimation from stereoscopic image pairs assuming piecewise continuos surfaces. In: Image Processing for Broadcast and Video Production: Proceedings of the European Workshop on Combined Real and Synthetic Image Processing for Broadcast and Video Production, Hamburg, 23–24 November 1994, pp 115–127. Springer, Cham (1995). https://doi.org/10.1007/978-1-4471-3035-2
7. Fu, H, Gong, M., Wang, C., Batmanghelich, K., Tao, D.: Deep ordinal regression network for monocular depth estimation. In: Proceedings of the IEEE Conference on Computer Vision and Pattern Recognition, pp 2002–2011 (2018)
8. Geiger, A., Lenz, P., Urtasun, R.: Are we ready for autonomous driving? The KITTI vision benchmark suite. In: 2012 IEEE Conference on Computer Vision and Pattern Recognition, pp 3354–3361 (2012)
9. Godard, C., Mac Aodha, O., Brostow, G.J.: Unsupervised monocular depth estimation with left-right consistency. In: Proceedings of the IEEE Conference on Computer Vision and Pattern Recognition, pp 270–279 (2017)
10. Godard, C., Mac Aodha, O., Firman, M., Brostow, G.J.: Digging into self-supervised monocular depth estimation. In: Proceedings of the IEEE/CVF International Conference on Computer Vision, pp 3828–3838 (2019)
11. Heikkila, J., Silvén, O.: A four-step camera calibration procedure with implicit image correction. In: Proceedings of IEEE Computer Society Conference on Computer Vision and Pattern Recognition, pp. 1106–1112. IEEE (1997)
12. Huang, G., Liu, Z., Van Der Maaten, L., Weinberger, K.Q.: Densely connected convolutional networks. In: Proceedings of the IEEE Conference on Computer Vision and Pattern Recognition, pp 4700–4708 (2017)
13. Jiao, J., Cao, Y., Song, Y., Lau, R.: Look deeper into depth: monocular depth estimation with semantic booster and attention-driven loss. In: Proceedings of the European Conference on Computer Vision (ECCV), pp 53–69 (2018)
14. Johari, M.M., Carta, C., Fleuret, F.: (2021) DepthInSpace: exploitation and fusion of multiple video frames for structured-light depth estimation. In: Proceedings of the IEEE/CVF International Conference on Computer Vision, pp 6039–6048
15. Liu, B., Gould, S., Koller, D.: Single image depth estimation from predicted semantic labels. In: 2010 IEEE Computer Society Conference on Computer Vision and Pattern Recognition, pp 1253–1260. IEEE (2010)
16. Ming, Y., Meng, X., Fan, C., Yu, H.: Deep learning for monocular depth estimation: a review. Neurocomputing **438**, 14–33 (2021)
17. Park, H., Van Messemac, A., De Neveac, W.: Box-Scan: an efficient and effective algorithm for box dimension measurement in conveyor systems using a single RGB-D camera. In: Proceedings of the 7th IIAE International Conference on Industrial Application Engineering, Kitakyushu, Japan, pp. 26–30 (2019)
18. Shuai, S., et al.: Research on 3D surface reconstruction and body size measurement of pigs based on multi-view RGB-D cameras. Comput. Electron. Agric. **175**, 105543 (2020)
19. Spencer, J., et al.: The monocular depth estimation challenge. In: Proceedings of the IEEE/CVF Winter Conference on Applications of Computer Vision, pp 623–632 (2023)
20. Tyagi, K., et al.: Regression analysis. In: Artificial Intelligence and Machine Learning for EDGE Computing, pp 53–63. Elsevier (2022)
21. Valentin, J., et al.: Depth from motion for smartphone AR. ACM Trans. Graph. (ToG) **37**, 1–19 (2018)

22. Yang, X., Luo, H., Wu, Y., Gao, Y., Liao, C., Cheng, K.-T.: Reactive obstacle avoidance of monocular quadrotors with online adapted depth prediction network. Neurocomputing **325**, 142–158 (2019)
23. Zhou, K., et al.: Devnet: Self-supervised monocular depth learning via density volume construction. In: Computer Vision–ECCV 2022: 17th European Conference, Tel Aviv, Israel, 23–27 October 2022, Proceedings, Part XXXIX, pp 125–142. Springer, Cham (2022). https://doi.org/10.1007/978-3-031-19842-7_8

Wave-Shaping Neural Activation for Improved 3D Model Reconstruction from Sparse Point Clouds

Georgios Triantafyllou, George Dimas, Panagiotis G. Kalozoumis, and Dimitris K. Iakovidis(✉)

Department of Computer Science and Biomedical Informatics, University of Thessaly, Volos, Greece
{gtriantafyllou,gdimas,pkalozoumis,diakovidis}@uth.gr

Abstract. The quality of a 3D model depends on the object digitization process, which is usually characterized by a tradeoff between volume resolution and scanning speed, *i.e.*, higher resolution scans require longer scanning times. Aiming to improve the quality of lower resolution 3D models, this paper proposes a novel approach to 3D model reconstruction from an initially coarse point cloud (PC) representation of an object. The main contribution of this paper is the introduction of a novel periodic activation function, named Wave-shaping Neural Activation (WNA), in the context of implicit neural representations (INRs). The use of the WNA function in a multilayer perceptron (MLP) can enhance the learning of continuous functions describing object surfaces given their coarse 3D representation. Then, the trained MLP can be regarded as a continuous implicit representation of the 3D representation of the object, and it can be used to reconstruct the originally coarse 3D model with higher detail. The proposed methodology is experimentally evaluated by two case studies in different application domains: a) reconstruction of complex human tissue structures for medical applications; b) reconstruction of ancient artifacts for cultural heritage applications. The experimental evaluation, which includes comparisons with state-of-the-art approaches, verifies the effectiveness and improved performance of the WNA-based INR for 3D object reconstruction.

Keywords: Machine Learning · Implicit Neural Representations · Reconstruction · 3D Modeling · Gastrointestinal Tract · Cultural Heritage

1 Introduction

In the last few decades, the reconstruction of three-dimensional (3D) models has been widely investigated. Nowadays, 3D models of real objects are typically created from image sequences depicting a target object from different perspectives, or by using 3D laser scanning techniques, *e.g.*, LIght Detection And Ranging (LIDAR). The quality of these methods can be hindered by various factors related to the digitization process, *e.g.*, low image or volume resolution, while there is usually a tradeoff between resolution and scanning speed, *i.e.*, higher resolution scans require longer scanning times, whereas faster scans result in coarser 3D models.

J. Blanc-Talon et al. (Eds.): ACIVS 2023, LNCS 14124, pp. 172–183, 2023.
https://doi.org/10.1007/978-3-031-45382-3_15

The significance of digital twins (DTs) has been recognized in a broad variety of domains, including industrial, medical, and cultural heritage applications. Recently, the use of DT models of human tissue structures has been identified as an impactful research topic in biomedicine [18]. DTs are employed to simulate the pathophysiology of human organs under different conditions. Hence, they can be of paramount importance for biomedical applications, since they can shorten the time required for the various trial phases, *e.g.*, clinical trials required for the development of novel biomedical devices. The accuracy of these simulations depends on the fidelity of the 3D model of a human organ. Such a 3D model can be reconstructed from magnetic resonance (MR) or computed tomography (CT) images. However, due to uncertainty factors introduced in this process, *e.g.*, imperfect segmentation of the tissue structures in the tomographic slices, the quality of the obtained 3D models varies, and a post processing stage is required to improve the overall model quality.

In the domain of cultural heritage, museums, cultural venues, and archeological sites can benefit from the creation of digital replicas of their exhibits. Such digital replicas can be used as part of a virtual experience for people that cannot physically visit these venues, or to generate physical 3D-printed replicas [12], *e.g.*, for the preservation of the original artifacts or for the creation of tactile exhibitions for visually impaired people (VIP) [37].

Considering the above, several deep learning (DL) approaches have been proposed for the reconstruction of 3D models. These approaches employ voxels [10, 40], meshes [2, 15], or point clouds (PCs) [32, 42] to train deep artificial neural networks (ANNs). Nevertheless, the interpolation capability of these networks is limited. Implicit neural representations (INRs) have exhibited the capacity of expressing shapes in the form of continuous functions with the use of multilayer perceptrons (MLPs) [7, 8, 27, 30] or convolutional neural networks (CNNs) [31]. These networks are tasked to approximate implicit functions based on raw data, PCs, or latent codes with or without supervision [27, 30, 36]. Signed distance functions (SDFs) have been recently utilized in INRs to infer different geometries [14]. Recently, the utilization of the sinusoidal (*sin*) function for enhancing the performance of INRs (SIREN) has been proposed in [36].

Methods related to the 3D reconstruction of complex human tissue structures include structure from motion (SfM) approaches [16] and their more recent variations, such as non-rigid-structure-from-motion (NRSfM) [13, 21, 34]. More recently, generative models for 3D organ shape reconstruction have been proposed using autoencoder architectures [3, 38]. In [43], a DL-based framework was utilized to reconstruct 3D colon structures based on a colon model segmented from CT images and monocular colonoscopy images. Similar methods have also been applied to the 3D reconstruction of cultural heritage artifacts, focusing mainly on obtaining 3D models based on SfM [26] and multiview stereo photogrammetry techniques [19] as an alternative to high-resolution laser scanners. Moreover, DL has been implemented for 3D reconstruction based on dense models [6]. Other studies have utilized non-DL approaches that require manual curation or use of 3D modelling-related software [4, 39]. Based on the above studies, it can be inferred that all current methods depend on acquiring an initially high-resolution 3D model based on traditional photogrammetry methods, *e.g.*, SfM. Thus, in cases where only a coarse 3D model of the object can be obtained, the quality of the extracted mesh is limited. In such scenarios, an INR-based post-processing step that can successfully reconstruct the extracted coarse 3D model at higher resolution without using computationally expensive methodologies would be extremely beneficial.

To this end, this paper proposes an MLP-based INR comprising a neural network that utilizes a novel activation function, called hereinafter Wave-shaping Neural Activation (WNA). The development and use of this novel activation function has been motivated by SIREN. WNA enables an MLP to provide better reconstruction results given just a sparse PC of an object. Subsequently, coarse 3D models can be properly restored to be used in various configurations, *e.g.*, virtual reality and DT applications. Finally, the capacity of MLPs of learning continuous INRs to reconstruct high-quality 3D meshes of the gastrointestinal (GI) tract and cultural heritage artifacts given sparse PC representations, is demonstrated. To the best of our knowledge WNA functions have not been previously considered in the context of INRs, and this is the first time that INRs have been used to reconstruct 3D models given sparse PCs in either the cultural or biomedical domains.

2 Method

The proposed methodology aims at reconstructing 3D coarsely retrieved models by employing a fully-connected MLP network tasked to learn an implicit continuous representation of that model. An overview of the proposed methodology describing its pre-processing, training, and reconstruction stages is illustrated in Fig. 1. Each hidden layer of the MLP model comprises neurons formally expressed as $\varphi(\boldsymbol{x};\boldsymbol{W},b,\omega)$, where $\varphi(\cdot)$ denotes the WNA function, $\boldsymbol{W}$ represents the weights of the neurons, b is the bias, and ω is the learnable parameter of the WNA function. The proposed methodology does not require training on large datasets since it focuses on a single 3D model. The MLP network, utilizes the proposed WNA function to learn an SDF [28], which describes efficiently the 3D model that it aims to reconstruct. The MLP receives as input a point $\boldsymbol{u} = (x, y, z)^T$ of a 3D model, and it outputs a value approximating the respective SDF response. That value describes the distance of a point from the surface of the 3D model. After the training process, the model is capable of reconstructing the 3D model, which was initially coarse, at a higher resolution by predicting through inference if a point in the defined 3D space belongs to the surface of the model, *i.e.*, its distance from the surface is 0.

2.1 Implicit Neural Representation

Let us consider a point $\boldsymbol{p} = (x, y, z)^T$, with normalized coordinates $x, y, z \in [-1, 1]$ of an oriented PC $\boldsymbol{P}$, *i.e.*, the surface normal of each $\boldsymbol{p} \in \boldsymbol{P}$ is known, with 3D points that lie on the surface of a 3D model. This surface is described by an SDF $s(\bullet)$, s: $\mathbb{R}^3 \to \mathbb{R}$ [28], which encodes the surface of the 3D model as the signed distance of a point $\boldsymbol{p}$ to the closest surface. The response of the SDF corresponds to a distance d that encodes the position of the input point. Points that are outside and inside of the object surface are denoted by a positive and negative distance d, respectively. A response $d = 0$ denotes a point $\boldsymbol{p}$ that lies on the surface of the 3D model. An SDF that describes a particular 3D model can be approximated and implicitly represented by the weights of an MLP.

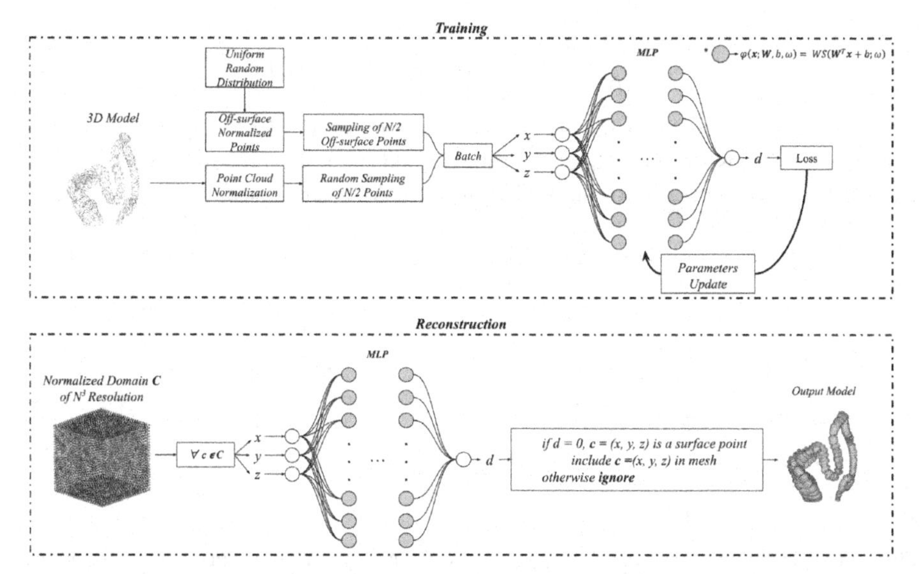

Fig. 1. Overview of the INR approach for 3D model refinement and restoration

An SDF describing a 3D model of an object can be approximated by an appropriately trained MLP $g(\bullet; \boldsymbol{\theta})$ parametrized by $\boldsymbol{\theta}$. The MLP $g(\bullet; \boldsymbol{\theta})$ should learn how the SDF responds for different points $\boldsymbol{p}$ and $\boldsymbol{q}$ that are on and off the surface of the model, respectively. In this paper, the loss function proposed in SIREN was used to train the MLP $g(\bullet; \boldsymbol{\theta})$. During the training of the network, the loss function leverages the properties of an SDF function by incorporating both the on and off surface points along with the respective surface normals. Each training batch is composed of N randomly sampled points $\boldsymbol{p}$ and $\boldsymbol{q}$, each of which comprising half the number ($N/2$) of points of the batch. Once g is trained, it can be used to reconstruct a mesh of an initially coarse model at higher resolutions. This is achieved by generating a cubic normalized PC $\boldsymbol{K}$ consisting of n^3 ($n \times n \times n$) points $\boldsymbol{k} = (x, y, z)^T$, $x, y, z \in [-1, 1]$ and inferring SDF values $\forall \boldsymbol{k} \in \boldsymbol{K}$ using g. For the reconstruction of the mesh only the points $\boldsymbol{k}$ that provide zero responses are considered. By determining the points belonging to the object surface the marching cubes algorithm is employed to generate the final refined 3D mesh [24].

2.2 WaveShaping Activation Function

The WNA function is introduced aiming at further improving the implicit representation capacity of MLPs regarding coarse 3D models. The implementation of the proposed WNA is achieved by applying the *tanh* function as a "wave-shaper" to the *sin* function. The *tanh* function is chosen since it is widely used as a wave-shaper in various applications in the context of signal processing [17, 22, 29]. The proposed WNA function and its derivative are:

$$WS(x; \omega) = tanh(sin(\omega x)) \quad (1)$$

$$\frac{dWS(x;\omega)}{dx} = \omega cos(\omega x)sech^2(sin(\omega x)) \tag{2}$$

where $\omega \in \mathbb{R}$ is a learnable parameter of WNA in contrast to SIREN, where it is applied manually as a constant throughout the training and inference processes. Equation (2) describes the first derivative of the WNA function. As it can be seen in Fig. 2(a), for $\omega = 1$, the WNA function can be regarded as a scaled approximation of *sin*.

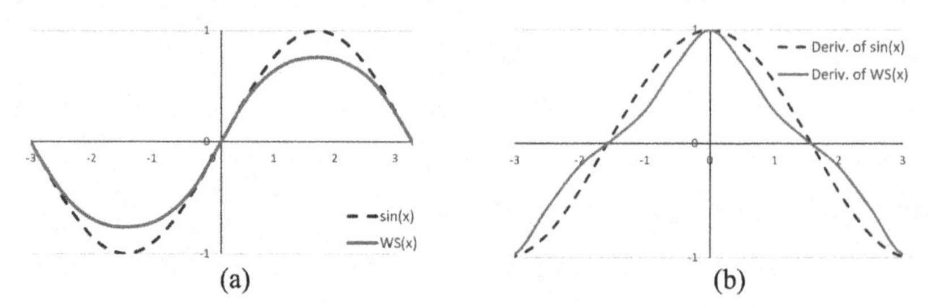

Fig. 2. Graphical representation of the proposed WNA and *sin* functions. (a) Responses and (b) derivatives of the *WNA* and *sin* functions.

This means that WNA maintains the properties of *sin*, *i.e.*, phase, periodicity, and it has upper and lower bounds. Nevertheless, it can be observed that the derivation of the WNA function produces a more complex expression (Fig. 2(b)); thus, it is expected that, during the training process, WNA will produce a different computation of gradients compared to *sin*. The following section shows that the gradients computed during the training of an MLP that utilizes the WNA function enable the network to efficiently reconstruct 3D models given only a small number of surface points.

3 Experiments and Results

3.1 Experimental Setup

The evaluation of the proposed method was performed with respect to the reconstruction accuracy of the high-resolution models given their coarse representations. For comparison, SIREN was incorporated in the evaluation. Additionally, the Screened Poisson (SP) surface reconstruction method [20] was used to generate 3D meshes as a baseline in the comparative study. More specifically, SP was selected due to its popularity and good performance amongst state-of-the-art surface reconstruction techniques [41, 44].

The reconstruction accuracy was assessed using the Chamfer Distance (CD) [4], Earth Mover's Distance (EMD) [23], and Mesh Completion (MC) [33] metrics. Given two different PCs, CD is used to evaluate the distances between their closest points. EMD is used for the comparison of two different data distributions and is approximated by the 1st Wasserstein Distance. Both CD and EMD are widely used for the evaluation of 3D reconstruction methods [1, 11]. These metrics were employed to assess the similarity of two PCs in terms of distance. MC is another metric that can be used to evaluate the quality of the generated mesh. MC indicates the degree that a generated mesh matches the ground truth mesh. This is achieved by selecting n samples of points ($n = 1000$) from the ground truth mesh and finding the fraction of points that are closer than a minimum distance Δd from the generated mesh. In this study, the value of Δd was selected according to the scale of each model.

In all the experiments, the architecture of the neural network comprised 4 hidden layers, with 256 neurons each, the proposed WNA function, and a linear output layer. This type of MLP architecture has been widely used in related works [35, 36]. All models used in the evaluation process were trained with the same number of epochs, batch size, and initialization parameters. In particular, the batch size was 5×10^3 on-surface points and 5×10^3 off-surface points, and the number of epochs for the experiments was 3×10^3; it was observed that training the network for more epochs did not yield better results. The Adam optimizer was utilized for the weights optimization with a learning rate of 10^{-4}. To simulate 3D models with a sparse 3D representation, the dense PC of each high-resolution 3D model was used to generate sparse PCs with densities of 0.5%, 1%, 5%, 10%, 20%, and 40% the original PC density. This was achieved by randomly sampling points from each original PC based on a uniform distribution.

3.2 Biomedical 3D Models

In the case study of biomedical 3D models, a set of 10 different high-resolution 3D models of the large and small intestines was used for evaluation. These models were reconstructed from CT data obtained from The Cancer Imaging Archive [9], and were sub-sampled based on the method described in Subsect. 3.1. To further assess the performance of the proposed method on data that resemble PCs extracted from sparse CT scans (CT slices with large thickness), a different sub-sampling method was additionally implemented. More specifically, the PC points were periodically sampled from the original PCs every 5 mm along the z axis, emulating PCs obtained from a CT scan with a slice thickness of 5 mm. This resulted in 10 more sub-sampled PCs that will be referred to as CT-sampled PCs. Thus, the dataset of the biomedical 3D models contained 70 different PCs of various densities.

Figure 3 qualitatively compares the reconstruction results obtained by different methodologies of a large intestine. The original PC in Fig. 3 comprised ~ 80k points. As it can be observed, the utilization of the WNA function improves substantially the refinement capacity of the MLP when the PC is highly sparse (Figs. 3(a)). As the PC of the model becomes denser (Figs. 3(c, d)), i.e., more detailed, the results produced by the proposed methodology are comparable to those produced by SIREN. Moreover, Figs. 3(b, c, d) show that the baseline model can reconstruct the shape of the target 3D geometry, but the quality is worse compared to the other two methods. Lastly, the models trained with the CT-sampled PCs appear to have a comparable quality for both the proposed method and SIREN (Figs. 3(d)).

Table 1 summarizes the quantitative results obtained from the comparative evaluation in terms of average CD, EMD, and MC. The best results are indicated in boldface typesetting. The quantitative results confirm the qualitative observations. It can be noticed that the proposed methodology results in a notably better performance with only 1% and 5% points of the original PC. According to the average MC results, the proposed method outperforms the other methods for each density level, except for the 40% one, indicating

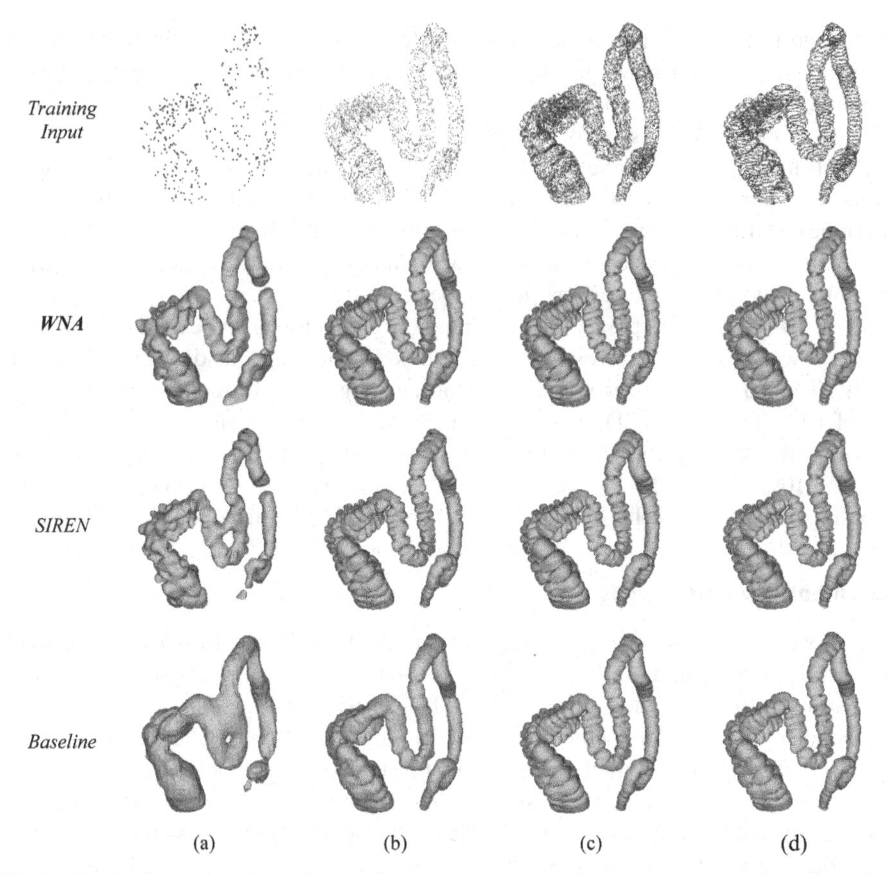

Fig. 3. Qualitative comparison of the reconstruction outcome for a representative large intestine reconstructed from PCs with various densities: (a) 0.5%; (b) 5%; (c) 20%; (d) CT-sampled.

that it can produce meshes with better quality. Moreover, the proposed method and SIREN needed 40% of the original PC to learn the represented shape and only 5% to generate high-quality meshes.

3.3 Cultural Heritage 3D Models

In the cultural heritage domain, a set of 10 high-resolution 3D models of exhibits displayed in the Archaeological Museum of Delphi, Greece, including the Charioteer of Delphi, a sarcophagus, bronze statuette of Kouros, chryselephantine statuette, clay figurine on tripod seat, and others, was used.

As in Subsect. 3.2, the original PCs of these models were subsampled resulting in a dataset of 60 PCs. The evaluation results of the reconstructed models agree well with those obtained for the GI tract models. Figure 4 shows the reconstruction results for a 3D mesh of a sarcophagus that originally comprised ~ 277k points. It can be observed that for low PC densities (Figs. 4(a, b)) the proposed method is able to reconstruct the mesh with coarse details, whereas SIREN produces a mesh with more artifacts and holes.

Table 1. Average CD, EMD, and MC of the proposed WNA function against other methods for GI tract 3D models with different PC densities. ↑ and ↓ means higher and lower is better, respectively.

PC Density (%)	Methodologies								
	WNA			*SIREN*			*Baseline (SP)*		
	CD↓	EMD↓	MC↑	CD↓	EMD↓	MC↑	CD↓	EMD↓	MC↑
0.5	**4.386**	**1.390**	**0.427**	5.015	1.722	0.372	9.443	3.942	0.004
1	**3.164**	**0.865**	**0.601**	3.343	0.982	0.566	6.314	3.655	0.023
5	**2.330**	**0.434**	**0.949**	2.343	0.446	0.926	3.487	1.509	0.157
10	**2.391**	0.413	**0.983**	2.290	**0.409**	0.958	2.496	0.956	0.405
20	2.201	**0.419**	**0.998**	**2.200**	0.427	0.995	2.245	1.263	0.513
40	**2.197**	0.399	**1.000**	2.199	**0.395**	**1.000**	2.265	1.750	0.904
CT-sample	**2.228**	0.423	**0.982**	2.233	**0.361**	0.971	2.289	1.834	0.529

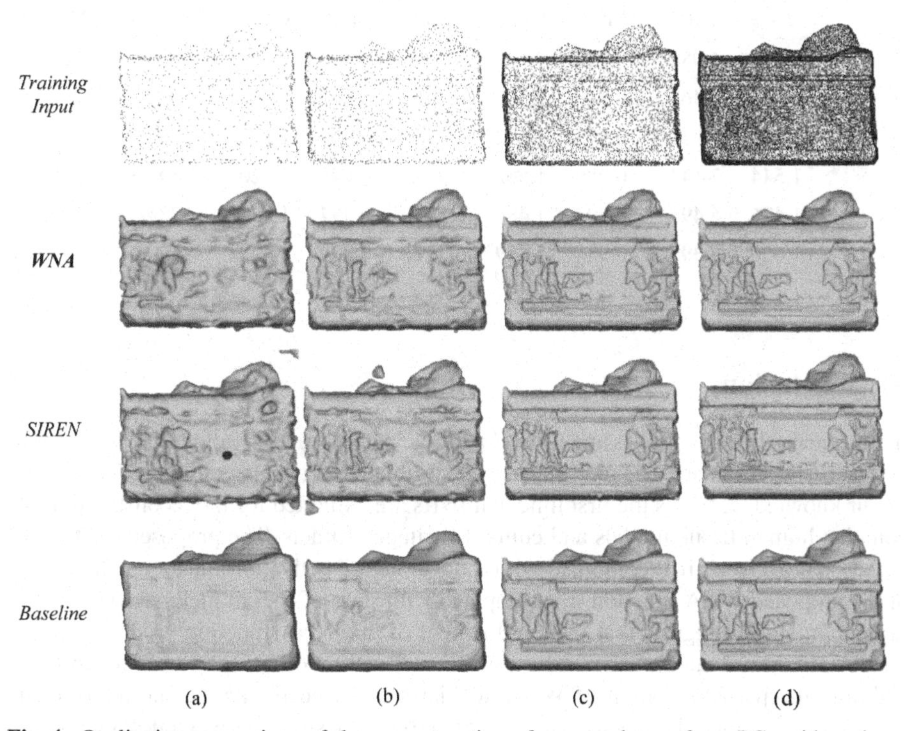

Fig. 4. Qualitative comparison of the reconstruction of a sarcophagus from PCs with various initial densities: (a) 0.5%; (b) 5%; (c) 10%; (d) 20%.

As in Figs. 3 (b, c), the quality of the meshes improves with increasing PC density. Moreover, it can be seen that the baseline model produces low-quality meshes even for higher densities (Figs. 4(c, d)).

Table 2 exhibits the quantitative results obtained from the comparative evaluation of the cultural objects. Based on the CD values, the proposed methodology outperforms the other methods in terms of the similarity between PCs, except from the 40% PC density. In addition, the MC values verify the capacity of the proposed method to generate more complete meshes given PCs with low density (Fig. 4(a)). Again, for higher PC densities the proposed method is on par with SIREN. These types of objects (Fig. 4) encompass challenging surface geometries, *i.e.*, both smooth and irregular. This may be the reason why both the proposed method and SIREN produce meshes with comparable EMD values.

Table 2. Average CD, EMD, and MC of the proposed WNA function against other methods for cultural heritage 3D models with different PC densities. ↑ and ↓ means higher and lower is better, respectively.

PC Density (%)	Methodologies								
	WNA			*SIREN*			*Baseline (SP)*		
	CD ↓	EMD↓	MC↑	CD ↓	EMD↓	MC↑	CD ↓	EMD↓	MC↑
0.5	**3.747**	**6.084**	**0.747**	7.533	7.065	0.700	9.459	6.799	0.055
1	**2.571**	6.465	**0.860**	2.590	6.891	0.851	5.863	**4.206**	0.166
5	**1.668**	6.941	**0.975**	1.692	6.778	0.969	2.862	**3.209**	0.610
10	**1.544**	7.144	0.989	1.565	6.909	**0.991**	2.120	**3.835**	0.889
20	**1.476**	**6.495**	0.993	1.486	6.523	**0.994**	1.870	7.683	0.943
40	1.650	**5.395**	0.984	**1.459**	6.371	**0.990**	1.699	7.414	0.963

4 Conclusions

In this paper, the use of INRs and a novel periodic parametric activation function for the reconstruction of coarse 3D models has been evaluated in two case studies. To the best of our knowledge, this is the first time that INRs are exploited for the reconstruction of complex human tissue models and cultural heritage artifacts. The proposed method is based on a 5-layer MLP combined with a novel neural activation function. The effect of the proposed WNA function was compared with state-of-the-art methodologies. The evaluation study suggests that the employment of the WNA function produces better results, when evaluated in the context of 3D reconstruction of sparse 3D models. In addition, the parametrization of WNA alleviates the need for its manual adjustment when the target model changes. The proposed 3D reconstruction methodology can be regarded as unsupervised since it is trained directly on a PC of a coarse 3D model without the need for a huge dataset of other models and any labeled data. Given the fact that periodic functions seem to have a positive effect on INRs, future work should attempt to explore other periodic activation functions, determine how different types of parameterization affect the performance of the network, and seek for ways to improve the reconstruction quality in scenarios with extremely sparse PCs.

Acknowledgement. We acknowledge support of this work by the project "Smart Tourist" (MIS 5047243) which is implemented under the Action "Reinforcement of the Research and Innovation Infrastructure", funded by the Operational Programme "Competitiveness, Entrepreneurship and Innovation" (NSRF 2014–2020) and co-financed by Greece and the European Union (European Regional Development Fund).

References

1. Achlioptas, P., Diamanti, O., Mitliagkas, I., Guibas, L.: Learning representations and generative models for 3D point clouds. In: International Conference on Machine Learning. PMLR, pp. 40–49 (2018)
2. Bagautdinov, T., Wu, C., Saragih, J., Fua, P., Sheikh, Y.: Modeling facial geometry using compositional VAEs. In: Proceedings of the IEEE Conference on Computer Vision and Pattern Recognition, pp 3877–3886 (2018)
3. Balashova, E., Wang, J., Singh, V., Georgescu, B., Teixeira, B., Kapoor, A.: 3D organ shape reconstruction from Topogram images. In: Chung, A.C.S., Gee, J.C., Yushkevich, P.A., Bao, S. (eds.) IPMI 2019. LNCS, vol. 11492, pp. 347–359. Springer, Cham (2019). https://doi.org/10.1007/978-3-030-20351-1_26
4. Ballarin, M., Balletti, C., Vernier, P.: Replicas in cultural heritage: 3D printing and the museum experience. Int. Arch. Photogramm. Remote. Sens. Spat. Inf. Sci. **42**, 55–62 (2018)
5. Chabra, R., et al.: Deep local shapes: learning local SDF priors for detailed 3D reconstruction. In: Vedaldi, A., Bischof, H., Brox, T., Frahm, J.-M. (eds.) ECCV 2020. LNCS, vol. 12374, pp. 608–625. Springer, Cham (2020). https://doi.org/10.1007/978-3-030-58526-6_36
6. Chen, X., et al.: A fast reconstruction method of the dense point-cloud model for cultural heritage artifacts based on compressed sensing and sparse auto-encoder. Opt. Quant. Electron. **51**, 1–16 (2019)
7. Chen, Z., Zhang, H.: Learning implicit fields for generative shape modeling. In: Proceedings of the IEEE/CVF Conference on Computer Vision and Pattern Recognition, pp 5939–5948 (2019)
8. Chibane, J., et al.: Neural unsigned distance fields for implicit function learning. In: Advances in Neural Information Processing Systems, vol. 33, pp. 21638–21652 (2020)
9. Clark, K., et al.: The cancer imaging archive (TCIA): maintaining and operating a public information repository. J. Digit. Imaging **26**, 1045–1057 (2013)
10. Dai, A., Ruizhongtai Qi, C., Nießner, M.: Shape completion using 3D-encoder-predictor CNNs and shape synthesis. In: Proceedings of the IEEE Conference on Computer Vision and Pattern Recognition, pp 5868–5877 (2017)
11. Deng, Z., Yao, Y., Deng, B., Zhang, J.: A robust loss for point cloud registration. In: Proceedings of the IEEE/CVF International Conference on Computer Vision, pp. 6138–6147 (2021)
12. Garcia Carrizosa, H., Sheehy, K., Rix, J., Seale, J., Hayhoe, S.: Designing technologies for museums: accessibility and participation issues. J. Enabl. Technol. **14**, 31–39 (2020)
13. Gómez-Rodrguez, J.J., Lamarca, J., Morlana, J., Tardós, J.D., Montiel, J.M.: SD-DefSLAM: Semi-direct monocular SLAM for deformable and intracorporeal scenes. In: 2021 IEEE International Conference on Robotics and Automation (ICRA), pp 5170–5177. IEEE (2021)
14. Gropp, A., Yariv, L., Haim, N., Atzmon, M., Lipman, Y.: Implicit geometric regularization for learning shapes. arXiv preprint arXiv:200210099 (2020)
15. Groueix, T., Fisher, M., Kim, V.G., Russell, B.C., Aubry, M.: A papier-mâché approach to learning 3D surface generation. In: Proceedings of the IEEE Conference on Computer Vision and Pattern Recognition, pp 216–224 (2018)

16. Hu, M., Penney, G., Edwards, P., Figl, M., Hawkes, D.J.: 3D reconstruction of internal organ surfaces for minimal invasive surgery. In: Ayache, N., Ourselin, S., Maeder, A. (eds.) MICCAI 2007. LNCS, vol. 4791, pp. 68–77. Springer, Heidelberg (2007). https://doi.org/10.1007/978-3-540-75757-3_9
17. Huovilainen, A.: Non-linear digital implementation of the Moog ladder filter. In: Proceedings of the International Conference on Digital Audio Effects (DAFx-04), pp 61–64 (2004)
18. Kalozoumis, P.G., Marino, M., Carniel, E.L., Iakovidis, D.K.: Towards the development of a digital twin for endoscopic medical device testing. In: Hassanien, A.E., Darwish, A., Snasel, V. (eds.) Digital Twins for Digital Transformation: Innovation in Industry. Studies in Systems, Decision and Control, vol. 423, pp. 113–145. Springer, Cham (2022). https://doi.org/10.1007/978-3-030-96802-1_7
19. Kaneda, A., Nakagawa, T., Tamura, K., Noshita, K., Nakao, H.: A proposal of a new automated method for SfM/MVS 3D reconstruction through comparisons of 3D data by SfM/MVS and handheld laser scanners. PLoS ONE **17**, e0270660 (2022)
20. Kazhdan, M., Hoppe, H.: Screened Poisson surface reconstruction. ACM Trans. Graph. (ToG) **32**, 1–13 (2013)
21. Lamarca, J., Parashar, S., Bartoli, A., Montiel, J.: DefSLAM: tracking and mapping of deforming scenes from monocular sequences. IEEE Trans. Rob. **37**, 291–303 (2020)
22. Lazzarini, V., Timoney, J.: New perspectives on distortion synthesis for virtual Analog oscillators. Comput. Music. J. **34**, 28–40 (2010)
23. Levina, E., Bickel, P.: The earth mover's distance is the mallows distance: some insights from statistics. In: Proceedings Eighth IEEE International Conference on Computer Vision. ICCV 2001, pp 251–256. IEEE (2001)
24. Lewiner, T., Lopes, H., Vieira, A.W., Tavares, G.: Efficient implementation of marching cubes' cases with topological guarantees. J. Graph. Tools **8**, 1–15 (2003)
25. Ma, B., Han, Z., Liu, Y.-S., Zwicker, M.: Neural-pull: learning signed distance functions from point clouds by learning to pull space onto surfaces. arXiv preprint arXiv:201113495 (2020)
26. Makantasis, K., Doulamis, A., Doulamis, N., Ioannides, M.: In the wild image retrieval and clustering for 3D cultural heritage landmarks reconstruction. Multimed. Tools Appl. **75**, 3593–3629 (2016)
27. Mescheder, L., Oechsle, M., Niemeyer, M., Nowozin, S., Geiger, A.: Occupancy networks: learning 3D reconstruction in function space. In: Proceedings of the IEEE/CVF Conference on Computer Vision and Pattern Recognition, pp 4460–4470 (2019)
28. Osher, S., Fedkiw, R.: Signed distance functions. In: Level Set Methods and Dynamic Implicit Surfaces, pp 17–22. Springer (2003)
29. Pakarinen, J., Yeh, D.T.: A review of digital techniques for modeling vacuum-tube guitar amplifiers. Comput. Music. J. **33**, 85–100 (2009)
30. Park, J.J., Florence, P., Straub, J., Newcombe, R., Lovegrove, S.: DeepSDF: learning continuous signed distance functions for shape representation. In: Proceedings of the IEEE/CVF Conference on Computer Vision and Pattern Recognition, pp 165–174 (2019)
31. Peng, S., Niemeyer, M., Mescheder, L., Pollefeys, M., Geiger, A.: Convolutional occupancy networks. In: Vedaldi, A., Bischof, H., Brox, T., Frahm, J.-M. (eds.) ECCV 2020. LNCS, vol. 12348, pp. 523–540. Springer, Cham (2020). https://doi.org/10.1007/978-3-030-58580-8_31
32. Qi, C.R., Su, H., Mo, K., Guibas, L.J.: Pointnet: Deep learning on point sets for 3D classification and segmentation. In: Proceedings of the IEEE Conference on Computer Vision and Pattern Recognition, pp 652–660 (2017)
33. Seitz, S.M., Curless, B., Diebel, J., Scharstein, D., Szeliski, R.: A comparison and evaluation of multi-view stereo reconstruction algorithms. In: 2006 IEEE Computer Society Conference on Computer Vision and Pattern Recognition (CVPR 2006), pp 519–528. IEEE (2006)
34. Sengupta, A., Bartoli, A.: Colonoscopic 3D reconstruction by tubular non-rigid structure-from-motion. Int. J. Comput. Assist. Radiol. Surg. **16**, 1237–1241 (2021)

35. Ben-Shabat, Y., Koneputugodage, C.H., Gould, S.: DiGS: divergence guided shape implicit neural representation for unoriented point clouds. In: Proceedings of the IEEE/CVF Conference on Computer Vision and Pattern Recognition, pp 19323–19332 (2022)
36. Sitzmann, V., Martel, J., Bergman, A., Lindell, D., Wetzstein, G.: Implicit neural representations with periodic activation functions. In: Advances in Neural Information Processing Systems, vol. 33, pp. 7462–7473 (2020)
37. Vaz, R., Freitas, D., Coelho, A.: Blind and visually impaired visitors' experiences in museums: increasing accessibility through assistive technologies. Int. J. Inclusive Mus. **13**, 57 (2020)
38. Wang, Z., et al.: A Deep Learning based Fast Signed Distance Map Generation. arXiv preprint arXiv:200512662 (2020)
39. Wilson, P.F., Stott, J., Warnett, J.M., Attridge, A., Smith, M.P., Williams, M.A.: Evaluation of touchable 3D-printed replicas in museums. Curator Mus. J. **60**, 445–465 (2017)
40. Wu, Z., et al.: 3D shapenets: a deep representation for volumetric shapes. In: Proceedings of the IEEE Conference on Computer Vision and Pattern Recognition, pp 1912–1920 (2015)
41. Xu, Z., Xu, C., Hu, J., Meng, Z.: Robust resistance to noise and outliers: screened Poisson surface reconstruction using adaptive kernel density estimation. Comput. Graph. **97**, 19–27 (2021)
42. Yuan, W., Khot, T., Held, D., Mertz, C., Hebert, M.: PCN: point completion network. In: 2018 International Conference on 3D Vision (3DV), pp 728–737. IEEE (2018)
43. Zhang, S., Zhao, L., Huang, S., Ma, R., Hu, B., Hao, Q.: 3D reconstruction of deformable colon structures based on preoperative model and deep neural network. In: 2021 IEEE International Conference on Robotics and Automation (ICRA), pp 1875–1881. IEEE (2021)
44. Zhou, L., Sun, G., Li, Y., Li, W., Su, Z.: Point cloud denoising review: from classical to deep learning-based approaches. Graph. Models **121**, 101140 (2022)

A Deep Learning Approach to Segment High-Content Images of the *E. coli* Bacteria

Dat Q. Duong[1,2,3], Tuan-Anh Tran[4], Phuong Nhi Nguyen Kieu[1,2,3], Tien K. Nguyen[2,3], Bao Le[1,2,3], Stephen Baker[5], and Binh T. Nguyen[1,2,3](✉)

[1] AISIA Research Lab, Ho Chi Minh City, Vietnam
ngtbinh@hcmus.edu.vn
[2] University of Science, Ho Chi Minh City, Vietnam
[3] Vietnam National University, Ho Chi Minh City, Vietnam
[4] Nuffield Department of Medicine, University of Oxford, Oxford, UK
[5] Department of Medicine, University of Cambridge, Cambridge, UK

Abstract. High-content imaging (HCI) has been used to study antimicrobial resistance in bacteria. Although cell segmentation is critical for accurately analyzing bacterial populations, existing HCI platforms were not optimized for bacterial cells. This study proposes a convolutional neural network-based approach utilizing transfer learning and fine-tuning to perform instance segmentation on fluorescence images of *E. coli*. The method uses the pre-trained EfficientNet as the encoder for feature extraction and U-Net for reconstructing the segmentation maps containing the cell cytoplasm and the cell instance boundary. Next, individual cells are separated using a marker-controlled watershed transformation. The EffNetB7-UNet yields the best performance with the highest F1-Score of 0.91 compared to other methods.

Keywords: Bacteria · Data set · Convolutional neural networks · Instance segmentation · Transfer learning

1 Introduction

Antimicrobial resistance (AMR) is a global health challenge [25]. A systematic review published in 2019 estimated that approximately 1.27 million deaths are attributable to AMR infections worldwide [22]. Remarkably, there were 33,000 deaths caused by resistant bacteria in Europe [7] and 23,000 deaths in the USA [12]. According to the WHO, the hospitalization costs for patients infected with AMR bacteria are significantly higher than that for patients infected with susceptible organisms [24]. The economic cost of AMR infections in Europe was estimated to be around 1.1 billion Euros [26]. In a broader view, without efficient intervention, AMR was predicted to directly or associatively cause 10 million deaths worldwide and will cost the global economy £55 trillion to treat infection with AMR pathogens [23].

J. Blanc-Talon et al. (Eds.): ACIVS 2023, LNCS 14124, pp. 184–195, 2023.
https://doi.org/10.1007/978-3-031-45382-3_16

Microscopic techniques have been at the heart of biology, especially microbiology. Advance in automation and image processing algorithms has enabled the development of HCI technology. Compared to conventional microscopes, HCI is able to analyze biological samples at the single-cell level, which produces an enormous amount of imaging data providing in-depth biological insights into the samples [4]. Contributing to the fight against AMR, HCI has been recently utilized to predict the modes of action (MoA) of antimicrobials [31] and infer the resistance to clinically significant antimicrobials [28].

The process of delineating cell instances is often the first and fundamental step for various biomedical applications to extract meaningful biological insights into the bacterial properties [5,21]. This task involves performing semantic segmentation or instance segmentation. While semantic segmentation typically involves assigning the image's pixel values to the object it belongs to (e.g., bacterial cells or background), instance segmentation requires computationally identifying each individual cell separately in the image and assigning each cell with unique values [15]. However, manual segmentation for bacterial images is time-consuming and labor-intensive, especially in dense bacterial populations. Built-in image-processing programs for HCI platforms, such as Harmony [20], are optimized on eukaryotic cell images and not optimal for bacterial images. In recent years, deep learning has been proven to be a prominent tool for microscopy image analysis [15], and many open-source applications and packages have been developed for segmentation and analysis of bacterial images [2,9,27,32,33]. Of these, SuperSegger [33] is a threshold-based method for single cell analysis, while Ilastik [2], MiSiC [27], and Omnipose [9] utilize deep learning approaches. The latter two methods provide instance segmentation with two different approaches. MiSiC is a watershed-based method, whereas the more recent method Omnipose, developed from the framework of Cellpose [32], is based on the gradient-flows tracking method.

Wei et al. [37] mentioned that CNN-based instance segmentation methods could be categorized into two groups: top-down and bottom-up. Top-down methods such as Mask-RCNN [13] directly predict the cell instances from the input image. In contrast, bottom-up techniques such as UNet-based methods usually involve predicting a binary mask and a boundary mask [8], or a binary mask and gradient flow maps [32], followed by a series of post-processing techniques such as watershed transformation [36], connected components labeling [11] or gradient-flow tracking [32] to separate the instances. Bottom-up methods, i.e., UNet-based methods, have shown better results in recent biological cell segmentation challenges [6,17] and can be adapted for 3D problems without having difficulties setting proper anchor sizes in some top-down methods [18]. Our approach followed the path of the bottom-up methods utilizing the marker-based watershed algorithm [3] and the transfer learning to our data set.

Transfer learning is an approach that focuses on transferring knowledge from a source domain to a target domain to improve the performance of target learners on target domains. Transfer learning offers many benefits, such as reduced training time and improved network performance compared to random initialization making it very useful in biomedical image analysis when annotated data is scarce [14,19]. Transfer learning and fine-tuning approaches have been used in recent research and competitions. Gihun and his colleagues [17] proposed a holistic pipeline that employs pretraining and finetuning, namely MEDIAR, for cell instance segmentation under a multi-modality environment. The approach achieved the highest score on the NeurIPS 2022 cell segmentation challenge.

In this study, we first develop a protocol to acquire high-content fluorescent images of the *E. coli* bacteria cells in densely populated environments. One hundred full-size images of 2160 $\times$ 2160 pixels are carefully annotated and will be released for public access to contribute to the biomedical community. Compared to other public data sets [9,30], our data set contains images at a higher resolution, with a dense distribution of bacterial cells in each field of view. We then employ a convolutional neural network approach, namely EffNet-UNet, to boost the segmentation network performance by utilizing transfer learning. The proposed method outperforms other methods in this study and can be considered a standardized benchmark for bacterial segmentation for high-content imaging analysis.

2 Methodology

This section presents our main segmentation problem, how we collected and annotated data for the experiments, the proposed deep learning approach, and the evaluation metrics.

2.1 Data Acquisition

One antimicrobial-resistant isolate was randomly selected for this study from a collection of twenty-six bloodborne *E. coli* [35]. The study was approved by the Oxford Tropical Research Ethic Committee (OxTREC, reference number 35–16) and the Ethic Committee of Children's Hospital 1 (reference number 73/GCN/BVND1). One laboratory isolate of ATCC 25922 was included in this study as a susceptible reference.

Image acquisition was performed on Opera Phenix high-content imaging platform (PerkinElmer, Germany); 100 images were selected randomly for annotation at full resolution (2160 $\times$ 2160) using the APEER Annotate platform[1]. An example of a raw image and its corresponding instance segmentation mask is shown in Fig. 1.

2.2 Modeling Approach

This study proposes a transfer-learning-based approach, namely, Efficient-UNet [1], to perform semantic segmentation of the *E. coli* cell body and cell boundary contour, followed by a marker-controlled watershed transform to separate

[1] https://www.apeer.com/annotate.

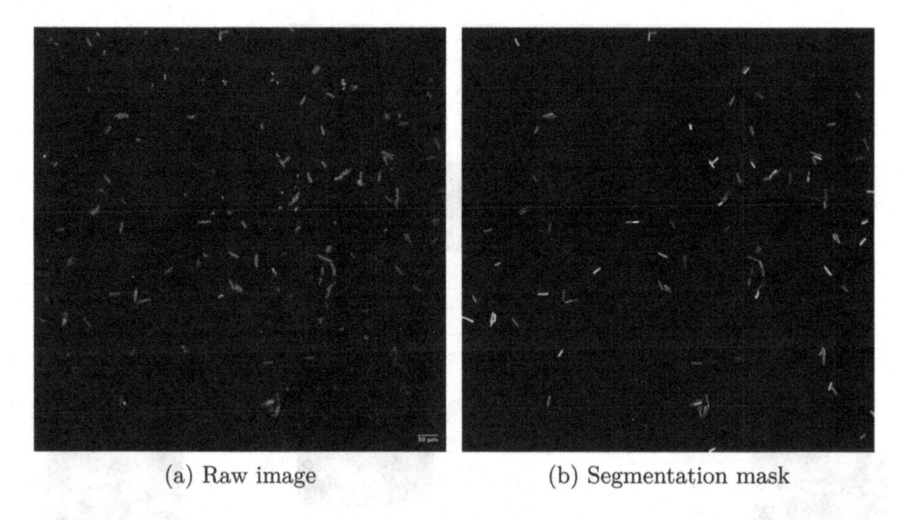

(a) Raw image (b) Segmentation mask

Fig. 1. A full-size *E. coli* fluorescence image and the corresponding segmentation mask.

the individual cells for the final instance segmentation result. EfficientNet is the convolutional neural network (CNN) architecture and compound scaling method [34] that effectively scales up the network depth, width, and resolution to achieve state-of-the-art results on the classification task of the ImageNet data set. There are eight variants in the EfficientNets family, namely EfficientNetB0 to EfficientNetB7. EfficientNets have been shown to consistently achieve higher accuracy on small new data sets with an order of magnitude fewer parameters than existing methods by employing transfer learning using the pre-trained model on the ImageNet [10] data set. We aim to combine the effectiveness of EfficientNet as a feature extractor in the encoder and the U-Net [29] decoder for generating fine segmentation maps.

Due to the nature of high resolution in microscopy images, it is difficult for neural network models to train on native-resolution data. Many approaches have been proposed to resolve this issue, such as resizing, random cropping, and creating smaller patches of the input images to feed into the model [6,8,17]. We adopt the approach of creating a patch data set to use as inputs for training the neural networks. Multiple patches of size (256×256) are extracted from the original data set of size (2160×2160); we choose a minimal overlapping size between patches and remove the patches with no bacteria cells or if the cell size is smaller than 100 pixels. During validation and inference, the full-size image will be predicted using a sliding-window process. An overview of our approach is shown in Fig. 2.

Network Structure. The network architecture is based on the renowned U-Net [29] with a modification to the encoder path for better feature extraction [1]. The EfficientNet is used as an encoder in the contracting path instead of a conventional set of convolution layers to capture low-level feature maps. And the

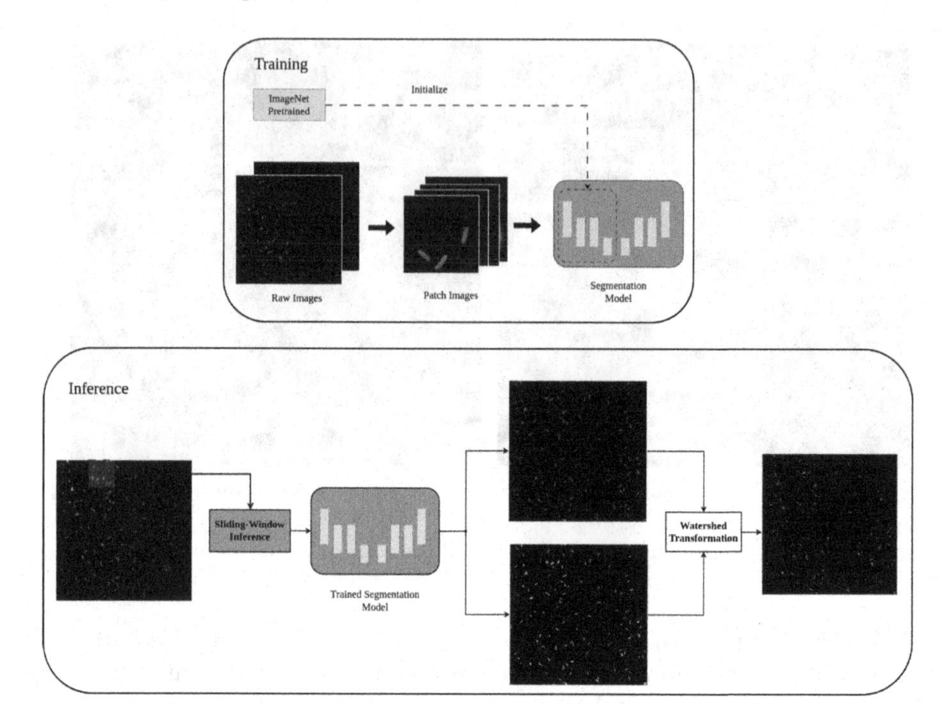

Fig. 2. An overview of our approach for *E.coli* cell instance segmentation.

decoder part from the original U-Net is reused. The input size is 256×256 patch images from the preprocessing step. In the decoder, transposed convolution was employed to restore the resolution of the feature maps. The output segmentation maps consist of two channels, one for the semantic mask of the cell cytoplasm and the other for the cell instance boundary. The pre-trained weights of EfficientNet on ImageNet are used as initialization in the encoder and then further fine-tuned to our data set.

Loss Function. For training the U-Net and EfficientNet-UNet models, we adopt a weighted Binary cross entropy (WBCE) loss function to account for the binary semantic mask y_{seg} and boundary contours mask y_{cont} [18], defined as:

$$\mathcal{L}_{WBCE}(x, y) = \lambda \mathcal{L}_{BCE}(\hat{y}_{seg}, y_{seg}) + (1 - \lambda)\mathcal{L}_{BCE}(\hat{y}_{cont}, y_{cont}), \tag{1}$$

where $\hat{y}_{seg}$ and $\hat{y}_{cont}$ are the decoder outputs that correspond to the segmentation mask and the boundary contours given the input data x. We set λ as 0.5 in our experiments to balance between optimizing the segmentation mask and the boundary contour.

Inference Method. To conduct inference on large-size input images, one can employ a sliding window process at a fixed overlap size between adjacent patches

[17]. In our experiments, we use a window size of (256 × 256) with a minimal overlap size of 8.4 pixels to reconstruct the final image.

Post-processing. We observe that the smallest cell in our data set has an area of 101 pixels. Thus, after obtaining the semantic mask and the instance boundary contours, small blobs with an area of fewer than 100 pixels are removed from the predictions by morphological masking operation. Then, a marker-controlled watershed transformation [3] is applied to the output probability maps. In our experiments, we set the threshold of 0.5, 0.1, and 0.3 for computing the seed map, boundary contour, and semantic mask from the output probability maps, respectively.

2.3 Quantitative Evaluation Metrics

Three metrics, namely, Precision, Recall, and F1-Score (Eq. 2), are used for evaluating instance segmentation. A true positive is counted when a single predicted object matches a ground truth object with an Intersection over Union (IoU) above a threshold τ. A false positive (FP) indicates a predicted object had no associated ground truth object. A false negative (FN) indicates a ground truth object had no associated predicted object. *Precision* is calculated as the total number of true positive samples divided by the predicted cell instances. *Recall* is the fraction of true positive samples among the ground truth cells. The F1-Score is defined as the harmonic mean of the model's *precision* and *recall.* The metrics are calculated at the threshold value $\tau = 0.5$.

$$\mathcal{P}_\tau = \frac{TP_\tau}{TP_\tau + FP_\tau}; \quad \mathcal{R}_\tau = \frac{TP_\tau}{TP_\tau + FN_\tau}; \quad \mathcal{F}1_\tau = \frac{2 * TP_\tau}{2 * TP_\tau + FP_\tau + FN_\tau} \tag{2}$$

3 Experiments

Due to the small size of the data set, we conducted a 5-fold cross-validation to evaluate the performance and stability of the proposed methods. The number of training, validation, and testing images for each fold is 78, 2, and 20, respectively. During training, the patch data set is generated for every image in the training folds; the validation and testing images are kept in full-size resolution. The number of images in every training fold is increased to around 3000 images after the generation of the patch data set.

3.1 Model Settings

Cellpose [32] and Omnipose [9] models are implemented from the open-sourced GitHub repositories using Pytorch. For Cellpose and Omnipose models, we initialized the weights using the pre-trained "cyto" model and "bact_omni_fluor" model, respectively, then trained the models on our data set. The U-Net and EffNet-UNet models are implemented using TensorFlow. Experiments are run using 2 RTX-2080Ti GPU cards.

Data Augmentation. During training, real-time data augmentation is employed with six transformation techniques: random rotations in the range (−180°, 180°), random horizontal and vertical flipping, and elastic transformation [29], random contrast, and brightness adjustment.

Hyper-Parameters. The Cellpose and Omnipose models are trained using the default hyperparameters from the papers [9,32]. The U-Net and EffNet-UNet models are trained using the Adam stochastic optimizer [16] with an initial learning rate of 0.0003. The batch size number is set to 8, and the number of epochs is set at 200 with an early stopping criteria based on the validation set loss for 50 epochs.

3.2 Results and Discussion

In this study, we assess the performance of the proposed transfer learning method, in which multiple pre-trained EfficientNets were used as backbones with increasing size, namely, EffNetB0, EffNetB3, and EffNetB7. In addition, experiments are conducted to compare the proposed methods with other approaches, including Harmony [20], Omnipose [9], Cellpose [32], and the original U-Net architecture [29]. The Cellpose and Omnipose models are trained using weights initialized from pre-trained models mentioned in Sect. 3.1. The EfficientNets models are trained with encoder weights initialized from pre-trained models on the ImageNet-1K data set. The comparison results are shown in the means and standard deviations of precision, recall, and F1-Score are given in Table 1 and Table 2.

The means and standard deviations are calculated across all test sets from the 5-fold cross-validation procedure. Due to the nature of fluorescence staining and cell immobilization method, images with dense bacterial communities contain a high level of noise and artifacts. Therefore, we also aim to investigate the capability to perform instance segmentation of cells in high-resolution fluorescent images with multiple dense microcolonies where bacterial cells are in tight contact. Figures 3 and 4 show a few example images of the ground truth and predict segmentation at full resolution with multiple dense microcolonies and artifacts.

For the Harmony software, we developed a pipeline using a combination of multiple thresholding and filtering methods. The Harmony pipeline displayed high precision equivalent to other deep learning methods. However, in scenarios where images contain multiple artifacts and noise, the pipeline could not capture the cell contexts resulting in a low recall value of 0.6 (Table 1). This low recall explains Harmony's inability to segment bacterial cells in Fig. 4b accurately. The program segmented only two of 23 cells in the cropped image (Fig. 4b). Overall, since Harmony software uses intensity-thresholding methods, selecting a universal threshold for various scenarios is challenging, especially for images with low contrast and over-crowded bacterial populations.

For other deep learning approaches, Cellpose and Omnipose displayed better results than the Harmony program, with an F1-score of 0.87 and 0.73, respectively. Interestingly, the Omnipose model did not perform as well as the Cellpose model after fine-tuning process in our data set, although Ominipose was developed for bacterial segmentation. These results can be explained by intrinsic variation between distinct data sets with different conditions of data acquisition. Cellpose possessed the highest precision among all methods tested but lower recall and F1-score compared to the original U-Net and EffNetB7-UNet (Table 1). The effects of these results were demonstrated in Fig. 4c. Though Cellpose's prediction was accurate, it missed out on detecting many bacteria cells that are close together and under the effects of artifacts leading to a lower recall score.

Compared to the Cellpose and Onipose methods which are based on gradient-flow tracking, all the marker-controlled watershed U-Net-based models performed well in capturing dense bacterial cells, yielding high precision, recall, and F1-score. Overall, the precisions of these models were over 0.87, while recalls exceeded 0.91 and F1-scores reached at least 0.89 (Table 1). Among these models, the EffNetB7-UNet showed the highest performance with precision, recall, and F1-Score of 0.89, 0.94, and 0.91, respectively (Table 1). As shown in Fig. 4f, the method could segment 23 out of 23 cells from cropped images, which was explained by the high recall and F1-score. In addition, we also investigate the transfer learning performance of the EffNetB7-UNet model with weights initialized from a pre-trained model on the "bact_fluor" data set, i.e., the data set of bacterial fluorescent images published along with the Omnipose method [9]. However, only precision increased to 0.90, while the F1-score remained unchanged (Table 2).

Table 1. The average results from the 5-Fold cross-validation procedure between the proposed methods and other methods.

Model	$\mathcal{P}_{0.5}$	$\mathcal{R}_{0.5}$	$\mathcal{F}1_{0.5}$
Harmony	0.88 ± 0.02	0.60 ± 0.04	0.70 ± 0.04
Cellpose	$\mathbf{0.91 \pm 0.03}$	0.84 ± 0.07	0.87 ± 0.04
Omnipose	0.72 ± 0.03	0.81 ± 0.04	0.75 ± 0.03
U-Net	0.87 ± 0.04	0.91 ± 0.01	0.89 ± 0.03
EffNetB0-UNet	0.87 ± 0.02	0.91 ± 0.02	0.89 ± 0.02
EffNetB3-UNet	0.88 ± 0.02	0.92 ± 0.01	0.90 ± 0.02
EffNetB7-UNet	0.89 ± 0.03	$\mathbf{0.94 \pm 0.01}$	$\mathbf{0.91 \pm 0.01}$

Table 2. Results of EfficientNetB7-UNet when training with different weights initialization scheme.

Model	$\mathcal{P}_{0.5}$	$\mathcal{R}_{0.5}$	$\mathcal{F}1_{0.5}$
EffNetB7-UNet w/ImageNet encoder	0.89 ± 0.03	0.93 ± 0.01	0.91 ± 0.01
EffNetB7-UNet pretrained bact_fluor	$\mathbf{0.90 \pm 0.01}$	0.93 ± 0.01	0.91 ± 0.01

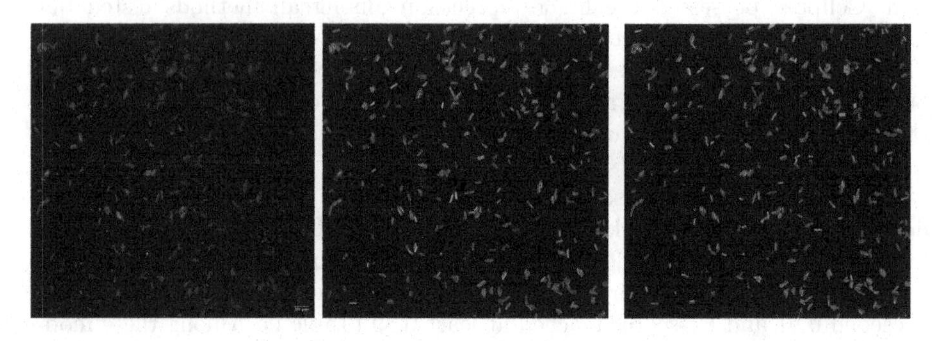

Fig. 3. *E. coli* cell instance segmentation on full-size images in dense microcolonies and multiple artifacts. The first column shows the raw input image, second and third columns show the ground truth annotation and predicted segmentation of the EffNetB7-UNet model, respectively. Segmentation maps are overlayed on the input images for better visualization.

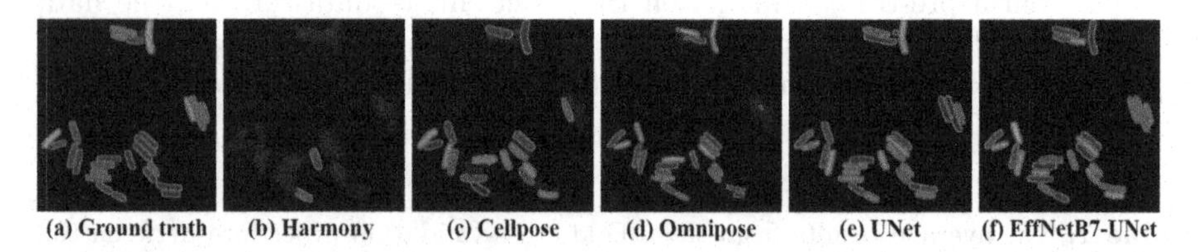

Fig. 4. Comparison between different segmentation methods. Images are zoomed in for clear visualization.

4 Conclusion and Future Works

Automatic delineation of bacterial cells is challenging, especially for dense populations. This study proposes a transfer learning approach using UNet-based deep convolution neural network models that incorporate knowledge from another domain into the biomedical domain to segment the *E. coli* bacteria from high-content images. Briefly, the EfficientNet [34] was used as an encoder for feature extraction combined with the original U-Net [29] decoder for generating segmentation maps containing the semantic mask of the cell cytoplasm and the cell instance boundary. Subsequently, a marker-controlled watershed transform was employed to create the instance segmentation of dense *E. coli* bacterial microcolonies from high-resolution fluorescent images. Quantitative evaluations between machine-generated and ground truth delineation are performed using a 5-fold cross-validation strategy for the proposed methods. The experimental results show that using transfer learning on a pre-trained EffNetB7-UNet architecture achieves the best performance with the highest F1-Score of 0.91 compared to other methods.

For future work, more annotations from the raw images of our data set will be collected to contribute to the biomedical community and perform a deeper analysis of the *E. coli* cells from HCI analysis.

Acknowledgment. This research is conducted under the Collaborative Research Agreement between the University of Oxford, the University of Cambridge, and the University of Science, Vietnam National University in Ho Chi Minh City, Viet Nam. We also want to acknowledge the support of the AISIA Research Lab during this paper.

References

1. Baheti, B., Innani, S., Gajre, S., Talbar, S.: Eff-UNet: a novel architecture for semantic segmentation in unstructured environment. In: 2020 IEEE/CVF Conference on Computer Vision and Pattern Recognition Workshops (CVPRW), Seattle, WA, USA, pp. 1473–1481. IEEE (2020)
2. Berg, S., et al.: Ilastik: interactive machine learning for (bio) image analysis. Nat. Methods **16**(12), 1226–1232 (2019)
3. Beucher, S., Meyer, F.: Segmentation: the watershed transformation. Mathematical morphology in image processing. Opt. Eng. **34**, 433–481 (1993)
4. Boutros, M., Heigwer, F., Laufer, C.: Microscopy-based high-content screening. Cell **163**(6), 1314–1325 (2015)
5. Caicedo, J., et al.: Data-analysis strategies for image-based cell profiling. Nat. Methods **14**(9), 849–863 (2017)
6. Caicedo, J.C., et al.: Nucleus segmentation across imaging experiments: the 2018 data science bowl. Nat. Methods **16**(12), 1247–1253 (2019)
7. Cassini, A., et al.: Attributable deaths and disability-adjusted life-years caused by infections with antibiotic-resistant bacteria in the EU and the European economic area in 2015: a population-level modelling analysis. Lancet. Infect. Dis **19**(1), 56–66 (2019)
8. Chen, H., Qi, X., Yu, L., Heng, P.A.: DCAN: deep contour-aware networks for accurate gland segmentation. In: 2016 IEEE Conference on Computer Vision and Pattern Recognition (CVPR), Las Vegas, NV, USA, pp. 2487–2496. IEEE (2016)
9. Cutler, K.J., et al.: Omnipose: a high-precision morphology-independent solution for bacterial cell segmentation. Nat. Methods **19**(11), 1438–1448 (2022)
10. Deng, J., Dong, W., Socher, R., Li, L., Li, K., Fei-Fei, L.: ImageNet: a large-scale hierarchical image database. In: 2009 IEEE Computer Society Conference on Computer Vision and Pattern Recognition Workshops (CVPR Workshops), Los Alamitos, CA, USA, pp. 248–255. IEEE Computer Society (2009)
11. Dillencourt, M.B., Samet, H., Tamminen, M.: A general approach to connected-component labeling for arbitrary image representations. J. ACM **39**(2), 253–280 (1992)
12. Hampton, T.: Report reveals scope of us antibiotic resistance threat. JAMA **310**(16), 1661–1663 (2013)
13. He, K., Gkioxari, G., Dollár, P., Girshick, R.: Mask R-CNN (2017)

14. Iman, M., Rasheed, K., Arabnia, H.R.: A review of deep transfer learning and recent advancements (2022)
15. Jeckel, H., Drescher, K.: Advances and opportunities in image analysis of bacterial cells and communities. FEMS Microbiol. Rev. **45**(4), fuaa062 (2021)
16. Kingma, D.P., Ba, J.: Adam: a method for stochastic optimization. In: Bengio, Y., LeCun, Y. (eds.) 3rd International Conference on Learning Representations, ICLR 2015, San Diego, CA, USA, 7–9 May 2015, Conference Track Proceedings, p. 80. ICLR, San Diego (2015)
17. Lee, G., Kim, S., Kim, J., Yun, S.Y.: Mediar: harmony of data-centric and model-centric for multi-modality microscopy (2022)
18. Li, M., Chen, C., Liu, X., Huang, W., Zhang, Y., Xiong, Z.: Advanced deep networks for 3D mitochondria instance segmentation (2021)
19. Maqsood, M., et al.: Transfer learning assisted classification and detection of Alzheimer's disease stages using 3D MRI scans. Sensors **19**, 2645 (2019)
20. Massey, A.J.: Multiparametric cell cycle analysis using the operetta high-content imager and harmony software with phenologic. PLoS ONE **10**(7), e0134306 (2015)
21. Mermillod, M., Bugaiska, A., Bonin, P.: The stability-plasticity dilemma: investigating the continuum from catastrophic forgetting to age-limited learning effects. Front. Psychol. **4**, 504 (2013)
22. Murray, C.J., et al.: Global burden of bacterial antimicrobial resistance in 2019: a systematic analysis. Lancet **399**(10325), 629–655 (2022)
23. O'Neill, J.: Tackling Drug-Resistant Infections Globally: Final Report and Recommendations. Review on Antimicrobial Resistance. Wellcome Trust and HM Government (2016)
24. World Health Organization, et al.: Antimicrobial resistance: global report on surveillance. World Health Organization, Geneva (2014)
25. World Health Organization, et al.: Ten threats to global health in 2019 (2019)
26. World Health Organization, et al.: Antimicrobial resistance surveillance in Europe 2022–2020 data. WHO: World Health Organization, Copenhagen (2022)
27. Panigrahi, S., et al.: Misic, a general deep learning-based method for the high-throughput cell segmentation of complex bacterial communities. Elife **10**, e65151 (2021)
28. Quach, D., Sakoulas, G., Nizet, V., Pogliano, J., Pogliano, K.: Bacterial cytological profiling (BCP) as a rapid and accurate antimicrobial susceptibility testing method for staphylococcus aureus. EBioMedicine **4**, 95–103 (2016)
29. Ronneberger, O., Fischer, P., Brox, T.: U-Net: convolutional networks for biomedical image segmentation. In: Navab, N., Hornegger, J., Wells, W.M., Frangi, A.F. (eds.) MICCAI 2015. LNCS, vol. 9351, pp. 234–241. Springer, Cham (2015). https://doi.org/10.1007/978-3-319-24574-4_28
30. Spahn, C., et al.: DeepBacs for multi-task bacterial image analysis using open-source deep learning approaches. Commun. Biol. **5**(1), 688 (2022)
31. Sridhar, S., et al.: High-content imaging to phenotype antimicrobial effects on individual bacteria at scale. Msystems **6**(3), e00028-21 (2021)
32. Stringer, C., Wang, T., Michaelos, M., Pachitariu, M.: Cellpose: a generalist algorithm for cellular segmentation. Nat. Methods **18**(1), 100–106 (2021)
33. Stylianidou, S., Brennan, C., Nissen, S.B., Kuwada, N.J., Wiggins, P.A.: SuperSegger: robust image segmentation, analysis and lineage tracking of bacterial cells. Mol. Microbiol. **102**(4), 690–700 (2016)

34. Tan, M., Le, Q.: EfficientNet: rethinking model scaling for convolutional neural networks. In: Chaudhuri, K., Salakhutdinov, R. (eds.) Proceedings of the 36th International Conference on Machine Learning, Long Beach, California, USA. Proceedings of Machine Learning Research, vol. 97, pp. 6105–6114. PMLR (2019)
35. Tuan-Anh, T., et al.: Pathogenic Escherichia coli possess elevated growth rates under exposure to sub-inhibitory concentrations of azithromycin. Antibiotics **9**(11), 735 (2020)
36. Wang, W., et al.: Learn to segment single cells with deep distance estimator and deep cell detector. Comput. Biol. Med. **108**, 133–141 (2019)
37. Wei, D., et al.: MitoEM dataset: large-scale 3D mitochondria instance segmentation from EM images. In: Martel, A.L., et al. (eds.) MICCAI 2020. LNCS, vol. 12265, pp. 66–76. Springer, Cham (2020). https://doi.org/10.1007/978-3-030-59722-1_7

Multimodal Emotion Recognition System Through Three Different Channels (MER-3C)

Nouha Khediri[1,2(✉)], Mohammed Ben Ammar[2], and Monji Kherallah[3]

[1] Faculty of Sciences of Tunis, University of Tunis El Manar, Tunis, Tunisia
[2] Faculty of Computing and IT, Northern Border University, Rafha, Kingdom of Saudi Arabia
Nouha.khediri@fst.utm.tn, {Nuha.khediri,Mohammed.Ammar}@nbu.edu.sa
[3] Faculty of Sciences, University of Sfax, Sfax, Tunisia
Monji.kherallah@fss.usf.tn

Abstract. The field of machine learning and computer science known as "affective computing" focuses on how to recognize and analyze human emotions. Different modalities can complement or enhance one another. This paper focuses on merging three modalities, face, text, and speech, for detecting a user's emotion. To do this and obtain a multimodal emotion recognition system, we use deep-learning techniques. Among these are (a) Convolutional Neural Network (CNN) which is used in our case to detect emotion from both modalities speech and face. Then, we adopt (b) Bidirectional Long Short-Term Memory (Bi-LSTM) for predicting emotions from text information. Our experiments show that our proposed decision fusion method gives a good accuracy of 93% on the three modalities.

Keywords: Multimodal emotion recognition system · affective computing · deep learning · decision fusion · text · face · speech

1 Introduction

Affective computing is a branch of artificial intelligence research and development that combines the disciplines of computer science, psychology, and cognitive science to create systems and tools that can identify, comprehend, and process human sensibility. According to Christine Lisetti [1], "it takes into account the role of emotions in cognition in order to improve human-machine interaction". The field of human-machine interaction has achieved huge success in many research areas [2–6]. Affective computing converts emotions into machines. Before the conversion, the emotion must be captured using different channels that contain affective information as shown in Fig. 1. These different channels can be ***Physiological Signals*** like respiration rate (RR) and Blood Volume Pulse (BVP). ***Vocal channel*** contains all verbal communication like speech

J. Blanc-Talon et al. (Eds.): ACIVS 2023, LNCS 14124, pp. 196–208, 2023.
https://doi.org/10.1007/978-3-031-45382-3_17

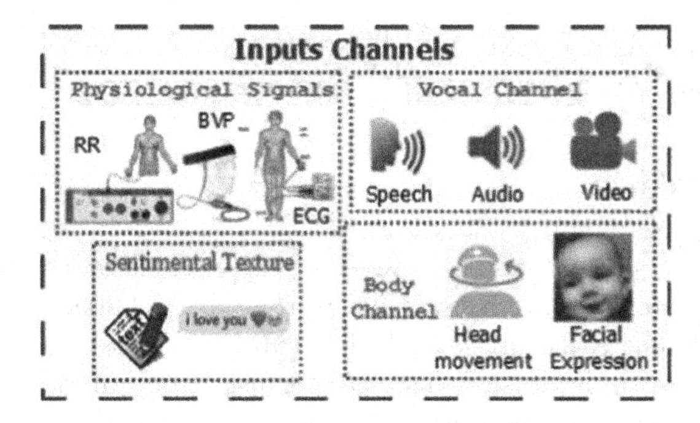

Fig. 1. Different Emotion's Input Channels

and video. ***Body channel*** contains non-verbal communication such as gestures, facial expressions, and head movement. ***Textual channel*** serves as the primary communication channel between people and computers through e-mails, text messages, etc. More information about the user and many external elements cannot be provided by unimodal systems. Instead of unimodal systems, multimodal systems have emerged due to this. However, the majority of research depends on two modalities which are the face and audio like [7–9]. But these two modalities generate a bi-modal system. Hillel et al. [10] emphasize the need for a multimodal system since it is more similar to the human sensory system and would be excellent for automatic human emotion recognition. This paper aims to create a machine that can automatically understand and recognize users' emotions by analyzing and processing multimodal inputs: face, text, and speech.

Fusion techniques are employed in multimodal systems to produce reliable results from several input channels. Various fusing methods include single-level fusion, feature-level fusion, decision-level fusion, and hybrid-level fusion. In this paper, we propose a decision-level fusion using the majority voting technique to better detect emotions from the three modalities (face, text, and speech).

Compared to machine learning techniques, deep learning techniques have two key advantages. First of all, they automatically extract different features from the raw data, avoiding the need for a separate extraction module. Second, using these clever techniques results in a shorter processing time for high-dimensional data sets. One of the most powerful and successful methods for applying massive volumes of data to pattern recognition applications and modeling complex processes is CNN, which we use to recognize emotions from faces and speech separately. From text modality, we apply another deep learning technique to recognize emotion, which is bidirectional long short-term memory (Bi-LSTM).

The paper is organized as follows: Sect. 2 summarizes the related work of emotion recognition from the three modalities (face, speech, and text). Section 3 describes the majority voting technique proposed for multi-modal fusion and then our Multimodal Emotion Recognition (MER-3C) Model. Section 4 presents the experiments and the results achieved by the proposed approach for each modality (face, speech, and text) as well as with their fusion (MER-3C). Finally, the conclusions and perspectives on future work are stated in the last section.

2 Related Work

In this section, we present a brief overview of previous works that focus on multimodal emotion recognition using facial, textual, and speech channels. Mittal et al. [11] present a feature-level fusion (M3ER) where the fusion layer is integrated into a memory fusion network to fuse input modalities which are face, text, and speech. Its focuses on the issue of perceived emotion recognition rather than actual emotional state recognition. M3ER is resistant to sensor noise by adding a check phase that employs Canonical Correlational Analysis to distinguish between ineffective and effective modalities. The mean classification accuracies (MA) on the two datasets, IEMOCAP and CMU-MOSEI are 82.7% and 89% respectively. Tripathi et al. [12] use data from speech, text, and motion capture from hand and face movements, rotations, and expressions to conduct multimodal emotion identification on the IEMOCAP dataset. They use a late fusion precisely at the last layer.

Zhang et al. [13] propose a multimodal fusion based on graph neural network (GNN) for emotion learning structure. This research named as Heterogeneous Hierarchical Message Passing Network (HMPN) is composed of four levels such as the Feature-to-Feature level where each extracted feature from each modality is represented as a node in the graph. Feature-to-Label level where messages are submitted from the interactive multi-modal embeddings to the label embeddings in order to update them. There are other levels, such as the Label-to-Label level and the Modality-to-Label level. The accuracy of this model using CMU-MOSEI is 45.9%.

A hierarchical fusion method was reported by Majumder et al. [14] to obtain four distinct emotions. They first use RNN to extract context-aware unimodal features for each utterance for all three modalities. Then, they combine the two in two modalities, and last they combine these bimodal vectors into trimodal vectors using fully connected layers. Lian et al. [15] propose a multimodal emotion recognition from text image and audio but all of them are taken from only one input, which is video. They use a sampling technique to eliminate subsets that are less likely to produce useful results. In each cycle, they chose the top K subsets from a beam search of size K. Multiple classifiers are investigated and the recognition accuracy obtained is 60.34%. Shamane et al. [16] present a multimodal emotion recognition system with transformer-based self-supervised feature fusion. The inputs for the suggested fusion process are Self Supervised Learning (SSL) embeddings from the three modalities. A distinct CLS token (which is the first token of the sequence) from one modality can attend to the whole embedding sequence of another modality and receive crucial cross-modal data thanks to each of the transformer blocks. Then, six Inter-Modality-Attention (IMA) based transformer blocks are used to extract the information across all modalities. Finally, six CLS tokens that have been enhanced with inter-modality data are produced by the fusion procedure. In a recent work [17], Heredia et al. propose a flexible emotion recognition architecture that can work with multiple modalities and can handle varying levels of data quality in order to help robots better understand human moods in a given environment. The partial results from each modality are then combined using a previously published fusion method called EmbraceNet+ [18], which is modified and added to our suggested framework. In order to categorize emotions into the four categories of happiness, neutral, sadness, and anger, three modalities are combined which are text, face, and audio.

Most related work takes the textual data from the transcripts of the videos and is not an independent modality as we propose in this study.

We synthesize the studied research works, as illustrated in Table 1. For each work, we highlight the following characteristics: modalities, fusion method, accuracy, the dataset used, and classes of emotion detected.

Table 1. Summary of Related Works of Multimodal Emotion Recognition

Work	Modalities	Fusion	Dataset	Accuracy	Emotions
[12] 2018	Speech, text, face (head, hand) movement	Late fusion	IEMOCAP	71.04%	4
[15] 2018	text, audio face	Beam search fusion	AFEW	60.34%	7
[14] 2018	text, audio video	hierarchical fusion	IEMOCAP	76.5%	4
[11] 2020	Face, speech text	feature level fusion	IEMOCAP	82.7%	4
			CMU-MOSEI	89%	6
[16] 2020	text vision audio	attention-based fusion	IEMOLAP	64.3%	4
			MELD	64.3%	7
			CMU-MOSI	55.5%	
			CMU-MOSEI	55.7%	
[13] 2021	Text Voice face	multimodal fusion based on graph neural network	CMU-MOSEI	45.9%	6
[17] 2022	Face, text, speech posture, context touch sensing	fusion method EmbracetNet+	IEMOCAP	76.9%	4

3 Multimodal Emotion Recognition System Through Three Different Channels (MER-3C)

3.1 Majority Voting

Recent research has shown that integrating the classification judgments of different classifiers might produce better recognition outcomes. Pattern recognition classifier results have been combined effectively using the majority vote method. Although majority voting [19] is by far the simplest method for implementing the combination, research has shown that it is just as effective as more intricate plans. Another significant factor to take into account in pattern recognition applications is having low error or substitution rates. When the judgments of the n classifiers are combined, the sample is given the classification for which there is an agreement, or when the identification is agreed upon by more than half of the classifiers. Although each classifier has the potential to be right or wrong. A combined decision is incorrect only when a majority of the votes are incorrect and they all make the same error because of the nature of consensus. This is a benefit of the combination approach since it reduces the chances that most people will commit the same error [19].

3.2 Our Multimodal Emotion Recognition (MER-3C) Model

As CNN is capable of overcoming the difficulties presented by uncontrollable situations like illumination and occlusion. So, we use the architecture of CNN to build the face and speech models, and we keep the LSTM architecture for the text model to develop a novel Multimodal Emotion Recognition System called **MER-3C**.

Our FER Sub-system. The proposed Facial Emotion Recognition used is named **CV-FER** and described in detail in [20]. Each table entry in the model's input CSV file was transformed into a vector and subsequently into an image. We rotate certain images with a range of 10 in the pre-processing step. The image is then turned horizontally, moved to the left or right with a width and height range of 0.1, and the fill mode is set to nearest. Then, we move on to data augmentation so that we can feed the network with more images. We define our model, which is based on the VGG16 model [21], after data production. After that, we train our model to monitor how much better CNN is at classifying emotions.

Our CV-FER model has five convolutional layer blocks, each of which is followed by a maximum pooling layer, as outlined in Fig. 2. Each convolution bloc's filter numbers range from 256 to 16 (256, 128, 64, 32, 16), and stride (1,1). For the activation function, we used Rectified Linear Unit (ReLU) [22]. For the loss function, we used Categorical Cross-entropy. For the optimizer, we use Adam [23]. And finally, for the classifier we used Softmax.

Fig. 2. Our CV-FER PLOT

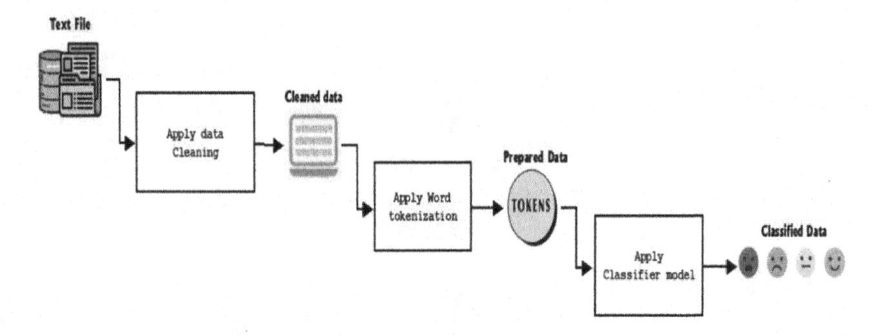

Fig. 3. Workflow chart of TER Process

Our TER Sub-system. The basic concepts behind textual emotion recognition (TER) involve the following phases to predict emotion shown in Fig. 3. The steps conducted for this study's pre-processing include "Case Folding" which converts all of the letters in the document to lowercase. "Data cleaning" is used to eliminate HTML tags, emails, special characters, and accents. "Tokenizing" is the method of breaking down a sentence into smaller bits that can be more readily given meaning, and "Stopword Removal" where Stop words are a collection of frequently used phrases (*e.g.* the, a, an, *etc*). The pre-processed data was used as the input for our model, which uses a Bi-LSTM architecture [24]. LSTMs have the ability to selectively retain patterns for a long time. A memory cell, as the name suggests, enables this. Four primary components make up its distinctive structure: an input gate, a neuron with a self-recurrent connection, a forget gate, and an output gate. It contains the following layers: Embedding, Bidirectional, and Dense layer. The embedding layer is an input layer that maps the words/tokenizers to a vector with input_dim dimensions. The bidirectional layer is an RNN-LSTM layer with an LSTM output. The dense is an output layer with nodes (indicating the number of classes) and softmax as an activation function. As we know from the softmax activation function, it helps to determine the probability of inclination a text will have in each class. The TER model takes categorical cross-entropy as a loss function and Adam as an optimizer.

Our SER Sub-system. In our SER sub-system, we use three features as Mel-Frequency Cepstral Coefficients (MFCC), Linear Predictive Coding (LPC), and Perceptual Linear Prediction (PLP). These features were separately classified using five different classifiers such as Gaussian Process Classifier (GPC), Random Forest Classifier (RFC), Convolutional Neural Network (CNN), Support Vector Machine (SVM), and K-Nearest Neighbor (KNN).

After that, we use a majority vote technique to select the best machine learning classifier among these options for each set of features. In theory, the majority voting method is similar to the electoral system. In the latter, the winner of an election is the candidate who obtains more than 50% of the vote. In our instance, the classifier that has the highest accuracy is deemed to be the best with regard to the relevant features. The architecture of our SER Sub-system is presented in Fig. 4.

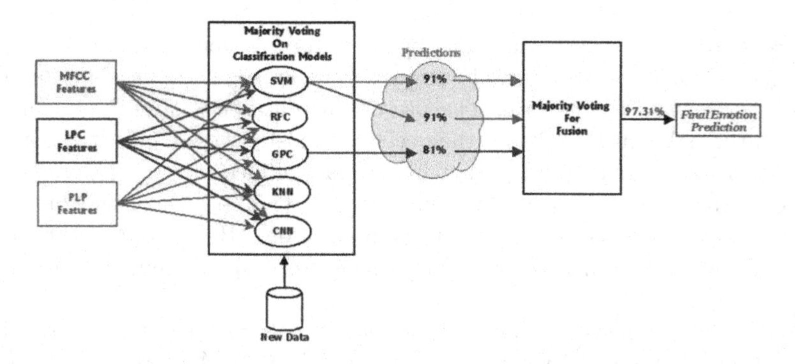

Fig. 4. SER Sub-system Architecture

Our MER-3C System. Our proposed architecture illustrated in Fig. 5 is demonstrated by a straightforward block diagram with four major blocks which are an input pre-processing block (for speech, image, and text) followed by the training block that uses deep learning algorithms to train separately the features extracted from each modality. We must understand that each possible emotion is affected differently by features from different modalities. For this, we proceed to a decision fusion block where a majority voting technique is applied along with the output classification block.

We assume that five classifiers are used for each input modality because we use a number of classifiers that should be greater than the number of final predicted classes. Every classifier generates a different result regarding the identity of the input channel.

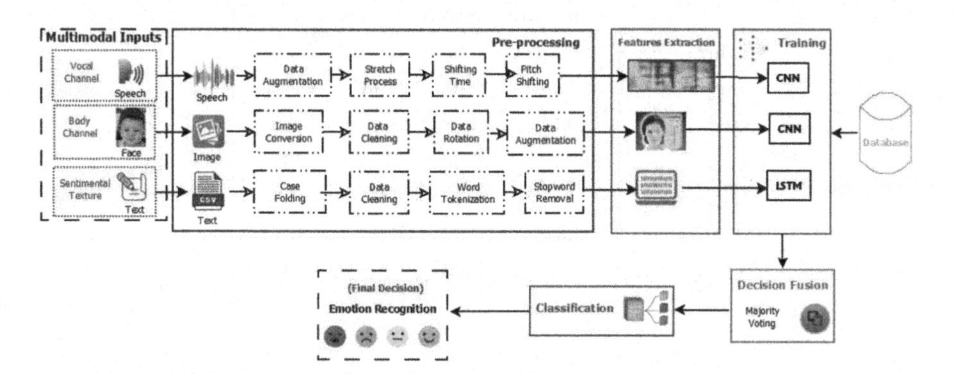

Fig. 5. MER-3C Architecture

4 Our MER-3C Results

We evaluate our proposed system on the three datasets:

- **FER-2013** [25] is used for face channel. It contains seven emotions mapped to integers as follows: (0) anger, (1) disgust, (2) fear, (3) happiness, (4) sadness, (5) surprise, and (6) neutral. The use of this dataset is already a challenge due to the imbalance problem, the occlusion mostly with hands, the contrast variations, and the eyeglasses that hide a part of faces.
- **Tweet Emotions**[1] is used for textual channel. This corpus consists of tweets. Each tweet expresses one of the thirteen emotions. But in our experiments, we will focus only on four emotions, which are sadness, neutral, happiness, and anger.
- **Ravdess** [26] is used for speech channel. It contains 1440 audio recordings that 24 trained actors pronounce with neutral North American accents and expresses seven emotions. We chose this dataset because of its excellent availability. In the second place, we attempt to employ a dataset with mixed gender. Last but not least, RAVDESS is widely utilized for comparative analysis in the SER sector and is quite well known.

But we train again our models with these datasets as we predict this time only four emotions which are calm, happy, sad, and angry.

From Table 2 achieves, we can conclude that our decision fusion achieves a good performance of 93%. This accuracy is 1% and 16% significantly higher than those achieved by our facial and textual emotion recognition baseline systems. The confusion matrix illustrated in Fig. 6 shows that our proposed system named **MER-3C** achieves good results in the detection of emotions but "Calm" is the toughest.

Table 2. Classification report of MER-3C

	Precision	*Recall*	*f1-score*	*Support*
Calm	0.93	0.95	0.94	350
Happy	0.93	0.93	0.93	344
Sad	0.93	0.88	0.91	352
Angry	0.95	0.98	0.96	331
Accuracy			0.93	1377
Macro avg	0.93	0.94	0.93	1377
Weighted avg	0.93	0.93	0.93	1377

We start our comparison figured in Table 3 by checking our results in the separated sub-systems compared with the result of the combined sub-systems after applying the decision fusion.

Table 3. Comparison of Uni-modal and Multi-modal ERS

Modality	*Accuracy*
Face Channel [20]	92%
Text Channel [24]	91%
Speech Channel	97.31%
Our Work (MER-3C)	93%

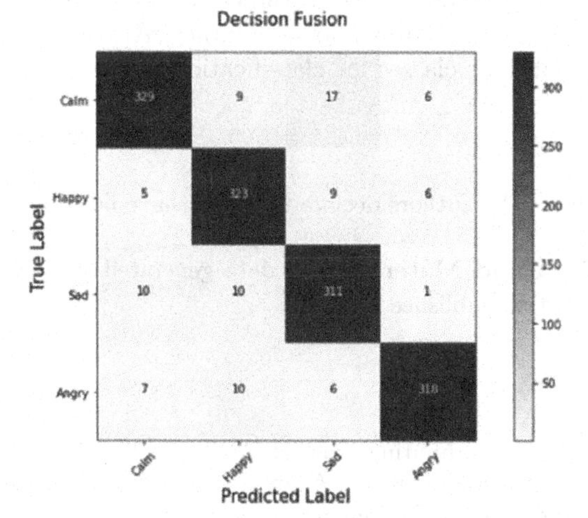

Fig. 6. Confusion Matrix of our MER-3C

We compare our MER-3C with some works that also predict the four emotions (happy, sad, angry, and neutral) using the fusion of three modalities face, text, and speech. Note that our emotion labeled "calm" is the same emotion labeled "neutral" in the other works. The comparison is presented in Table 4 for further validating. When compared to the other models, our model is the most efficient.

Table 4. Comparison of MER-3C with other works

Works	*Fusion method*	*Accuracy*
Mittal et al. [11]	MLP with multiplicative fusion	82.7%
Heredia et al. [17]	MLP with a single linear activation function	76.9%
Tripathi et al. [12]	Late fusion	71.04%
Majumder et al. [14]	Hierarchical fusion	76.5%
Siriwardhana et al. [16]	Attention-based fusion	64.3%
Our Work (MER-3C)	Decision fusion using majority voting	93%

[1] https://www.kaggle.com/code/hanifkurniawan/nlp-with-tweet-emotions/data.

5 Conclusion and Perspectives

The fast development of artificial intelligence in the modern world has increased the demand for more effective and natural interaction between humans and machines.

Applications involving emotion recognition systems are numerous. In the future, we plan to incorporate our **MER-3C** in an e-learning platform to detect the student's emotion and help the teacher in taking decisions depending on the detected emotion. As part of future studies, we would also investigate more complex fusion methods that can enhance accuracy. Also, it would be beneficial to consider more than four classes for classification in the future.

Declaration

Conflict of Interest. The authors declare that they have no conflict of interest.

Availability of Data and Materials. All data generated or analyzed during this study are included in this published article.

References

1. Lisetti, C.L.: Affective computing. Pattern Anal. Appl. **1**, 71–73 (1998)
2. Nikita, J., Vedika, G., Shubham, S., Agam, M., Ankit, C., Santosh, K.C.: Understanding cartoon emotion using integrated deep neural network on large dataset. Neural Comput. Appl. (2021). https://doi.org/10.1007/s00521-021-06003-9
3. Andres, J., Semertzidis, N., Li, Z., Wang, Y., Floyd Mueller, F.: Integrated exertion-understanding the design of human-computer integration in an exertion context. ACM Trans. Comput.-Hum. Interact. **29**(6), 1–28 (2023)
4. Fischer, F., Fleig, A., Klar, M., Müller, J.: Optimal feedback control for modeling human-computer interaction. ACM Trans. Comput.-Hum. Interact. **29**(6), 1–70 (2022)
5. Kosch, T., Welsch, R., Chuang, L., Schmidt, A.: The placebo effect of artificial intelligence in human-computer interaction. ACM Trans. Comput.-Hum. Interact. **29**, 1–32 (2022)
6. Glenn, A., LaCasse, P., Cox, B.: Emotion classification of Indonesian tweets using bidirectional LSTM. Neural Comput. Appl. **35**, 9567–9578 (2023). https://doi.org/10.1007/s00521-022-08186-1
7. Tang, K., Tie, Y., Yang, T., Guan, L.: Multimodal emotion recognition (MER) system. In: 2014 IEEE 27th Canadian Conference on Electrical and Computer Engineering (CCECE), pp. 1–6. IEEE (2014)
8. Veni, S., Anand, R., Mohan, D., Paul, E.: Feature fusion in multimodal emotion recognition system for enhancement of human-machine interaction. In: IOP Conference Series: Materials Science and Engineering, vol. 1084, no. 1, p. 012004. IOP Publishing (2021)

9. Luna-Jiménez, C., Griol, D., Callejas, Z., Kleinlein, R., Montero, J.M., Fernández-Martínez, F.: Multimodal emotion recognition on RAVDESS dataset using transfer learning. Sensors **21**(22), 7665 (2021)
10. Aviezer, H., Trope, Y., Todorov, A.: Body cues, not facial expressions, discriminate between intense positive and negative emotions. Science **338**(6111), 1225–1229 (2012)
11. Mittal, T., Bhattacharya, U., Chandra, R., Bera, A., Manocha, D.: M3ER: multiplicative multimodal emotion recognition using facial, textual, and speech cues. In: Proceedings of the AAAI Conference on Artificial Intelligence, vol. 34, no. 02, pp. 1359–1367 (2020)
12. Tripathi, S., Tripathi, S., Beigi, H.: Multi-modal emotion recognition on IEMOCAP with neural networks. arXiv (2018). arXiv preprint arXiv:1804.05788
13. Zhang, D., et al.: Multi-modal multi-label emotion recognition with heterogeneous hierarchical message passing. In: Proceedings of the AAAI Conference on Artificial Intelligence, vol. 35, no. 16, pp. 14338–14346 (2021)
14. Majumder, N., Hazarika, D., Gelbukh, A., Cambria, E., Poria, S.: Multimodal sentiment analysis using hierarchical fusion with context modeling. Knowl.-Based Syst. **161**, 124–133 (2018)
15. Lian, Z., Li, Y., Tao, J., Huang, J.: Investigation of multimodal features, classifiers and fusion methods for emotion recognition. arXiv preprint arXiv:1809.06225 (2018)
16. Siriwardhana, S., Kaluarachchi, T., Billinghurst, M., Nanayakkara, S.: Multimodal emotion recognition with transformer-based self supervised feature fusion. IEEE Access **8**, 176274–176285 (2020)
17. Heredia, J., et al.: Adaptive multimodal emotion detection architecture for social robots. IEEE Access **10**, 20727–20744 (2022)
18. Heredia, J., Cardinale, Y., Dongo, I., Díaz-Amado, J.: A multi-modal visual emotion recognition method to instantiate an ontology. In: 16th International Conference on Software Technologies, pp. 453–464. SCITEPRESS-Science and Technology Publications (2021)
19. Lam, L., Suen, C.Y.: A theoretical analysis of the application of majority voting to pattern recognition. In: Proceedings of the 12th IAPR International Conference on Pattern Recognition, vol. 3-Conference C: Signal Processing (Cat. No. 94CH3440-5), vol. 2, pp. 418–420. IEEE (1994)
20. Khediri, N., Ben Ammar, M., Kherallah, M.: Deep learning based approach to facial emotion recognition through convolutional neural network. In: International Conference on Image Analysis and Recognition, ICIAR (2022)
21. Simonyan, K., Zisserman, A.: Very deep convolutional networks for large-scale image recognition. arXiv preprint arXiv:1409.1556 (2014)
22. Nair, V., Hinton, G.E.: Rectified linear units improve restricted Boltzmann machines. In: Proceedings of the 27th International Conference on Machine Learning (ICML 2010), pp. 807–814 (2010)
23. Kingma, D.P., Ba, J.: Adam: a method for stochastic optimization. arXiv preprint arXiv:1412.6980 (2014)
24. Khediri, N., BenAmmar, M., Kherallah, M.: A new deep learning fusion approach for emotion recognition based on face and text. In: Nguyen, N.T., Manolopoulos, Y., Chbeir, R., Kozierkiewicz, A., Trawiński, B. (eds.) ICCCI 2022. LNCS, vol. 13501, pp. 75–81. Springer, Cham (2022). https://doi.org/10.1007/978-3-031-16014-1_7

25. Lucey, P., Cohn, J.F., Kanade, T., Saragih, J., Ambadar, Z., Matthews, I.: The extended Cohn-Kanade dataset (CK+): a complete dataset for action unit and emotion-specified expression. In: 2010 IEEE Computer Society Conference on Computer Vision and Pattern Recognition-Workshops, pp. 94–101. IEEE (2010)
26. Livingstone, S.R., Russo, F.A.: The Ryerson audio-visual database of emotional speech and song (RAVDESS): a dynamic, multimodal set of facial and vocal expressions in North American English. PLoS ONE **13**(5), e0196391 (2018)

Multi-modal Obstacle Avoidance in USVs via Anomaly Detection and Cascaded Datasets

Tilen Cvenkel, Marija Ivanovska, Jon Muhovič, and Janez Perš(✉)

Faculty of Electrical Engineering, University of Ljubljana, Tržaska 25, Ljubljana, Slovenia
janez.pers@fe.uni-lj.si, https://lmi.fe.uni-lj.si/en/home/

Abstract. We introduce a novel strategy for obstacle avoidance in aquatic settings, using anomaly detection for quick deployment of autonomous water vehicles in limited geographic areas. The unmanned surface vehicle (USV) is initially manually navigated to collect training data. The learning phase involves three steps: learning imaging modality specifics, learning the obstacle-free environment using collected data, and setting obstacle detector sensitivity with images containing water obstacles. This approach, which we call *cascaded datasets*, works with different image modalities and environments without extensive marine-specific data. Results are demonstrated with LWIR and RGB images from river missions.

Keywords: unmanned vehicles · USV · obstacle avoidance · anomaly detection

1 Introduction

Obstacle detection is crucial for autonomous vehicles. Unmanned robotic surface vehicles (USVs) can use various sensors, like RGB cameras [37], sometimes in stereo depth configuration [32], RADAR [35], LIDAR [23,41], and SONAR [20]. Cameras are appealing for their cost and superficial similarity to human perception, but require substantial image processing, which falls in the domain of *computer vision*.

Cameras' main drawback is their sensitivity to environmental variations. However, lately data-driven algorithms and deep neural networks (DNN) improved obstacle detection in marine environments [5,29,33], but require extensive annotations [46].

We propose *semi-supervised learning* for water obstacle detection, specifically *anomaly detection in a one-class learning setting* [3,39,42]. This approach trains on normal data and detects anomalies as non-conforming samples. Many, (but

This work was financed by the Slovenian Research Agency (ARRS), research projects [J2-2506] and [J2-2501 (A)] and research programs [P2-0095] and [P2-0250 (B)].

J. Blanc-Talon et al. (Eds.): ACIVS 2023, LNCS 14124, pp. 209–221, 2023.
https://doi.org/10.1007/978-3-031-45382-3_18

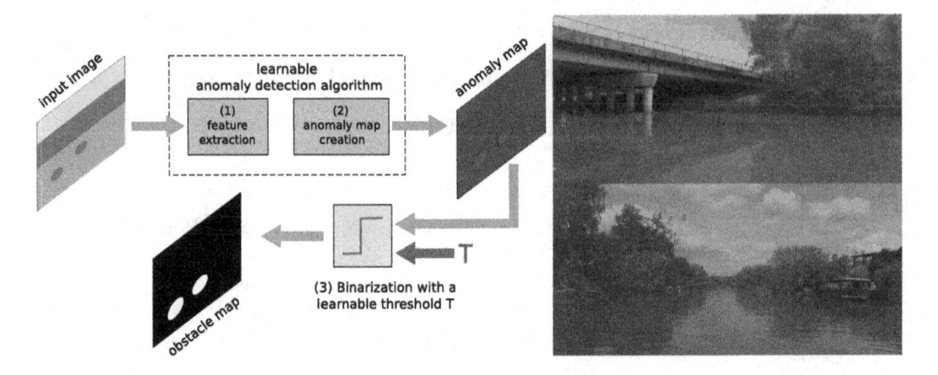

Fig. 1. Left: Obstacle avoidance through anomaly detection has three learnable components [(1), (2), (3)]. This approach allows separate learning using datasets without marine environments (1), obstacles (2), and with minimal detailed obstacle annotations (3). Right: Obstacle detection examples using one-class learning anomaly detector CS-Flow [39]: in the right-upper image correctly detects a bridge, while in the right-lower image detects a floating dock and a riverbank house.

not all) USVs are expected to operate in limited geographical domains, allowing one-class learning and anomaly detection.

Our contributions include: *i)* a novel strategy for water obstacle detection with limited annotated data, *ii)* a new approach to training with limited marine data, *iii)* evaluation of recent SOTA anomaly detection algorithms in realistic USV scenarios, and *iv)* comparison of a fully-supervised SOTA segmentation algorithm [6] with our approach.

2 Related Work

We review works using image data and computer vision for water obstacle detection, followed by literature on semi-supervised learning for anomaly detection, which avoids laborious annotation.

2.1 Detection of Water Obstacles

Detecting water obstacles under varying weather conditions is challenging. Early research used hand-crafted methods [22,49], while SOTA approaches leverage deep learning for more discriminative models. Many works [28,31,51] propose CNNs for vessel detection and classification. However, these don't generalize well and fail to recognize arbitrary objects.

Recent algorithms use neural networks for semantic segmentation. Cane *et al.* [10] and Bovcon *et al.* [7] applied segmentation models to marine data, while Kim *et al.* [26], Zhan *et al.* [55], and Steccanella *et al.* [43] proposed adapted segmentation architectures. The SOTA method [6] estimates water horizon location, fusing inertial information with RGB data. A time-efficient model is presented in [52]. Segmentation-based models require massive annotated data, posing a limitation, especially for less popular image modalities [33].

2.2 Anomaly Detection in One–class Learning Setting

Anomaly detection in industrial applications motivated one-class learning algorithms, trained using only normal data samples [1,40,42]. Reconstruction-based deep learning models use auto-encoder architectures, improved by pretext learning objectives [2,15,19,21,34,45,54].

Recent one-class anomaly detectors utilize pre-trained classification networks like VGG16, ResNet, and EfficientNet as backbone models. Defard *et al.* [13] proposed modeling pre-extracted features using a multivariate Gaussian distribution. Others pursued shallow approaches [11,38], knowledge distillation [4,30,47], or invertible neural networks [18,39,53] to construct flexible probability distributions of non-anomalous data.

2.3 Multi-modal Data

Autonomous navigation needs multi-modal data to perform well in diverse conditions, but the research into this problem is not as advanced as in RGB modality. Nirgudkar and Robinette [33] use LWIR modality. Datasets like nuScenes [9], Cityscapes [12], KITTI [16], and Waymo Open Dataset [50] include multiple RGB, depth cameras, lidar, and GPS. Pedestrian detection datasets like LLVIP [24] and CVC-14 [17] use LWIR and RGB cameras for better low-light performance. Wang et al. [48] propose a method for anomaly detection in hyperspectral satellite images.

3 Extension of Anomaly Detection to Water Environment

The presented solution to the obstacle detection problem as shown in Fig. 1 is based on a cascade of training datasets.

3.1 Cascaded Datasets

Our assumption is that the following non-overlapping image datasets are at our disposal:

- **Modality adaptation dataset:** Used for training a deep CNN architecture using a proxy task (e.g. general object detection on ImageNet). No pixel-wise annotations needed. Used in stage (1) of our approach, as in Fig. 1.
- **Environment adaptation dataset:** Trains the obstacle detector to the obstacle-free environment. No data annotations needed. Images must match the modality of the image acquisition hardware. No pixel-wise annotations needed. Used in stage (2) of our approach, shown in Fig. 1.
- **Tuning dataset:** Fine-tunes the model's sensitivity to obstacles using pixel-wise annotated images. We used 32 image items. The detection threshold T balances true and false positives. Tuning dataset is part of the stage (3) in Fig. 1.

3.2 Tuning Process and Tuning Targets

Anomaly detection algorithms yield prediction maps $f(u,v)$, with higher values indicating a higher certainty of anomalies. Binarization threshold T determines the operating point of the whole system and makes the it more or less sensitive to anomalies.

Evaluation and T tuning metrics depend on the detection model's functionality. While F1-score is widely used for comparing algorithms [8,25], FNR (the false negative rate – the probability that the USV runs over an obstacle) and FPR ction algorithms, FNR (the false negative rate – the probability that the USV runs over an obstacle) and FPR (the false positive rate – the probability that USV is unnecessarily stopped) are more intuitive. Detection goals can change based on the USV's location. In our approach, varying threshold T according to a function depending on the vehicle's GPS position could achieve different mission goals.

4 Experiments

We based our research on the LWIR and RGB image data, that was acquired using our own USV multi-sensor system [36], attached to a river boat. RGB and LWIR images were recorded.[1] The image acquisition took place on a stretch of Ljubljanica river[2], that represents predominantly natural (river,bush) environment with some urban elements.

4.1 Dataset

Data was gathered in June 2021 and September 2021. In June, only RGB images were taken under good weather conditions, while in September, both RGB and LWIR images were collected under cloudy conditions.

Data was organized into three main datasets: *LWIRSept*, *RGBJune*, and *RGBSept*, which were further divided into smaller subsets. These three datasets were then further divided into 5 smaller, non-overlapping subsets, i.e. *LWIRSeptemberNoObs*, *RGBJuneNoObs*, both consisting of obstacle–free images and *LWIRSeptObs*, *RGBJuneObs*, *RGBSeptObs*, containing images with obstacles. These three datasets were then further divided into 5 smaller, non-overlapping subsets, i.e. *LWIRSeptemberNoObs*, *RGBJuneNoObs*, both consisting of obstacle–free images and *LWIRSeptObs*, *RGBJuneObs*, *RGBSeptObs*, containing images with obstacles. Tuning datasets, *LWIRSeptObs32* and *RGBJuneObs32*, were created using selected images (see Subsect. 3.1). An overview of the datasets is presented in Table 1. Images in some subsets don't depict the same geographical location; subsets with obstacles and without obstacles were obtained on different river stretches.

[1] Stereolabs ZED stereo camera (only the left frame) and Device A-lab SmartIR384L thermal camera.

[2] Data was sampled from a section between 46.0402°N, 14.5125°E and 46.0234°N, 14.5079°E.

Table 1. Structure of our experimental dataset

Subset	Images	Annotated	Obstacles	Purpose
LWIRSeptNoObs	3263	No	No	Environment adaptation
LWIRSeptObs	436	Yes	Yes	Testing
LWIRSeptObs32	32	Yes	Yes	Parameter tuning
RGBJuneNoObs	600	No	No	Environment adaptation
RGBJuneObs	501	Yes	Yes	Testing
RGBJuneObs32	32	Yes	Yes	Parameter tuning
RGBSeptObs	521	Yes	Yes	Testing
RGBSeptObs33	33	Yes	Yes	Parameter tuning

In our experiments, we use *LWIRSeptNoObs* and *RGBJuneNoObs* in the role of an environment adaptation dataset (see Subsect. 3.1), and *LWIRSeptObs32*, *RGBJuneObs32* and *RGBSeptObs33* as tuning datasets. and *LWIRSeptObs*, *RGBJuneObs* and *RGBSeptObs* for the final evaluation of the trained obstacle detection model. In the testing phase of the trained model, *RGBJuneObs* was used to evaluate the algorithm under same weather conditions. *RGBSeptObs* was on the other hand obtained under very different weather conditions, so this dataset is used to evaluate the detection accuracy of the model in a much more challenging scenario. Note, that our acquired data (described in Table 1) does not provide any modality adaptation datasets. Since such images are not necessarily domain–specific, any publicly available dataset, representing the imaging modality of interest, can be used in this role. In our experiments ImageNet [14] was used as a modality adaptation dataset for the obstacle detection in RGB images, while Teledyne FLIR Thermal Dataset [44] was used for obstacle detection in LWIR images.

4.2 Evaluation

Our evaluation protocol is based on several assumptions, that are realistic for USV environments and commonly used in such scenarios. We consider the performance of the algorithm in the upper part of the image entirely irrelevant, as this part of the image contains the sky, and possibly the distant shore [27]. The demarcation line between the upper and bottom part of the image can be inferred from the inertial sensor in the vehicle, which has been shown to help with image segmentation in marine environment before [6]. Due to the slow dynamic of the vessel used for image acquisition (imperceptible pitch and roll), we evaluate our algorithms using a fixed horizontal line, as shown in Fig. 2.

In the testing phase, only the part of the image below the water edge is used for the quantitative evaluation of the trained model. Interesting structures, such as bridges and riverbank houses, appearing above the water edge are evaluated only qualitatively. Two such examples are presented in the Fig. 1. To exclude

Fig. 2. Left: Simplified water edge annotation, straight line. Right: precise polygon-based annotation of the water edge. The former could be easily inferred from USV's inertial sensor, as proposed by [6]. Our method is evaluated using both kinds of annotations.

all detections on the riverbank, we perform additional evaluation, using more detailed polygon annotations.

We evaluated two different recent anomaly detection algorithms, i.e. PaDiM [13] and CS-Flow [39]. PaDiM was tested on both, LWIR and RGB images, whereas CS-Flow was evaluated only on RGB images.

4.3 Training

The modality adaptation stage for the RGB images was skipped in the publicly available models of PaDiM and CS-Flow, since they both use a backbone classification CNN, pre-trained on RGB images. The modality adaptation was thus performed only for LWIR images. We trained feature extraction CNN from scratch, using the Teledyne FLIR Thermal Dataset for Algorithm Training [44] in object detection task as a proxy.

For the environment adaptation stage, PaDiM was trained on *RGBJuneNoObs* and *LWIRSeptNoObs* subsets, to adapt it to the target river environment for both (RGB and LWIR) camera modalities, respectively, resulting in two different models, one for each imaging modality. CS-Flow was trained on *RGBJuneNoObs* only, to adapt it to target river environment in the RGB images.

Finally, in the tuning stage, thresholds T were obtained using MODS evaluation scheme [8] on *RGBJuneObs32* for RGB models, and *LWIRSeptObs32* for the PaDiM model, which is adapted to LWIR images. Final threshold values were selected based on the highest $F1$ score.

Comparison to State-of-the-Art. To provide a comparison of our method with the fully–supervised SOTA segmentation algorithms, we used the publicly available version of WaSR [6], pre–trained on RGB images. The algorithm was used as an out–of–the–box method, without any modifications. Since [33] shows that WaSR does not work on LWIR images without retraining, we did not use it on LWIR.

5 Results

Each evaluation was performed twice, using first the straight line as annotation of water boundary, and then for the second time, using a more accurate polygon to delimit the water area where the evaluation is performed, as shown in Fig. 2.

5.1 Testing in Similar Weather Conditions

Both RGB models, PaDiM and CS-Flow, were tested on the *RGBJuneObs* subset, while LWIR PaDiM was tested on *LWIRSeptObs*. Obtained results are reported in Tables 2 and 3.

Table 2. PaDiM results on water edge annotated as polygon.

Environment adaptation	Tuning	Testing	TP	FP	FN	F1
LWIRSeptNoObs	LWIRSeptObs32	LWIRSeptObs	496	286	399	59.2%
RGBJunNoObs	RGBJunObs32	RGBJunObs	614	270	272	69.4%
RGBJunNoObs	RGBJunObs32	RGBSeptObs*	38	40	1529	4.6%

* denotes different weather conditions

Table 3. PaDiM results on water edge annotated with fixed horizontal line.

Environment adaptation	Tuning	Testing	TP	FP	FN	F1
LWIRSeptNoObs	LWIRSeptObs32	LWIRSeptObs	692	267	279	71.7%
RGBJunNoObs	RGBJunObs32	RGBJunObs	1116	613	565	65.5%
RGBJunNoObs	RGBJunObs32	RGBSeptObs*	52	40	2260	4.3%

* denotes different weather conditions

Table 4. CS-Flow results on water edge annotated as polygon.

Environment adaptation	Tuning	Testing	TP	FP	FN	F1
RGBJunNoObs	RGBJunObs32	RGBJunObs	875	1438	11	54.7%
RGBJunNoObs	RGBJunObs32	RGBSeptObs*	1013	1069	554	55.5%

* denotes different weather conditions

Table 5. CS-Flow results on water edge annotated with fixed horizontal line.

Environment adaptation	Tuning	Testing	TP	FP	FN	F1
RGBJunNoObs	RGBJunObs32	RGBJunObs	1433	924	248	71.0%
RGBJunNoObs	RGBJunObs32	RGBSeptObs*	1627	861	685	67.8%

* denotes different weather conditions

Fig. 3. Comparison of obstacle detection using anomaly detection by PaDiM (left) and WaSR (right) in the same river scene. PaDiM performs better in detecting obstacles not fully surrounded by water, like the paddleman. WaSR's misclassification can cause issues for tracking or motion prediction

5.2 Testing in Significantly Different Weather Conditions

Both RGB adapted algorithms, were then also evaluated on *RGBSeptObs*, where the weather conditions differ significantly from the *RGBJuneNoObs* training dataset (see Subsesct. 4.1 for more details). As expected, these results are worse than the results obtained under similar weather conditions. Obtained metrics values are reported in Tables 2, 3, 4 and 5.

For each of the previously described testing scenarios we report the number of true positives (TP), true negatives (TN), false positives (FP), and F1-score. Finally, in Table 6 we provide results of the SOTA segmentation–based algorithm WaSR, [6], for comparison.

Table 6. WaSR results for both types of water edge annotation.

	Fixed line water edge				Polygon water edge			
Testing	TP	FP	FN	F1	TP	FP	FN	F1
RGBJunObs	1679	179	2	94.9%	884	0	2	99.9%
RGBSeptObs	2185	352	127	90.1%	1433	1381	134	65.4%

From the results it can be concluded that WaSR still outperforms both anomaly detection methods in terms of the chosen evaluation metrics. If we compare these two methods to one another, we can see, that the CS-Flow [39] outperforms PaDiM [13], where the distribution of data features is modelled using a strong statistical prior. The difference is especially striking in situation where the training data is captured in sunny weather, while the testing data is captured in the rainy weather (Tables 2, 3, 4, 5).

However, this comparison is not entirely fair due to different data that has gone into training WaSR on RGB images. It needed 1320 accurately pixel-wise annotated training images [7], and efforts to improve its generalization with further training with diverse images from all around the world are still ongoing. This will be difficult to repeat with many other, less widespread modalities, such as near-infrared (NIR), LWIR (done only on geographically limited area so far), and especially various multi-spectral cameras.

To complete the insight into the performance of the anomaly detection methods on our data, we present visual evaluation on two images, containing above-water obstacles, as shown in Fig. 1. As can be seen, these particular examples show, that the methods successfully recognize any objects, that were not part of domain adaptation subset, as an anomaly, which is the key advantage of the one–class learning approach.

6 Discussion and Conclusion

To summarize, an important advantage of our cascaded training approach is the ability to split the training into three different phases, where each of them can be served with easily obtainable dataset. In limited geographical domains, common for many USV tasks, we simplify obstacle detection by modeling the obstacle-free environment, redefining obstacle detection as *detection of non-permanent scene items*. Our strategy is viable if *environment adaptation datasets* accurately represent operating conditions. Anomaly detection algorithms require varied training data, but such images can be obtained semi-automatically without labels. Models can be retrained with new data to maintain performance, unlike supervised segmentation models needing precise annotations.

Visual examination shows that performance of some methods might be affected by overshoots in true positive detections (as seen in Fig. 2). Further development in anomaly detection methods and more detailed anomaly maps (requiring more powerful hardware) can address this issue, localizing inconsistencies more precisely.

Different Nature of Detected Obstacles. Our framework uses anomaly detection for obstacle detection, which is of a *different nature* than discriminative methods like WaSR [6]. Riverbank features aren't considered obstacles; USVs need accurate GPS/DGPS and updated maps to avoid them. This approach, however, offers segmentation between riverbank and obstacle (unlike WaSR, as in Fig. 3), crucial for further analysis. A failure mode occurs when visually similar obstacles appear in unusual places (tree in the middle of the river), and this has to be handled by other sensors (e.g. inexpensive, one beam LIDAR).

References

1. Akcay, S., Atapour-Abarghouei, A., Breckon, T.P.: GANomaly: semi-supervised anomaly detection via adversarial training. In: Jawahar, C.V., Li, H., Mori, G., Schindler, K. (eds.) ACCV 2018. LNCS, vol. 11363, pp. 622–637. Springer, Cham (2019). https://doi.org/10.1007/978-3-030-20893-6_39
2. Bergman, L., Hoshen, Y.: Classification-based anomaly detection for general data. In: International Conference on Learning Representations (ICLR) (2020)
3. Bergmann, P., Fauser, M., Sattlegger, D., Steger, C.: MVTec AD – a comprehensive real-world dataset for unsupervised anomaly detection. In: IEEE Conference on Computer Vision and Pattern Recognition (CVPR), pp. 9584–9592 (2019)
4. Bergmann, P., Fauser, M., Sattlegger, D., Steger, C.: Uninformed students: student-teacher anomaly detection with discriminative latent embeddings. In: IEEE Conference on Computer Vision and Pattern Recognition (CVPR), pp. 4183–4192 (2020)
5. Bovcon, B., Kristan, M.: A water-obstacle separation and refinement network for unmanned surface vehicles. In: IEEE International Conference on Robotics and Automation (ICRA) (2020)
6. Bovcon, B., Kristan, M.: WaSR—a water segmentation and refinement maritime obstacle detection network. IEEE Trans. Cybern. **52**, 12661–12674 (2021)
7. Bovcon, B., Muhovic, J., Perš, J., Kristan, M.: The MaSTr1325 dataset for training deep USV obstacle detection models. In: IEEE International Conference on Intelligent Robots and Systems (IROS), pp. 3431–3438 (2019)
8. Bovcon, B., Muhovič, J., Vranac, D., Mozetič, D., Perš, J., Kristan, M.: MODS-A USV-oriented object detection and obstacle segmentation benchmark. IEEE Trans. Intell. Transp. Syst. **23**, 13403–13418 (2021)
9. Caesar, H., et al.: NuScenes: a multimodal dataset for autonomous driving. In: IEEE Conference on Computer Vision and Pattern Recognition (CVPR), pp. 11621–11631 (2020)
10. Cane, T., Ferryman, J.: Evaluating deep semantic segmentation networks for object detection in maritime surveillance. In: IEEE International Conference on Advanced Video and Signal Based Surveillance (AVSS), pp. 1–6 (2018)
11. Cohen, N., Hoshen, Y.: Sub-image anomaly detection with deep pyramid correspondences. arXiv preprint arXiv:2005.02357 (2020)
12. Cordts, M., et al.: The cityscapes dataset for semantic urban scene understanding. In: IEEE Conference on Computer Vision and Pattern Recognition (CVPR), pp. 3213–3223 (2016)
13. Defard, T., Setkov, A., Loesch, A., Audigier, R.: PaDiM: a patch distribution modeling framework for anomaly detection and localization. In: International Conference on Pattern Recognition (ICPR), vol. 12664, pp. 475–489 (2020)
14. Deng, J., Dong, W., Socher, R., Li, L.J., Li, K., Fei-Fei, L.: ImageNet: a large-scale hierarchical image database. In: IEEE Conference on Computer Vision and Pattern Recognition (CVPR), pp. 248–255 (2009)
15. Fei, Y., Huang, C., Jinkun, C., Li, M., Zhang, Y., Lu, C.: Attribute restoration framework for anomaly detection. IEEE Trans. Multimed. **24**, 116–127 (2021)
16. Geiger, A., Lenz, P., Urtasun, R.: Are we ready for autonomous driving? The KITTI vision benchmark suite. In: IEEE Conference on Computer Vision and Pattern Recognition (CVPR), pp. 3354–3361 (2012)
17. González, A., et al.: Pedestrian detection at day/night time with visible and FIR cameras: a comparison. Sensors **16**(6), 820 (2016)

18. Gudovskiy, D., Ishizaka, S., Kozuka, K.: CFLOW-AD: real-time unsupervised anomaly detection with localization via conditional normalizing flows. In: IEEE Winter Conference on Applications of Computer Vision (WACV), pp. 1819–1828 (2022)
19. Haselmann, M., Gruber, D.P., Tabatabai, P.: Anomaly detection using deep learning based image completion. In: International Conference on Machine Learning and Applications (ICMLA) (2018)
20. Heidarsson, H.K., Sukhatme, G.S.: Obstacle detection and avoidance for an autonomous surface vehicle using a profiling sonar. In: IEEE International Conference on Robotics and Automation (ICRA), pp. 731–736 (2011)
21. Hendrycks, D., Mazeika, M., Kadavath, S., Song, D.: Using self-supervised learning can improve model robustness and uncertainty. In: Advances in Neural Information Processing Systems (NIPS) (2019)
22. Hermann, D., Galeazzi, R., Andersen, J., Blanke, M.: Smart sensor based obstacle detection for high-speed unmanned surface vehicle. In: IFAC Conference on Manoeuvring and Control of Marine Craft (MCMC), vol. 48, pp. 190–197 (2015)
23. Jeong, M., Li, A.Q.: Efficient LiDAR-based in-water obstacle detection and segmentation by autonomous surface vehicles in aquatic environments. In: IEEE International Conference on Intelligent Robots and Systems (IROS), pp. 5387–5394 (2021)
24. Jia, X., Zhu, C., Li, M., Tang, W., Zhou, W.: LLVIP: a visible-infrared paired dataset for low-light vision. In: IEEE International Conference on Computer Vision (ICCV), pp. 3496–3504 (2021)
25. Kiefer, B., et al.: 1st workshop on maritime computer vision (MaCVi) 2023: challenge results. In: IEEE Winter Conference on Applications of Computer Vision Workshops (WACVW), pp. 265–302 (2023)
26. Kim, H., et al.: Vision-based real-time obstacle segmentation algorithm for autonomous surface vehicle. IEEE Access **7**, 179420–179428 (2019)
27. Kristan, M., Kenk, V.S., Kovačič, S., Perš, J.: Fast image-based obstacle detection from unmanned surface vehicles. IEEE Trans. Cybern. **46**(3), 641–654 (2016)
28. Lee, S.J., Roh, M.I., Lee, H.W., Ha, J.S., Woo, I.G.: Image-based ship detection and classification for unmanned surface vehicle using real-time object detection neural networks. In: International Ocean and Polar Engineering Conference (ISOPE) (2018)
29. Ma, L., Xie, W., Huang, H.: Convolutional neural network based obstacle detection for unmanned surface vehicle. Math. Biosci. Eng. **17**, 845–861 (2020)
30. Mai, K.T., Davies, T., Griffin, L.D.: Brittle features may help anomaly detection. In: Women in Computer Vision Workshop at CVPR (2021)
31. Moosbauer, S., Konig, D., Jakel, J., Teutsch, M.: A benchmark for deep learning based object detection in maritime environments. In: IEEE Conference on Computer Vision and Pattern Recognition Workshops (CVPRW) (2019)
32. Muhovič, J., Mandeljc, R., Bovcon, B., Kristan, M., Perš, J.: Obstacle tracking for unmanned surface vessels using 3-D point cloud. IEEE J. Oceanic Eng. **45**, 786–798 (2019)
33. Nirgudkar, S., Robinette, P.: Beyond visible light: usage of long wave infrared for object detection in maritime environment. In: International Conference on Advanced Robotics (ICAR), pp. 1093–1100 (2021)
34. Noroozi, M., Favaro, P.: Unsupervised learning of visual representations by solving jigsaw puzzles. In: Leibe, B., Matas, J., Sebe, N., Welling, M. (eds.) ECCV 2016. LNCS, vol. 9910, pp. 69–84. Springer, Cham (2016). https://doi.org/10.1007/978-3-319-46466-4_5

35. Onunka, C., Bright, G.: Autonomous marine craft navigation: on the study of radar obstacle detection. In: International Conference on Control Automation Robotics & Vision (ICARCV), pp. 567–572 (2010)
36. Perš, J., et al.: Modular multi-sensor system for unmanned surface vehicles. In: International Electrotechnical and Computer Science Conference (ERK) (2021)
37. Qiao, D., Liu, G., Li, W., Lyu, T., Zhang, J.: Automated full scene parsing for marine ASVs using monocular vision. J. Intell. Robot. Syst. **104**(2) (2022)
38. Rippel, O., Mertens, P., Merhof, D.: Modeling the distribution of normal data in pre-trained deep features for anomaly detection. In: International Conference on Pattern Recognition (ICPR) (2020)
39. Rudolph, M., Wehrbein, T., Rosenhahn, B., Wandt, B.: Fully convolutional cross-scale-flows for image-based defect detection. In: IEEE Winter Conference on Applications of Computer Vision (WACV), pp. 1829–1838 (2022)
40. Ruff, L., et al.: Deep one-class classification. In: International Conference on Machine Learning (ICML), pp. 4393–4402 (2018)
41. Ruiz, A.R.J., Granja, F.S.: A short-range ship navigation system based on ladar imaging and target tracking for improved safety and efficiency. IEEE Trans. Intell. Transp. Syst. **10**(1), 186–197 (2009)
42. Schlegl, T., Seeböck, P., Waldstein, S.M., Schmidt-Erfurth, U., Langs, G.: f-AnoGAN: fast unsupervised anomaly detection with generative adversarial networks. Med. Image Anal. **54**, 30–44 (2019)
43. Steccanella, L., Bloisi, D., Castellini, A., Farinelli, A.: Waterline and obstacle detection in images from low-cost autonomous boats for environmental monitoring. Robot. Auton. Syst. **124**, 103346 (2020)
44. Teledyne: Teledyne FLIR thermal dataset for algorithm training. https://www.flir.eu/oem/adas/adas-dataset-form/
45. Tsai, C.C., Wu, T.H., Lai, S.H.: Multi-scale patch-based representation learning for image anomaly detection and segmentation. In: IEEE Winter Conference on Applications of Computer Vision (WACV), pp. 3992–4000 (2022)
46. Žust, L., Kristan, M.: Learning maritime obstacle detection from weak annotations by scaffolding. In: IEEE Winter Conference on Applications of Computer Vision (WACV), pp. 955–964 (2022)
47. Wang, G., Han, S., Ding, E., Huang, D.: Student-teacher feature pyramid matching for unsupervised anomaly detection. In: British Machine Vision Conference (BMVC) (2021)
48. Wang, M., Wang, Q., Hong, D., Roy, S.K., Chanussot, J.: Learning tensor low-rank representation for hyperspectral anomaly detection. IEEE Trans. Cybern. **53**(1), 679–691 (2022)
49. Wang, W., Gheneti, B., Mateos, L.A., Duarte, F., Ratti, C., Rus, D.: Roboat: an autonomous surface vehicle for urban waterways. In: IEEE International Conference on Intelligent Robots and Systems (IROS), pp. 6340–6347 (2019)
50. Waymo: An autonomous driving dataset (2019)
51. Yang, J., Li, Y., Zhang, Q., Ren, Y.: Surface vehicle detection and tracking with deep learning and appearance feature. In: International Conference on Control, Automation and Robotics (ICCAR), pp. 276–280 (2019)
52. Yao, L., Kanoulas, D., Ji, Z., Liu, Y.: ShorelineNet: an efficient deep learning approach for shoreline semantic segmentation for unmanned surface vehicles. In: IEEE International Conference on Intelligent Robots and Systems (IROS), pp. 5403–5409 (2021)
53. Yu, J., et al.: FastFlow: unsupervised anomaly detection and localization via 2D normalizing flows. arXiv preprint arXiv:2111.07677 (2021)

54. Zavrtanik, V., Kristan, M., Skočaj, D.: Reconstruction by inpainting for visual anomaly detection. Pattern Recogn. **112**, 1–9 (2021)
55. Zhan, W., et al.: Autonomous visual perception for unmanned surface vehicle navigation in an unknown environment. Sensors **19**(10) (2019)

A Contrario Mosaic Analysis for Image Forensics

Quentin Bammey(✉)

Université Paris-Saclay, ENS Paris-Saclay, CNRS, Centre Borelli, Gif-sur-Yvette, France
quentin.bammey@ens-paris-saclay.fr

Abstract. With the advent of recent technologies, image editing has become accessible even without expertise. However, this ease of manipulation has given rise to malicious manipulation of images, resulting in the creation and dissemination of visually-realistic fake images to spread disinformation online, wrongfully incriminate someone, or commit fraud. The detection of such forgeries is paramount in exposing those deceitful acts. One promising approach involves unveiling the underlying mosaic of an image, which indicates in which colour each pixel was originally sampled prior to demosaicing. As image manipulation will alter the mosaic as well, exposing the mosaic enables the detection and localization of forgeries. The recent introduction of positional learning has facilitated the identification of the image mosaic. Nevertheless, the clues leading to the mosaic are subtle and frail against common operation such as JPEG compression. The pixelwise estimation of the mosaic is thus often imprecise, and a comprehensive analysis and aggregation of the results are necessary to effectively detect and localize forged areas. In this work, we propose MIMIC: Mosaic Integrity Monitoring for Image *a Contrario* forensics. an *a contrario* method to analyse a pixelwise mosaic estimation. We show that despite the weakness of these traces, the sole analysis of mosaic consistency is enough to beat the state of the art in forgery detection and localization on uncompressed images. Moreover, results are promising even on slightly-compressed images. The *a contrario* framework ensures robustness against false positives, and the complementary nature of mosaic consistency analysis to other forensic tools makes our method highly relevant for detecting forgeries in high-quality images.

Keywords: Positional learning · Image Forgery Detection · Demosaicing · a contrario · Media forensics

1 Introduction

The once-reliable status of photographic images as evidence is now uncertain, owing to the proliferation of digital photography and the development of sophisticated photo editing tools. Although image modifications are frequently intended to enhance an image's aesthetic appeal, they can alter the meaning of the image.

J. Blanc-Talon et al. (Eds.): ACIVS 2023, LNCS 14124, pp. 222–234, 2023.
https://doi.org/10.1007/978-3-031-45382-3_19

The addition, modification, or concealment of objects can give an image a completely new and potentially misleading meaning, especially as the modifications can now appear convincingly authentic (Fig. 1).

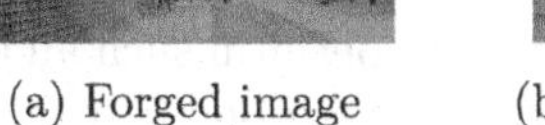

(a) Forged image

(b) Detected grid offsets

(c) Detected forgery

Fig. 1. MIMIC automatically detects forgeries based on the analysis of the underlying mosaic of an image. An *a contrario* detection automatically filters the grid estimates in search for significant inconsistencies, while keeping false positives under control.

However, images contain traces and artefacts left by the various operations of the image signal processing pipeline (ISP), from the camera sensors to the compressed version of the image. Those traces act as a signature to the image; as modifications made to an image will alter the original traces. As such, the resulting inconsistencies can be detected to show that the image has been forged. One such trace that can be analysed is the image mosaic.

Most cameras do not capture colours directly, instead, a colour filter array (CFA) is used to sample each pixel's value in a single colour. By applying filters of different colours to adjacent sensors, the pixels are sampled in different colours. The missing colours are interpolated with a demosaicing algorithm to provide a true colour image. We focus on the Bayer CFA, used in nearly all commercial cameras. This matrix samples half of the pixels in green, a quarter in red, and a quarter in blue, in a quincunx pattern. Depending on the offset, the image can be sampled in one of four patterns: $\begin{smallmatrix}R&G\\G&B\end{smallmatrix}$, $\begin{smallmatrix}B&G\\G&R\end{smallmatrix}$, $\begin{smallmatrix}G&R\\B&G\end{smallmatrix}$, or $\begin{smallmatrix}G&B\\R&G\end{smallmatrix}$. These patterns are phases of the 2-periodic CFA, offset by 0 or 1 in both directions. As demosaicing involves the reconstruction of missing data, no demosaicing method can be considered perfect, and each method introduces artefacts of some degree. As a result, these artefacts can reveal the image mosaic.

When an image is forged, the underlying mosaic of the image is altered as well. Copy-move forgeries, for instance, will displace the mosaic and might induce dephasing. Common operations in photo editing software, such as cloning and healing, often consist of multiple small copy-moves from smooth regions, the underlying mosaic of the resulting image will thus feature many small blobs of the original mosaic. Splicing from JPEG-compressed or resampled sources will make the underlying mosaic harder to detect, and might even alter its periodicity. Overall, revealing the underlying mosaic of an image provides important clues to the presence of image forgeries.

Of course, this mosaic is not explicitly known. Revealing the sampling colour of each pixel is a difficult enterprise, as the mosaic traces are hidden deep in the highest frequencies of an image. The slightest JPEG compression can dampen or even erase said traces [19], making it even more difficult to unconceal.

The recent advent of positional learning [5,6], coupled with internal fine-tuning, enabled the analysis of demosaicing traces on an image, even after a slight compression. However, even these methods remain locally inaccurate; reliable information on the mosaic can only be obtained when aggregating the method's output over a larger scale. Simply revealing the estimated mosaic is consequently no longer enough to detect forgeries. To provide reliable detections, a method must be able to analyse its own estimation so as to distinguish true mosaic inconsistencies from regions where the analysis is not accurate enough.

A contrario detection theory [13,14] provides a way to perform such an analysis. Based on the non-accidentalness principle, this theory proposes to detect data based on their unlikelihood under a background hypothesis, by thresholding the results based on a tolerated limit on the number of false alarms (NFA) under the hypothesis. This paradigm has seen successful applications in varied detection tasks [1,20–23,27–29], including forensics [3,7,8,16,18,28,31].

In this article, we propose MIMIC: Mosaic Integrity Monitoring for Image *a Contrario* forensics, an *a contrario* method to analyse a mosaic estimate and reliably detect image forgeries. Taking as only input the pixelwise mosaic estimate from an existing demosaicing analysis algorithm, the proposed method detects regions in which the estimate is significantly incoherent due to a shifted or even locally erased mosaic. MIMIC beats the forgery detection SOTA on high-quality images, even against a slight compression.

2 Related Works

Demosaicing Analysis. focused at first on linear estimation or using filters to detect inconsistencies [6,7,33]. Popescu and Farid jointly estimated a linear model and detected sampled pixels [33]. Ferrara looked for the local absence of demosaicing traces by comparing the variance between sampled and interpolated pixels [17]. Kirchner and Milani identified the sampling pattern by performing demosaicing with multiple algorithms [24,30]. Choi compared the counts of intermediate values in each lattice to estimate the correct pattern [4,10].
A contrario analysis is more than twenty years old [14], and has recently seen use in image forensics, primarily to attempt to detect inconsistencies in the JPEG [31] or demosaicing patterns [7,8] while controlling the risks of false positives. Blocks are made to vote for the most likely pattern, then the algorithm look for regions where one of the votes is significant enough to void the background hypothesis of an absence of demosaicing or JPEG compression. If two different patterns are significantly detected in different places, then the methods conclude to a forgery. This paradigm, however, is unable to detect regions without demosaicing traces, as is often possible in forged images, and also fails to reliably detect many forgeries even when they are visible in the vote maps.

Positional Learning. In AdaCFA [5], it is noted that due to translation invariance, convolutional neural networks cannot inherently detect the position of pixels, or information thereon. However, they can do so if the input image itself contains cues on the position of each pixel, as they will then learn to infer the positional information from these cues. In the case of demosaicing, the sampling mosaic of an image is 2-periodic, and demosaicing artefacts thus feature a strong 2-periodic component. If a CNN is trained on demosaiced images to detect information such as the modulo $(2, 2)$ position (horizontally and vertically) of each pixel, it will naturally rely on the demosaicing artefacts to use them as clue to the position of the pixel. Of course, the actual position of the pixel is already known, there is thus no need to actually detect it. What matters is that the network mimics the underlying mosaic of an image. Being trained on authentic images with integrate mosaics, the network indeed detects the correct positions of pixels in the training set. When used on a real image, however, the correctness of the output depends on the integrity of the image mosaic. If an image is authentic with an integrate mosaic, the network should correctly detect the position of pixels. More interestingly, if the tested image is locally forged, its mosaic will likely be locally altered as well. As a consequence, the network will yield incorrect outputs. If the mosaic has simply been shifted, for instance due to an internal copy-move, the model output will likewise be locally shifted on the forged area. If the mosaic is locally destroyed, for instance due to blurring or the insertion of a compressed and/or resampled object, whose mosaic is no longer visible, the network will simply render noise-like output instead of the actual position. In both cases, the fact that the network yields locally erroneous results is the very proof of a local forgery. This positional learning is introduced with AdaCFA [5] and refined with 4Point [6]. Combined with internal retraining on the very tested image, the internal mosaic of an image can be revealed even on slightly-compressed images.

3 Proposed Method: MIMIC

MIMIC extends on 4Point [6], where positional learning is coupled with internal fine-tuning on tested image. A network is trained on demosaiced images to detect two features on each pixel: its diagonal offset, which corresponds to whether the pixel was initially sampled in green or in another colour, and the evenness of the line and column of pixels that are on the main diagonal, which corresponds to whether the pixel (which is at this point known not to be sampled in green) was sampled in red or blue. This formulation is equivalent to detecting the modulo 2 position of the pixels horizontally and vertically, but is more natural in terms of demosaicing. From there, 4Point introduces a simple scheme to give a confidence over the authenticity of regions in the image. The mosaic estimated by the network is satisfying even on slightly-compressed images. However, as artefacts are not always visible and are often dampened or destroyed by even slight JPEG compression, the detected positions is rarely perfect. The result of the network only lead to an estimation of the underlying mosaic, and the estimation itself is not error-free. A thorough analysis of the result is thus needed to distinguish true clues of forgeries from mere estimation errors of the network.

Starting from the same base network and training, we introduce a more robust way to analyze the output of the network. Using an *a contrario* framework, we look for regions where the estimate is significantly more erroneous than in the rest of the image. Indeed, while the mosaic may be harder to detect on some images due to the processing of an image, this difficulty should not naturally vary within an image, as such, a locally erroneous estimate is a sign of forgery.

3.1 Block Votes

The network outputs two features for each pixel, namely its diagonal offset O_d and the evenness of its line offset O_l, both between 0 and 1 with extremes signifying more confidence in the result.

To bypass differences that may arise between differently-sampled pixels, the results are aggregated into 2×2 blocks, which correspond to a mosaic tile. The estimations on the four pixels of each block are averaged and compared to the midpoint, so as to return three binary outputs per block:

- $\delta_{\substack{\cdot\, G \,|\, G\, \cdot \\ G\, \cdot \,|\, \cdot\, G}}$ represents the diagonal of the block, more precisely whether the underlying mosaic tile of the block has green-sampled pixels in the top-right and bottom-left corners ($\substack{\cdot\, G \\ G\, \cdot}$ kind, with $\delta_{\substack{\cdot\, G \,|\, G\, \cdot \\ G\, \cdot \,|\, \cdot\, G}} = 0$) or in the top-left and bottom-right corners ($\substack{G\, \cdot \\ \cdot\, G}$ kind, with $\delta_{\substack{\cdot\, G \,|\, G\, \cdot \\ G\, \cdot \,|\, \cdot\, G}} = 1$)
- $\delta_{\substack{R\, G \,|\, B\, G \\ G\, B \,|\, G\, R}}$ (resp. $\delta_{\substack{G\, R \,|\, G\, B \\ B\, G \,|\, R\, G}}$) refine on the diagonal to estimate the full pattern. They represent whether, assuming the block tile is of the $\substack{\cdot\, G \\ G\, \cdot}$ kind (resp. $gxxg$), whether it is more likely to be on a $\substack{R\, G \\ G\, B}$ or $\substack{B\, G \\ G\, R}$ tile (resp. $\substack{G\, R \\ B\, G}$ or $\substack{G\, B \\ R\, G}$).

We now know the detected diagonal and pattern of each 2×2 block in the image. The detected diagonal and pattern of the whole image is then defined as the mode of the blocks' diagonals and patterns. Let $D_g \in \{\substack{\cdot\, G \\ G\, \cdot}, \substack{G\, \cdot \\ \cdot\, G}\}$ and $P_g \in \{\substack{R\, G \\ G\, B}, \substack{B\, G \\ G\, R}, \substack{G\, R \\ B\, G}, \substack{G\, B \\ R\, G}\}$ denote the global diagonal and pattern of the image.

3.2 *A Contrario* Automatic Forgery Detection

Algorithm 1: Error map computation

1 function **compute_errormap(P)**
  Input **P**: patterns or diagonal of each block, size (X, Y).
  Output **E**: Error map of P, same size.
  Output P_g: Global detected pattern or diagonal.
2 $P_g \coloneqq \texttt{mode}(P)$
3 $E \coloneqq \mathbf{0}_{X,Y}$
4 for x from 0 to X and y from 0 to Y do
5   if $P_{x,y} \neq P_g$ then
6     $E_{x,y} \coloneqq 1$
7 return **E**, P_g

In each 2×2 block of the image, we have derived an estimation of the diagonal tile and full pattern of the underlying mosaic. These estimations can be compared to the global estimations to look for forgeries. So as to avoid false positives that are solely due to misinterpretation of the detected mosaic map, we propose an *a contrario* framework to automatically detect significantly deviant regions.

As proposed in [7], we could focus on regions that present a significant grid that is different from the grid of the global image. Yet, this would not enable us to detect areas with multiple small patches of different grids (as is frequently the case on inpainted images); nor would we see the localised absence of demosaicing.

Instead, we detect regions where the detection is significantly erroneous, i.e. where the network makes more mistakes than in the rest of the image. We apply the method separately on the detected diagonals and patterns. Let E_d (resp. E_p) be a binary map which equals 1 for each block whose detected diagonal (resp. pattern) is different from D_g (resp. P_g). This is a map of the "wrong" blocks. The computation of those maps is described in Algorithm 1.

For the rest of the subsection, E represents either E_d or E_p. The empirical probability of any block on the image being wrong is denoted by p_0, and is computed as the mean of E. We want to find regions in which the error density is significantly higher than p_0.

Algorithm 2: NFA computation

```
1 function get_rectangle_nfa(E, d, x0, x1, y0, y1)
      Input E: 1 if the block is erroneous, 0 if it is correct. Size (X, Y).
      Input d: Downsampling coefficient, given by the radius of the linear
               estimation filters.
      Input x0, x1, y0, y1: Coordinates of the rectangle whose NFA to compute
      Output ε: NFA of the rectangle
```

2 $p_0 \coloneqq \frac{1}{X \cdot Y} \sum_{x=0}^{X} \sum_{y=0}^{Y} E_{xy}$

3 $n_{\text{tests}} \coloneqq 2 * (X \cdot Y)^2$

4 $k \coloneqq \sum_{x=x_0}^{x_1} \sum_{y_0}^{y_1} E_{xy}$

5 $n \coloneqq (x_1 - x_0) \cdot (y_1 - y_0)$

6 $NFA \coloneqq n_{\text{tests}} \cdot I_{p0} \left(\frac{k}{d^2} - 1, \frac{n-k}{d^2} \right)$

7 return NFA

Let us assume that, in a given rectangle, k out of the n blocks contained in the rectangle are incorrect. Under the background hypothesis that the probability of error is p_0, and assuming that the blocks are independent, the probability of having at least k wrong blocks in the area is the survival function of the binomial distribution $\text{Binom}_{sf}(k, n, p_0)$. Yet a first obstacle to this simple strategy arises, as the grid values of different blocks are not independent, as the neural network uses inputs that overlap between neighbouring blocks. To achieve independence, we simulate down-sampling and divide k and n by d^2, where d is the distance between two independent outputs. We set d to 17, the radius of the CNN. To account for the fact that in the binomial integers are then replaced by floating values, we use the Beta distribution to interpolate the binomial. The probability of having at least k wrong blocks in this area is thus evaluated by

$$p_{k,n,p_0} = I_{p_0}\left(\frac{k}{d^2}+1, \frac{n-k}{d^2}\right),$$

where I_x is the regularized incomplete Beta function. Under the *a contrario* framework, the number of tests is the possible number of rectangles in the image, that is, the number of blocks squared, multiplied by a factor 2 since we work separately on the patterns and diagonal. The NFA associated with the detection, whose computation is described in Algorithm 2, is consequently defined by

$$\mathrm{NFA}_{k,n,p_0} = 2n_{\mathrm{blocks}}^2 I_{p_0}\left(\frac{k}{d^2}-1, \frac{n-k}{d^2}\right). \tag{1}$$

Algorithm 3: Forgery detection

```
1  function get_forgery_mask(E, d, s)
       Input P: Patterns or diagonals of each block, size (X, Y).
       Input p_b: 0.5 if P represents diagonal, 0.25 if it represents patterns.
       Input d: Downsampling coefficient, given here by the size of the filters.
       Input s: Stride at which to search for rectangles. Here, s = 16.
       Output D: Forgery mask, each pixel represents the NFA detection score of
                 its corresponding 2 × 2 block.
2      D ≔ +∞
3      E, P_g ≔ compute_errormap(P)
4      E_c ≔ morphological_closing(E, disk_2)
5      labels ≔ label_connected(E)
6      for label from 0 to max(labels) do
7          M = labels_{x,y} = label
8          if (1/|E_c|) Σ_{x∈E_c} 1_{E_C}(x) > 1 − p_b then
9              ε ≔ 0 for R rectangle within the bounding box of M at
               step s do
10                 ε_R ≔ get_rectangle_nfa(E, d, R)
11                 ε ≔ min(ε, ε_R)
12             M ≔ morphological_closing(M, disk_8)
13             D_{|M} ≔ min(D_{|M}, ε)
14     return D
```

Ideally, to detect forgeries in an image, we would compute the NFA of all the rectangles in the image. The score of a pixel would be the minimal score among the rectangles containing that pixel, and the score of an image would be

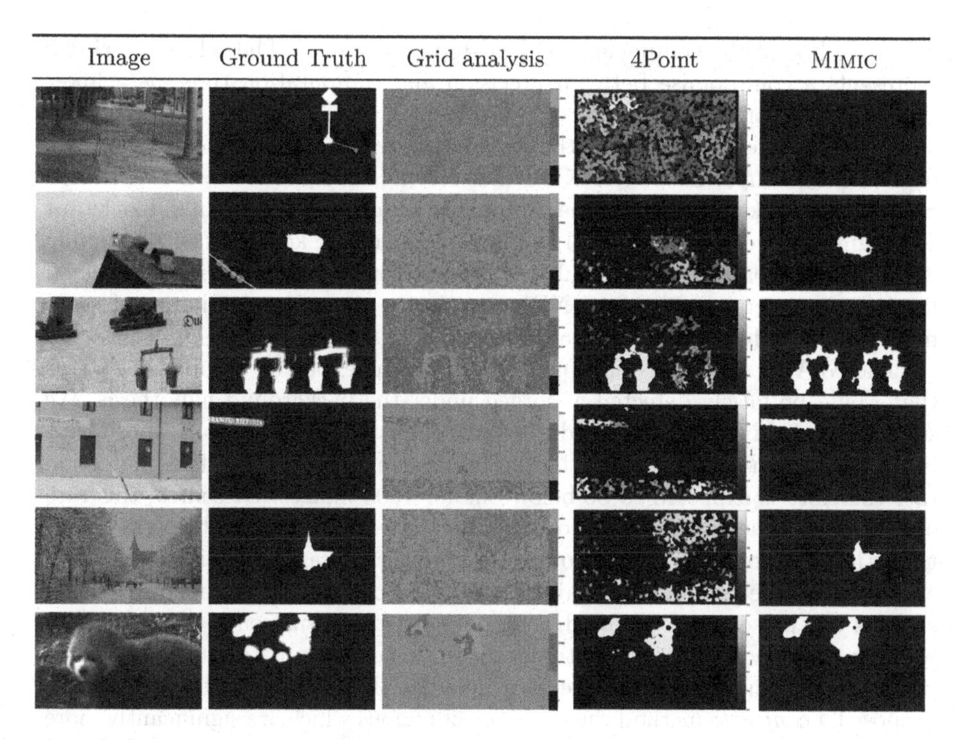

Fig. 2. Visual results on Korus forged images. In most cases, MIMIC detects the forgeries, even with inaccurate grid estimates. Although it uses the same grid estimate as 4Point, its detections are much more precise, in addition to being automatic. Its main caveat is that it misses some detections when they are too thin or diagonal, such as in the second and last columns.

that of the most significant rectangle. However, the number of rectangles scales quadratically with the number of pixels in an image. Hence, checking all possible rectangles is not possible. Even if a forgery is detected, some rectangles bigger than the forgery itself may still be significant, and the detection will therefore be too large; conversely, if part of a forgery is detected, we should detect nearby parts of the same forgery as well, even if they are not as significant as the detected part. As a consequence, we propose to first detect and separate all potential forgeries, and then to decide on their significance, so as to improve the localization of the forgeries. The method used is described in Algorithm 3.

Still separately on the diagonal and full patterns, we use the map E of 2×2 blocks whose diagonal/pattern is erroneous. We apply a morphological closing to this map with a disk of size 2 to connect inconsistent blocks, and segment the resulting map into connected components. Components where the global pattern (respectively diagonal) represent more than 25% (respectively 50%) of the blocks are immediately rejected and not tested for forgeries.

Each of the remaining components is tested to determine whether it is a forgery. On each component, we test all the rectangles contained within the bounding box of the component, with a step of 16 pixels. The selected striding represents a compromise between precision and computation time, as a lower stride means more rectangles need to be checked.

Finally, we keep the NFA of the most significant rectangle. We set the score of the whole component to this NFA, thus solving the final two issues addressed above: only blocks that were in the component are given this NFA, and blocks out of a significant rectangle but still in the component are kept. Forgery detection is performed separately on the full pattern and on diagonals, then the detected forgeries are merged. The final NFA map is the pointwise minimum of score maps of the patterns and diagonals NFA.

The NFA of a region is an upper bound on the expected number of regions that would be falsely detected as forged under the background hypothesis. We set the threshold to $\epsilon = 10^{-3}$, and the final, binary map keeps pixels whose NFA is below this threshold. Under the background hypothesis, the false detection rate therefore is expected to be below one for 1000 images. Of course, this rate only concerns the risk of false positives that are due to a misinterpretation of the estimated mosaic, and does not provide further guarantees against significant errors within the estimated mosaic, which could be due to the image structure, such as the presence of textured areas, or post-processing such as resampling which would modify the mosaic traces. Still, this enables us to filter out regions which are only marginally less precisely detected than the rest of the image. The proposed *a contrario* method thus only select regions which are significantly more erroneous than the rest of the image, regardless of the reason, and provides us with a mathematically rigorous way to automatically interpret the estimated mosaic to yield an automatic detection.

4 Results

We test MIMIC on the Trace CFA Grid [9] and on the Korus [25,26] datasets. The Trace CFA Grid dataset, on which results are shown in Table 1 contains 1000 forged images that can only be detected by their demosaicing traces, thus enabling a comparison between demosaicing analysis methods and evaluation of the sensitivity to demosaicing artefacts of more generic methods. The Korus dataset, on which results are shown in Table 2 and visually in Fig. 2, features 220 forged images from four cameras.

Results are presented using Matthew's Correlation Coefficient (MCC), a metric ranging from 1 (perfect detection) to -1 (opposite detection). Any input-independent method has a zero MCC expectation. The MCC can only be measured on binary detections, which is only the case for MIMIC. For the other methods, we threshold the output on the threshold that maximizes the score, giving those methods a slight advantage compared to a real case scenario, where adjusting the threshold to the data would not be possible.

Experiments on the Trace dataset show that MIMIC beats the state of the art even when the images are JPEG-compressed at quality levels 95 and 90, while generic methods are shown to be insensitive to demosaicing artefacts and

Table 1. Results on the CFA Grid exomask (Grid) dataset of the Trace database, on uncompressed images and after compression with quality factors 95 and 90. The methods are grouped, after the proposed method MIMIC are methods based on demosaicing analysis, then more generic methods that do not specifically target demosaicing artefacts. Our analysis of the mosaic improves on the results of 4Point, especially on stronger compression ($Q = 90$). As already established in the literature [5,6,8,9], generic methods that do not specifically target demosaicing artefacts are entirely blind to shifts in the mosaic.

Method	Uncompressed	JPEG $Q = 95$	JPEG $Q = 90$
MIMIC	**0.724**	**0.311**	**0.196**
4Point [6]	0.709	0.307	0.151
AdaCFA [5]	0.692	0.005	0.003
DDem [8]	0.401	0.129	0.093
Shin [34]	0.104	0.001	0.001
Choi [4,10]	0.603	0.156	0.070
Ferrara [17,36]	0.071	0.000	0.000
Dirik [15,36]	−0.002	0.000	0.001
Park [32]	0.116	0.001	0.000
Noiseprint [12]	−0.001	0.004	0.001
Splicebuster [11]	0.003	0.004	0.001
ManTraNet [2,35]	0.000	−0.001	0.002

Table 2. Results on the Korus dataset of forged images. No demosaicing artefacts are found on the Canon 60D images by any of the demosaicing-based method, thus we can safely conclude they do not feature demosaicing artefacts. On all the other images, as well as overall, the proposed method significantly improves over the existing state of the art on this dataset.

Method	Overall	Canon 60D	Nikon D7000	Nikon D90	Sony α57
Proposed	**0.472**	0.000	**0.595**	**0.630**	**0.662**
4Point [6]	0.353	0.00	0.401	0.378	0.624
AdaCFA [5]	0.167	0.002	0.049	0.044	0.574
Shin [34]	0.143	0.021	0.003	0.012	0.511
Choi [4,10]	0.238	0.004	0.176	0.251	0.251
Ferrara [17,36]	0.321	−0.016	0.498	0.461	0.339
Dirik [15,36]	0.153	0.036	0.241	0.275	0.062
Park [32]	0.338	0.018	0.540	0.491	0.302
Noiseprint [12]	0.202	**0.153**	0.322	0.236	0.148
Splicebuster [11]	0.238	0.153	0.329	0.222	0.155
ManTraNet [2,35]	0.169	0.121	0.229	0.193	0.143

inconsistencies. On the Korus dataset, one quarter of the images do not feature traces of demosaicing, as validated by all tested methods. The proposed method is thus unable to detect forgeries on this part of the dataset. Despite that, MIMIC still yields the best overall score on this dataset. We further note that on the images without traces of demosaicing, MIMIC does not make any false detection.

5 Conclusion

In this paper, we proposed MIMIC (Mosaic Integrity Monitoring for Image *a Contrario* forensics), an *a contrario* method that extends on the demosaicing analysis network 4Point [6] to refine its analysis and automatically detect image forgeries. MIMIC identifies forgeries as regions where the network output is significantly more erroneous than in the rest of the image. An *a contrario* framework helps limit the risk of false positives.

Demosaicing artefacts are frail and subtle, yet they can provide highly-valuable information to detect forgeries. On high-quality images, their sole analysis yields better results on forgery detection than any other state-of-the-art method. These results are furthermore fully complementary to non-demosaicing-specific methods, which are not sensible to demosaicing artefacts.

Acknowledgment. This work has received funding by the European Union under the Horizon Europe vera.ai project, Grant Agreement number 101070093, and by the ANR under the APATE project, grant number ANR-22-CE39-0016. Centre Borelli is also a member of Université Paris Cité, SSA and INSERM. I would also like to thank Jean-Michel Morel and Rafael Grompone von Gioi for their insightful advice regarding this work.

References

1. Aguerrebere, C., Sprechmann, P., Muse, P., Ferrando, R.: A-contrario localization of epileptogenic zones in SPECT images. In: 2009 IEEE International Symposium on Biomedical Imaging: From Nano to Macro (2009)
2. Bammey, Q.: Analysis and experimentation on the ManTraNet image forgery detector. Image Processing On Line **12** (2022)
3. Bammey, Q.: Jade OWL: JPEG 2000 forensics by wavelet offset consistency analysis. In: 8th International Conference on Image, Vision and Computing (ICIVC). IEEE (2023)
4. Bammey, Q., Grompone von Gioi, R., Morel, J.M.: Image forgeries detection through mosaic analysis: the intermediate values algorithm. IPOL (2021)
5. Bammey, Q., von Gioi, R.G., Morel, J.M.: An adaptive neural network for unsupervised mosaic consistency analysis in image forensics. In: CVPR (2020)
6. Bammey, Q., von Gioi, R.G., Morel, J.M.: Forgery detection by internal positional learning of demosaicing traces. In: WACV (2022)
7. Bammey, Q., von Gioi, R.G., Morel, J.M.: Reliable demosaicing detection for image forensics. In: 2019 27th European Signal Processing Conference (EUSIPCO), pp. 1–5 (2019). https://doi.org/10.23919/EUSIPCO.2019.8903152

8. Bammey, Q., Grompone Von Gioi, R., Morel, J.M.: Demosaicing to detect demosaicing and image forgeries. In: 2022 IEEE International Workshop on Information Forensics and Security (WIFS), pp. 1–6 (2022). https://doi.org/10.1109/WIFS55849.2022.9975454
9. Bammey, Q., Nikoukhah, T., Gardella, M., von Gioi, R.G., Colom, M., Morel, J.M.: Non-semantic evaluation of image forensics tools: methodology and database. In: WACV (2022)
10. Choi, C.H., Choi, J.H., Lee, H.K.: CFA pattern identification of digital cameras using intermediate value counting. In: MM&Sec (2011)
11. Cozzolino, D., Poggi, G., Verdoliva, L.: Splicebuster: a new blind image splicing detector. In: WIFS (2015)
12. Cozzolino, D., Verdoliva, L.: Noiseprint: a CNN-based camera model fingerprint. IEEE TIFS (2020)
13. Desolneux, A., Moisan, L., Morel, J.: From gestalt theory to image analysis. Interdisc. Appl. Math. **35** (2007)
14. Desolneux, A., Moisan, L., Morel, J.M.: Meaningful alignments. IJCV (2000)
15. Dirik, A.E., Memon, N.: Image tamper detection based on demosaicing artifacts. In: ICIP (2009)
16. Ehret, T.: Robust copy-move forgery detection by false alarms control. arXiv preprint arXiv:1906.00649 (2019)
17. Ferrara, P., Bianchi, T., De Rosa, A., Piva, A.: Image forgery localization via fine-grained analysis of CFA artifacts. IEEE TIFS **7** (2012)
18. Gardella, M., Musé, P., Morel, J.M., Colom, M.: Noisesniffer: a fully automatic image forgery detector based on noise analysis. In: IWBF. IEEE (2021)
19. Gardella, M., Nikoukhah, T., Li, Y., Bammey, Q.: The impact of jpeg compression on prior image noise. In: 2022 IEEE International Conference on Acoustics, Speech and Signal Processing (ICASSP), ICASSP 2022, pp. 2689–2693 (2022). https://doi.org/10.1109/ICASSP43922.2022.9746060
20. Grompone von Gioi, R., Jakubowicz, J., Morel, J.M., Randall, G.: LSD: a fast line segment detector with a false detection control. IEEE Trans. Pattern Anal. Mach. Intell. **32** (2010)
21. Grompone von Gioi, R., Jakubowicz, J., Morel, J.M., Randall, G.: LSD: a line segment detector. IPOL **2** (2012)
22. Grompone von Gioi, R., Randall, G.: Unsupervised smooth contour detection. Image Process. On Line **6** (2016)
23. Gómez, A., Randall, G., Grompone von Gioi, R.: A contrario 3D point alignment detection algorithm. Image Process. On Line **7** (2017)
24. Kirchner, M., Fridrich, J.: On detection of median filtering in digital images. In: MFS II (2010)
25. Korus, P., Huang, J.: Evaluation of random field models in multi-modal unsupervised tampering localization. In: IEEE WIFS (2016)
26. Korus, P., Huang, J.: Multi-scale analysis strategies in PRNU-based tampering localization. IEEE Trans. Inf. Forensics Secur. (2017)
27. Lezama, J., Randall, G., Morel, J.M., Grompone von Gioi, R.: An unsupervised point alignment detection algorithm. Image Process. On Line **5** (2015)
28. Li, Y., et al.: A contrario detection of h.264 video double compression. In: 2023 IEEE International Conference on Image Processing (ICIP) (2023)
29. Lisani, J.L., Ramis, S.: A contrario detection of faces with a short cascade of classifiers. Image Process. On Line **9** (2019)
30. Milani, S., Bestagini, P., Tagliasacchi, M., Tubaro, S.: Demosaicing strategy identification via eigenalgorithms. In: ICASSP (2014)

31. Nikoukhah, T., Anger, J., Ehret, T., Colom, M., Morel, J.M., Grompone von Gioi, R.: JPEG grid detection based on the number of DCT zeros and its application to automatic and localized forgery detection. In: CVPRW (2019)
32. Park, C.W., Moon, Y.H., Eom, I.K.: Image tampering localization using demosaicing patterns and singular value based prediction residue. IEEE Access **9**, 91921–91933 (2021). https://doi.org/10.1109/ACCESS.2021.3091161
33. Popescu, A.C., Farid, H.: Exposing digital forgeries in color filter array interpolated images. IEEE Trans. Signal Process. **53**, 3948–3959 (2005)
34. Shin, H.J., Jeon, J.J., Eom, I.K.: Color filter array pattern identification using variance of color difference image. J. Electron. Imaging (2017)
35. Wu, Y., AbdAlmageed, W., Natarajan, P.: Mantra-net: manipulation tracing network for detection and localization of image forgeries with anomalous features. In: IEEE CVPR (2019)
36. Zampoglou, M., Papadopoulos, S., Kompatsiaris, Y.: Large-scale evaluation of splicing localization algorithms for web images. Multimed. Tools Appl. **76** (2017)

IRIS Segmentation Technique Using IRIS-UNet Method

M. Poovayar Priya[1(✉)] and M. Ezhilarasan[2]

[1] Department of Computer Science and Engineering, Pondicherry Engineering College, Puducherry 605014, India
poovayarpriya25@gmail.com

[2] Department of Information Technology, Pondicherry Engineering College, Puducherry 605014, India
mrezhil@pec.edu

Abstract. Precise segmentation of the iris from the eye images is an essential task in iris diagnosis. Most of the predictions fail due to improper segmentation of iris images, which results in false predictions of the patient's disease. However, the traditional methods for segmenting the iris are not suitable for iris diagnosis applications. In the field of medical purposes. The iris of the eye could be separated from the eye using deep learning techniques. To overcome this issue, we design a model called Iris-UNet which can effectively segment the limbic and pupillary boundary from the eye images. Using the Iris-UNet model, high-level features are extracted in the encoder path, and segmentation of limbic and pupillary boundaries takes place in the decoder path. We have evaluated our Iris-UNet model on real patient datasets: CASIA, MMU, and PEC datasets. Our Iris-UNet model shows an outperforming solution through the experimental results compared with other traditional methods.

Keywords: Iris segmentation · deep learning · Iris-UNet · iris diagnosis · medical purposes

1 Introduction

Image Segmentation is the approach where a digital image is split into different parts known as image segments which makes it reduces the complexity of an image for further processing or analysing the image. Image segmentation is mainly used to detect the objects and boundaries of an image such as lines, curves, etc., in an image. The limitations of image segmentation are over-segmentation when the image suffers from noise or has intensity variations. To overcome this, many new algorithms have been implemented. Many applications are involved in image segmentation such as medical image segmentation, audio and video surveillance, iris detection, object detection, traffic control system, recognition tasks, etc.

The process of separating the iris portion from the ocular image is known as iris segmentation. Sometimes, it is referred to as iris localization, an important step in iris recognition. The thin, annular organ known as the iris, which is located inside the eye,

J. Blanc-Talon et al. (Eds.): ACIVS 2023, LNCS 14124, pp. 235–249, 2023.
https://doi.org/10.1007/978-3-031-45382-3_20

is in charge of regulating the pupil's size and diameter. Five layers of fibre-like tissues make up the iris' extremely fine structure. These tissues are extremely complex and can be seen as reticulation, thread-like, linen-like, burlap-like, etc. The iris' surface is covered in a variety of intricate texture patterns, including crystals, thin threads, spots, concaves, radials, furrows, strips, etc. (Fig. 1).

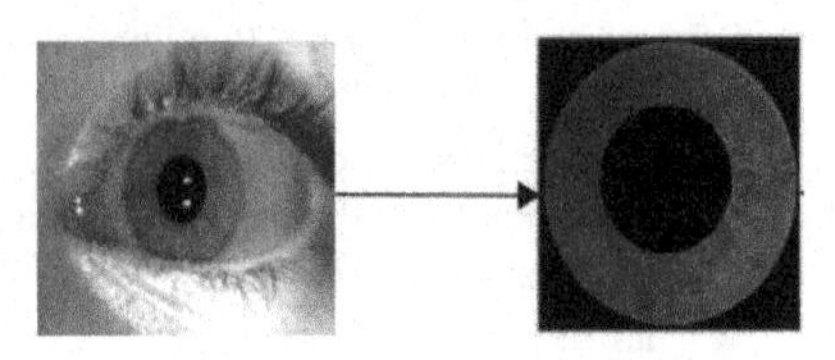

Fig. 1. Iris segmentation process

2 Literature Survey

2.1 Memristive Network-Based Genetic Algorithm (MNGA)

By fusing the memristive network and the genetic algorithm, Yu Yongbin [1] proposed a network model known as memristive network-based genetic algorithm (MNGA). Edge detection using MNGA was suggested. This new edge detection approach uses filtering technique and a fitness function to assess each pixel in order to overcome the limitations of the existing methods, such as noise and the difficulty in designing an individual's fitness and edge information.

$$FF(x, y) = \sum_{\substack{i=-1,0,1 \\ j=-1,0,1}} |P(x, y) - P(x + i, y + j)| \tag{1}$$

where P(x, Y) stands for the image's pixel value for the xth row and yth column. The fitness of a pixel is indicated by FF(x,y). (x, y). Based on FoM, which produced a value of 62.7%, the performance was assessed.

2.2 B-Edge Approach Based on Deep Learning

Mamta Mittal [2] proposed a robust edge detection algorithm using multiple threshold approaches(B-Edge). The edge connectivity and edge thickness are the two main restrictions encountered. With minimal noise, the suggested approach successfully recognises strong and thin edges. Better edge continuity and entropy value are both provided by it.

$$\psi = \frac{\left(\sum_{\alpha=1}^{n} \sum_{\beta=1}^{m} ara\right)}{m \cdot n} \tag{2}$$

In this case, m and n are the pixel sizes, and area is an array representing the input image. With each image input, Graythresh calculates a unique average of the inputted images.

$$Phi = \frac{\psi \cdot 20}{8.33} \tag{3}$$

The performance is evaluated based on PSNR which gave 62.47% as result. This recommended solution does not work for blur images properly. Time calculations need to be made more accurate. A deep learning strategy can be used to get the best outcomes.

2.3 Self Attention Deep Neural Network

Davood Karimi [3], an alternative deep neural network that does not use convolutional operations has been proposed to provide more accurate segmentation than FCNs. Which is based on self-attention between close picture patches. This model, which was developed using self-attention between nearby 3D patches rather than the traditional convolution building block, produced an accurate outcome. Starting with the position- and embedded-encoded patch input sequence, X^0 described above, the kth stage of the network performs the following operations to map X^k to X^{k+1}.

With 89.2% accuracy, the proposed approach was assessed using the DSC (Dice Similarity Coefficient). Future study will involve using this model for other medical image analysis tasks, like classifying and detecting anomalies.

2.4 Multi-atlas Based on Soft Segmentation

Iman Aganj [4], a brand-new atlas-based technique for supervised soft image segmentation was introduced that, even when only one valid label is discovered, nonetheless in the new image, it assigns an anticipated label at each position. They computed the anticipated label value map by performing a straightforward convolution with the key after a fast Fourier transform(FFT). When N atlases with manual labels are available and are aligned in the same space, for instance, the equation can be expressed as follows:

$$E := \frac{1}{N} \sum_{i=1}^{N} \mathbf{E}[L_i \circ \boldsymbol{T} \mid I, J_i], \tag{4}$$

where L_i and J_i are i^{th} pair of manual label and atlas images, respectively. The performance is evaluated based on DSC showed the result as 92%. Still, the segmentation accuracy can be improved by using an ELV map.

2.5 SR-Guided Threshold Maneuvering and Window Mapping

Deepak Dhillon [5] introduced an enhanced Canny edge detection method that works similarly to standard Canny but creates better-connected edges and lowers noise. This method uses Stochastic Resonance guided threshold manoeuvring and window mapping

[5]. Noisy structures and discontinuous edges are the paper's two primary shortcomings. Along with SR, a comprehensive examination of CED is also provided.

The suggested algorithm significantly outperformed CED. With several measures, ODS = 0.79, OIS = 0.81, and AP = 0.50, it produced performance outcomes. However, the algorithm's performance has been constrained by the lack of any pre- or post-processing. Future research will concentrate on developing deep neural networks for enhanced edge detection methods to overcome these constraints.

2.6 SpineParseNet-Based Semantic Segmentation

To achieve automatic spine parsing for volumetric MR images, Shumao Pang [6] suggested a brand-new two-stage system called SpineParseNet. SpineParseNet employs 2D segmentation refinement for 2D residual U-Net (ResUNet) and 3D coarse segmentation for segmenting 3D graphs. Memory costs are decreased during the training and testing phases of our two-stage segmentation scheme. The main benefit of the suggested strategy is that GCN is used to enhance the distinction of various spinal structures. This is how the semantic graph is produced:

$$F^S = \sigma\left(A^e \sigma\left(A^e \sigma\left(A^e F^G W_1^e\right) W_2^e\right) W_3^e\right) \tag{5}$$

where $W_1^e \in \mathbb{R}^{m \times m}$, $W_2^e \in \mathbb{R}^{m \times m}$, and $W_3^e \in \mathbb{R}^{m \times m}$ are three trianable weight matrices.

SpineParseNet successfully completed accurate spine parsing for volumetric MR images as a result. DSC was used as a metric, and it obtained 0.87. But the top structure of the image cannot be segmented by SpineParseNet. In the future, regions with high levels of uncertainty can be segmented using a region-specific classifier to get better segmentation results.

2.7 Self-ensembling-Based Semi-supervised Segmentation

According to a recent semi-supervised technique for medical image segmentation by Xiaomeng Li [7], the weighted combination of a simply labelled inputs for common supervised loss and of both labelled and unlabeled data for regularisation loss is maximised in this network. This improved the regularisation effect for pixel-level predictions by introducing a transformation-consistent approach into the self-ensembling model for semi-supervised segmentation. The lack of labelled data is what drives research into techniques that can be learned with less supervision, such as semisupervised learning.

One restriction is the assumption that same distribution of labelled and unlabelled data. However, labelled and unlabelled data may have different distributions in the field of medical applications. This technique is all-encompassing and is applicable to many different semi-supervised medical imaging issues. Based on JA and DICE, the performance is rated as 0.75 and 0.93, respectively. An investigation into domain adaptability can be expanded in the future.

2.8 Deep Symmetric Adaptative Network

Xiaoting Han [8] presented the novel deep symmetric architecture of unsupervised domain adaptation consisting of two symmetric target and source domain translation sub-networks and a segmentation sub-network. Existing UDA techniques concentrate by reducing discrepancies in source and target domain distributions resulting from feature alignment or picture translation. Additionally, this model examined the semantic information from various image formats and provided a method for alignment over the source and destination translation sub-networks. The Translation sub-network adversarial loss concept is as follows:

$$\begin{aligned}\mathcal{L}^{s}_{adv}(E, U_s, D_s) = &\mathbb{E}_{x^z\sim\mathcal{X}^s}\left[\log D_s(x^s)\right]\\ &+\mathbb{E}_{x^t\sim\mathcal{X}^t}\left[\log\left(1 - D_s\left(U_s\left(E\left(x^t\right)\right)\right)\right)\right].\end{aligned} \tag{6}$$

It is possible to quantify optimisation loss for the target and source domain translation sub-network as follows:

$$\begin{aligned}\mathcal{L}^{t}_{gen} &= \lambda_{rec}\mathcal{L}^{t}_{rec} + \lambda_{adv}\mathcal{L}^{t}_{adv},\\ \mathcal{L}^{s}_{gen} &= \lambda_{rec}\mathcal{L}^{s}_{rec} + \lambda_{adv}\mathcal{L}^{s}_{adv},\end{aligned} \tag{7}$$

Comparing this procedure to cutting-edge techniques, it demonstrates a significant advantage. It achieved a Dice score of 78.50%. This method can be further improved to better generalize in the target domain.

2.9 SIMCVD Segmentation Method

Chenyu You [9] proposed SimCVD, a straightforward contrastive distillation framework, a semi-supervised segmentation algorithm. Previous semi-supervised learning techniques had significant problems with suboptimal performance, geometric information loss, and generalisation. To solve these issues: Boundary-aware representations that take into account detailed information about the geometry of the object and a signed distance map, a segmentation map, and other outcomes from multi-task learning and a pair-wise distillation goal all at once are two features of SimCVD. For training on labelled data, supervised loss can be calculated as follows:

$$\mathcal{L}_{\text{sup}} = \frac{1}{N}\sum_{i=1}^{N}\mathcal{L}_{\text{seg}}\left(\mathbf{Q}_i^s, \mathbf{Y}_i\right) + \frac{\alpha}{N}\sum_{i=1}^{N}\mathcal{L}_{\text{mse}}\left(\mathbf{Q}_i^{s,\text{sdm}}, \mathbf{Y}_i^{\text{sdm}}\right) \tag{8}$$

InfoNCE loss is defined by,

$$\mathcal{L}\left(\mathbf{h}^t_{i,j}, \mathbf{h}^s_{i,j}\right) = -\log\frac{\exp\left(\mathbf{h}^t_{i,j}\cdot\mathbf{h}^s_{i,j}/\tau\right)}{\sum_{k,l}\exp\left(\mathbf{h}^t_{i,j}\cdot\mathbf{h}^s_{k,l}/\tau\right)} \tag{9}$$

SimCVD produced brand-new state-of-the-art outcomes. The Dice score for this approach is 89.03%. The approach may also be expanded to address multi-class medical image segmentation challenges.

2.10 Two-Stream Graph Convolutional Network

Yue Zhao [10] created a teeth segmentation model using a two-stream graph convolutional network (TSGCN). The nave concatenation of many raw qualities during the input phase causes unneeded complexity when describing and differentiating between mesh cells. Modern deep learning-based systems, in contrast, present completely different geometric data and experience various raw properties. This TSGCN model can successfully handle inter-view confusion between them in order to combine the complementary information from various raw attributes and create discriminative multi-view geometric representations. The intra-oral scanning image's basic topology is extracted by the C-stream from each cell's coordinates. The information from each neighbourhood is combined and sent to each centre as follows:

$$\mathbf{f}_i^{l+1} = \sum_{m_{ij} \in \mathcal{N}(i)} \alpha_{ij}^l \odot \hat{\mathbf{f}}_{ij}^l, \tag{10}$$

For all the cells, the normal vectors are assigned to a canonical space as inputs to N-stream via an input transformer module before the hierarchical extraction of higher-level feature representations. The following formulation describes the boundary representation for the relevant centre:

$$\mathbf{f}_i^{l+1} = maxpooling\left\{\hat{\mathbf{f}}_{ij}^l, \forall m_{ij} \in \mathcal{N}(i)\right\}. \tag{11}$$

This approach achieves an overall accuracy of 96.69%. Future work can be expanded to implement a large number of training samples since current method is only appropriate for a small number of training samples.

2.11 Semi-supervised Approach for Image Segmentation

Kai Han [11] proposed a deep semi-supervised method that incorporates a pseudo-labeling technique for segmenting liver CT images. The fundamental obstacle to developing a deep segmentation model continues to be the volume of training data. A liver image segmentation based on a semi-supervised framework was introduced, using the direction from labeled photos to produce pseudo-unlabelled images in a high-quality manner.

Dice score for this approach was 86.83%. In order to normalize, direct the formation of labels, future work will focus on creating more reliable class representations that are produced by the network architecture and the quantity of output channels.

2.12 SSL-ALPNet

Cheng Ouyang [12] proposed SSL-ALPNet, a brand-new self-supervised few-shot segmentation framework for medical pictures. In order to acquire picture representations that generalise to unknown testing classes, conventional FSS approaches for segmenting images rely on a large amount of annotated data from training classes. However, in medical imaging circumstances when annotations are rare, such training is impractical. The suggested solution leverages pseudo-labels based on superpixels to deliver supervision signals. Additionally, a powerful pooling module based on adaptive local prototype was suggested to increase segmentation precision. The suggested method is made to only segment one label class at a time, or one-way segmentation. The support feature map that spatially averages beneath the object's whole binary mask, specifically:

$$p_l^g\left(c^{\hat{j}}\right) = \frac{\sum_h \sum_w \mathbf{y}_l^s\left(c^{\hat{j}}\right)(h, w)\mathbf{z}_l^s(h, w)}{\sum_h \sum_w \mathbf{y}_l^s\left(c^{\hat{j}}\right)(h, w)}. \tag{12}$$

The local similarity score is defined as:

$$S_{k_j}\left(c^j\right)(h, w) = \alpha \operatorname{sim}\left(p_{k_j}\left(c^j\right), \mathbf{z}^q(h, w)\right), \tag{13}$$

Channel dimension for softmax function is calculated as:

$$\hat{\mathbf{y}}^q(h, w) = \operatorname*{softmax}_{c^j}\left(\left\{S'\left(c^j\right)(h, w)\right\}\right). \tag{14}$$

The mean value of this method is 80.16%. The future work can be extended to segment more than one class i.e., multi-way segmentation.

2.13 Deep Multi-task Attention Model

IrisParseNet is a deep learning based iris segmentation method with high efficiency that Caiyong Wang [13] suggested. Since iris images are frequently taken in unfavorable conditions, unfavorable noise complicates many of the approaches currently used for iris segmentation. The suggested method offers a comprehensive iris segmentation, including parameterized outer and inner iris boundaries and iris masking, to address this issue. On newly annotated iris databases, the suggested method is compared to conventional iris segmentation algorithms and exhibits superior performance on numerous benchmarks. This is how the segmentation error rate is calculated:

$$E1 = \frac{1}{n \times c \times r} \sum_{c'} \sum_{r'} G(c', r') \otimes M(c', r') \tag{15}$$

The F1 score for the IrisParseNet technique is 83.05%. To enhance segmentation performance, future work might investigate more effective ways to explicitly leverage the spatial link between the outer and the inner boundaries and the iris mask.

2.14 Deep Convolutional Neural Network (DCNN)

Zhiyong Wang created a deep convolutional neural network (DCNN) that automatically extracts the iris and pupil pixels of each eye from the input image [14]. UNet and SqueezeNet were merged to develop a potent convolutional neural network for image categorization in this network. With just a single RGB camera, our method advances the state-of-the-art in 3D eye gaze tracking. The cross-entropy loss to evaluate the segmentation outcome and the ground truth is expressed as follows:

$$E = \arg\min_{\theta} - \sum_{i \in \Omega} w_i log(P_i) + (1 - w_i) log(1 - P_i) \tag{16}$$

By using this method, a mean inaccuracy of 8.42° in eye gaze directions was obtained. This method works well since it is quick, completely automatic, and precise. Both PCs and cell phones can use this technology in real-time.

2.15 Iris Segmentation Using Optimized U-Net

By increasing the size of the convolutional kernels, for example, Sabry Abdalla M et al. suggested some new, straightforward improvements to U-Net that can enhance segmentation outcomes in comparison to the original U-Net design. U-Net, a well-known deep network architecture, has been adjusted to provide models for iris segmentation that are more accurate and run more quickly. The accuracy rate of this approach, which used hyperparameters, was 0.96%.

3 Material and Method

3.1 Data and Pre-processing

The data consists of the CASIA database, MMU database, and PEC database.

3.1.1 CASIA Datasets

4,035 iris images from more than 70 participants make up the version of CASIA-IrisV3 that was used. All of the iris photographs were captured using a near-infrared light source and are 8-bit grey-level JPEG files. The circular NIR LED array that was created with an appropriate luminous flux for iris imaging is the most impressive aspect of this iris camera. 20% of the images are for testing, while the remaining 80% are for training.

3.1.2 MMU Datasets

Images of the eyes are available in the Multimedia University (MMU1) public database and can be used to develop IRIS-based biometric attendance models. Since each person has unique IRIS patterns for each Eye, it is simple to identify a specific person. For a total of 46 individuals, this dataset contains 460 images, including a few empty bmp files and five pictures of each person's left and right IRIS. Using the retained data, IRIS segmentation can be utilised to identify and classify each unique IRIS image.

3.1.3 PEC Datasets

PEC datasets are a college database made up of student iris scans for study. This collection of 450 photos includes iris images for each subject's left and right eyes as png files. Using an L1 solution sensor and near-infrared illumination, all of the images are captured. Our iris camera can take incredibly clear iris photographs thanks to this creative design. It has 80% of the iris pictures needed for testing and 20% for training (Table 1).

Table 1. Details of different datasets

Parameters	CASIA	MMU	PEC
No. of subjects	70	46	140
No. of images	4035	460	450
Image size	320 * 280	320 * 240	640 * 480
File format	.jpeg	.bmp	.png

3.2 Structure of Iris-UNet Model

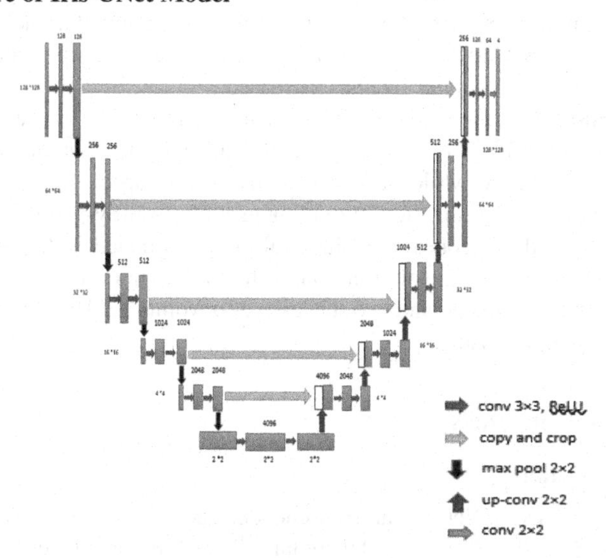

Fig. 2. Iris-UNet architecture

The architecture includes two paths:

A. Encoder: The encoder path is first. Encoder/contraction path refers to the side of the "U"-shaped design that faces left. A max pooling layer and a stack of convolutional layers make up this path. This serves to preserve the image's context.
B. Decoder: The decoder path is second. The decoder/expansion path is the term for the right-hand side of the "U"-shaped design. Convolutional layers that have been transposed make up the path. This helps to enable more accurate localization.

This network is fully convolutional from beginning to end. (FCN). Since it lacks fully connected dense layers, it can accept any image size. The size of the supplied image is 128 * 128 * 3. The left contractor path and right expansion path are depicted in the image above. Convolution layers and max pooling layers are used by Contractor Path to extract information from the image. As we need to do image segmentation, which is a pixel-by-pixel classification of images, the correct path aids in the exact localization of images. This is accomplished on the right side by combining transposed convolutional layers with standard convolutions.

The image is doubled in height and breadth for each transpose convolution. While the number of channels is reduced by half. One iteration on the left uses two convolutions (which expand the filter's size to obtain more features) and one maxpooling. (Decreasing the size of the image, thus keeping relevant information only). Up Sampling with Transpose Convolutional Layer, Skip Connection (obtaining data from the parallel convolutional side of the encoder), and 2 Convolutions are the only iterations shown on the right-hand side. (so that the model can learn to assemble a more precise output).

The output tensors for convolutional layers are c1–c13. The Maxpooling layer's outputs range from p1 to p6. The outputs of the transpose convolutional layer, which was used for upsampling, are u8–u13. The image is down-sampled in the encoder from 128 * 128 * 3 to 8 * 8 * 2048. This indicates that the image's height and width have been decreased while the channel size has increased.

The information about the image is contained in this output. The image is up-sampled in the decoder from 8 * 8 * 2048 to 128 * 128 * 1. This indicates that the image's height and breadth have increased while the channel size has dropped. This output includes details about the image's localization, providing us with pixel-by-pixel classification.

The output from the encoder convolutional layers is added at the same level as the output of the transpose convolutional layer to enable more accurate and detailed localization. Using this skip link and two successive convolutional layers, we can provide an output that is more precise.

3.3 Implementation Details

3.3.1 Network Details

As shown in Fig. 2, Iris-UNet contains an encoder and decoder path. The size of the input image is 128 * 128 * 3. This model contains 7 convolutional layers in the encoder side which is a down-sampling path. Max-pooling layers are used for each patch of a feature map to compute the maximum value and use it to make down-sampled features. In this model, 6 max-pooling layers were used. In the encoder path, convolutional layers and max-pooling layers functions take place. There are 6 convolutional layers and 6 transpose convolutional layers in the decoder path. The functionality of the decoder path is to segment the features from the encoder path using convolutional layers and transposed layers.

3.3.2 Details of Training Data

Our Iris-UNet technique was learned with 50 training epochs on NVIDIA GTX 1080 GPUs by minimising the cross-entropy loss. The mini-batch size was set to 4 for the Adam optimizer.

4 Practical Details

4.1 Exporational Setup

In this paper, an automatic separation of the iris part from the eye image is shown in Fig. 2. Training data for 20 participants and testing data for 5 subjects were randomly selected from the datasets. The outputs are visualized using seismic and salt plots in Fig. We use three different datasets for IRIS segmentation: CASIA, MMU, and PEC database (Fig. 3).

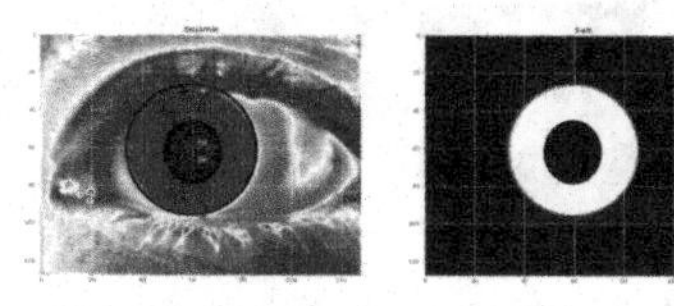

Fig. 3. Iris segmented using the Iris-UNet model

Our Iris-UNet was correlated with two different models which are traditional U-Net and LinkNet. The three datasets are used for comparison as shown in the table. The overall segmentation performance was computed by the metrics' overall accuracy is calculated as:

$$Accuracy = \frac{Number\ of\ Correct\ Predictions}{Total\ Number\ of\ Predictions}$$

Table 2. Comparison of our method - different three datasets with state-of-the-art methods

Dataset	Model	Accuracy
CASIA	U-Net [15]	0.88
	LinkNet	0.42
	Proposed	0.97
MMU	U-Net [15]	0.89
	LinkNet	0.79
	Proposed	0.98
PEC	U-Net [15]	0.91
	LinkNet	0.60
	Proposed	**0.99**

4.2 Correlating with Other Methods

The Table 2 provides a summary of the segmentation results from each competing approach using the various datasets. We can make at least two observations from the

Table 2: 1) Among all three datasets, the PEC database achieves higher accuracy than all other commercially available datasets which is student data taken from Puducherry Technological University. 2) Our model Iris-UNet achieves higher in all three dataset comparisons.Fig. 4 represents the segmentation results with different plots as seismic, salt, salt predicted and salt predicted binary from which we can have two observations: 1) In Fig. 4, our Iris-UNet segments the outer part of the iris i.e., it separates the limbic boundary from the eye image. 2) In Fig. 5, it segments the inner part from the pupils by separating the pupillary boundary from the iris part.

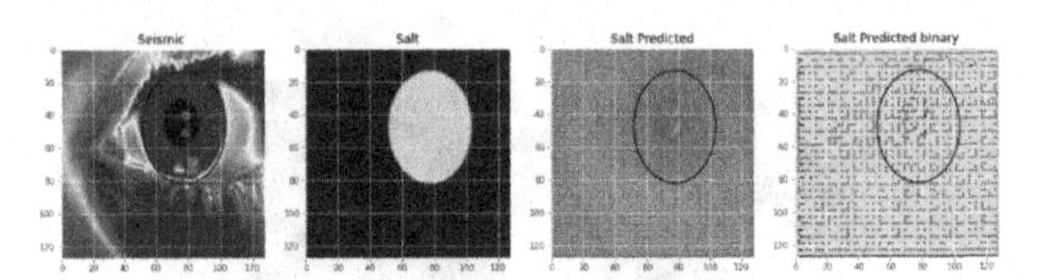

Fig. 4. Limbic boundary segmentation

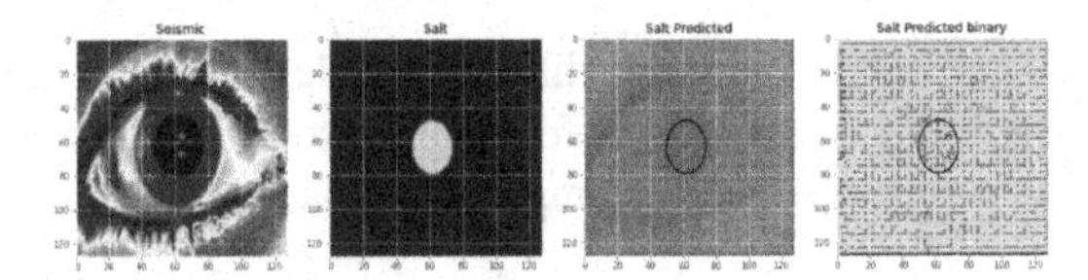

Fig. 5. Pupillary boundary segmentation

4.3 Discussion

4.3.1 Effectiveness of Iris-UNet

Rather than a simple UNet model, this Iris-UNet model adopts two more convolutional layers which will increase the depth of the model so that the capacity of the model increases. The traditional UNet model contains the output tensors of the convolutional layers from c1–c9, but this model has four layers extra i.e., c10–c13. The outputs of the Maxpooling layer also differ from the simple UNet having two more layers, i.e., p5–p6.

In the encoder, increasing the number of layers in the convolutional layers increases the accuracy of the model. In the decoder also, the outputs of the transpose convolutional layers which are used for up-sampling uses layers from u8–u13.

4.3.2 Graphical Representation of the Model

In Fig. 6, the segmented output results are visualized in graph representation. The different datasets CASIA, MMU, and PEC datasets are represented as blue, green, and red lines respectively. From this, we can easily conclude that the PEC database achieves 0.99% when compared with other methods.

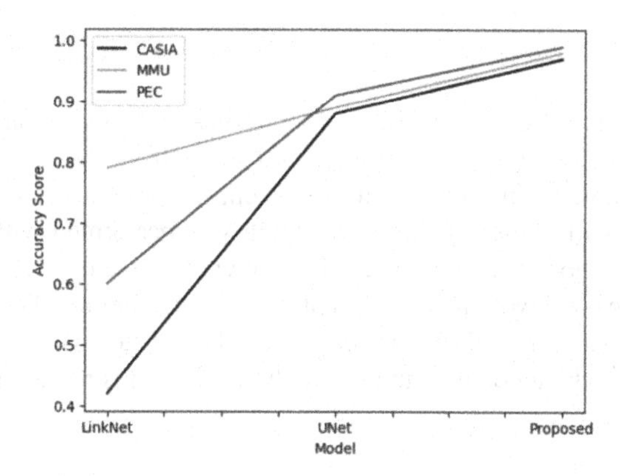

Fig. 6. Graphical representation of the methods

4.3.3 Evaluation Metrics

In this paper, we used precision-recall metrics to evaluate using three different datasets i.e., CASIA, MMU, and PEC datasets. Being evaluated using this, our method is also compared with the standard U-Net model as well as other conventional methods. The precision-recall curve is shown in Fig. 7. When compared with all the databases, the PEC database achieves a 0.99 score. The performance when compared to other methods, our proposed methods achieve better results.

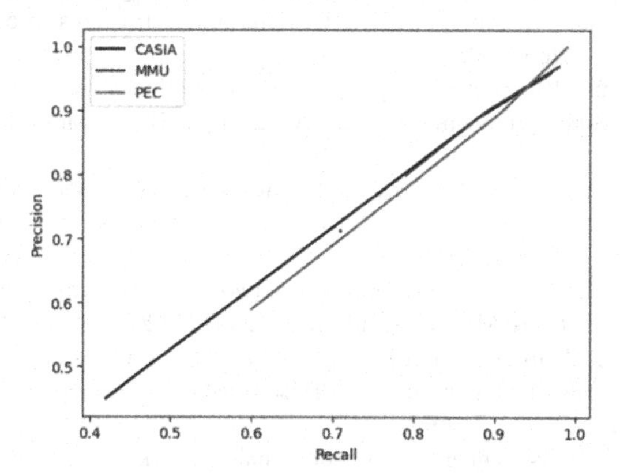

Fig. 7. Precision-Recall curve

4.3.4 Limitations

Although our method and the dataset achieve higher performance, compared with all other methods. Our method fails in eye occlusion cases where the human closes their eye by their upper or lower lid while capturing the photo of the iris through sensors. In those cases, our Iris-UNet cannot be able to detect the limbic and pupillary boundary from the eye image.

5 Conclusion

In this paper, the Iris-UNet approach has been developed to segment the limbic and pupillary from the eye image i.e., separation of the iris from the eye. First, the limbic boundary was separated from the eye image. Then, the pupillary boundary was separated from the iris part. An extensive comparison was performed with three different datasets and two methods. The outcomes show how effective our suggested strategies are, particularly in cases involving real patients. Eye occlusions are the major concern to be focussed on so that all the iris images can be easily detected particularly for medical diagnosis purposes, which could be the future direction in this research.

References

1. Yongbin, Y.U., Chenyu, Y.A.N.G., Quanxin, D.E.N.G., Tashi, N., Shouyi, L., Chen, Z.: Memristive network-based genetic algorithm and its application to image edge detection. J. Syst. Eng. Electron. **32**(5), 1062–1070 (2021)
2. Mittal, M., et al.: An efficient edge detection approach to provide better edge connectivity for image analysis. IEEE Access **7**, 33240–33255 (2019)
3. Karimi, D., Dou, H., Gholipour, A.: Medical image segmentation using transformer networks. IEEE Access **10**, 29322–29332 (2022)
4. Aganj, I., Fischl, B.: Multi-atlas image soft segmentation via computation of the expected label value. IEEE Trans. Med. Imaging **40**(6), 1702–1710 (2021)
5. Dhillon, D., Chouhan, R.: Enhanced edge detection using SR-guided threshold maneuvering and window mapping: handling broken edges and noisy structures in canny edges. IEEE Access **10**, 11191–11205 (2022)
6. Pang, S., et al.: SpineParseNet: spine parsing for volumetric MR image by a two-stage segmentation framework with semantic image representation. IEEE Trans. Med. Imaging **40**(1), 262–273 (2020)
7. Li, X., Yu, L., Chen, H., Fu, C.W., Xing, L., Heng, P.A.: Transformation-consistent self-ensembling model for semisupervised medical image segmentation. IEEE Trans. Neural Netw. Learn. Syst. **32**(2), 523–534 (2020)
8. Han, X., et al.: Deep symmetric adaptation network for cross-modality medical image segmentation. IEEE Trans. Med. Imaging **41**(1), 121–132 (2021)
9. You, C., Zhou, Y., Zhao, R., Staib, L., Duncan, J.S.: SimCVD: simple contrastive voxel-wise representation distillation for semi-supervised medical image segmentation. IEEE Trans. Med. Imaging **41**, 2228–2237 (2022)
10. Zhao, Y., et al.: Two-stream graph convolutional network for intra-oral scanner image segmentation. IEEE Trans. Med. Imaging **41**(4), 826–835 (2021)
11. Han, K., et al.: An effective semi-supervised approach for liver CT image segmentation. IEEE J. Biomed. Health Inform. **26**(8), 3999–4007 (2022). https://doi.org/10.1109/JBHI.2022.3167384
12. Ouyang, C., Biffi, C., Chen, C., Kart, T., Qiu, H., Rueckert, D.: Self-supervised learning for few-shot medical image segmentation. IEEE Trans. Med. Imaging **41**(7), 1837–1848 (2022). https://doi.org/10.1109/TMI.2022.3150682
13. Wang, C., Muhammad, J., Wang, Y., He, Z., Sun, Z.: Towards complete and accurate iris segmentation using deep multi-task attention network for non-cooperative iris recognition. IEEE Trans. Inf. Forensics Secur. **15**, 2944–2959 (2020)

14. Wang, Z., Chai, J., Xia, S.: Realtime and accurate 3D eye gaze capture with DCNN-based iris and pupil segmentation. IEEE Trans. Vis. Comput. Graph. **27**(1), 190–203 (2019)
15. Sabry Abdalla, M., Omelina, L., Cornelis, J., Jansen, B.: Iris segmentation based on an optimized U-Net. In: BIOSIGNALS, pp. 176–183 (2022)

Image Acquisition by Image Retrieval with Color Aesthetics

Huei-Fang Lin[1] and Huei-Yung Lin[2(✉)]

[1] Department of Electrical Engineering, National Chung Cheng University, Chiayi 621, Taiwan
glcopj1359@gmail.com
[2] Department of Computer Science and Information Engineering, National Taipei University of Technology, Taipei 106, Taiwan
lin@ntut.edu.tw

Abstract. The objective of this work is to obtained aesthetically pleasing images related based on the location information. To achieve this, we develop a system that can acquire photos using the geographic information and cloud data by image retrieval. This system comprises an insertion module and a search module. The insertion module recognizes the weather conditions, reads the image information and stores the image data in the cloud database. On the other hand, the search module reads the location information of the onboard sensors, searches for images with the similar location from the cloud database using the proposed optimal image selection algorithm. The search module then provides multiple photos with similar geographic information, selects the best image, and provides feedback suggestions. In addition to the software development, we also implement the proposed system on a hardware device that can directly retrieve the outdoor images for display and storage. The experiments with real scenes have demonstrated the feasibility of the proposed system.

Keywords: Image Acquisition · Image Retrieval · Color Aesthetics

1 Introduction

Due to the technological advancements in the past few decades, capturing well-composed photos is no longer restricted to personal expertise. Independent proxy cameras can produce stunning photographs, and photographers also share their work on the Internet [2]. Consequently, desired and high quality images can be possibly found through online searches. However, with an unlimited number of images available on the Internet, it would be time-consuming to perform the search one by one to find the image of interest. In this paper, we propose a technique that leverages location information and cloud data to obtain the photos without an actual image capture operation. The idea is to create a cloud database, and used for image retrieval under certain circumstances based on an IoT framework [10]. Our method consists of two parts, an insertion module and

J. Blanc-Talon et al. (Eds.): ACIVS 2023, LNCS 14124, pp. 250–261, 2023.
https://doi.org/10.1007/978-3-031-45382-3_21

a search module. The insert system selects the images to be stored, as well as shows the related information such as acquisition time, location, and weather conditions, etc. It can store photos and their information in the cloud database. The search system retrieves the images from the database based on the location information. Our system uses an image selection algorithm to recommend the best photos to the users for consideration. The overall flowchart is illustrated in Fig. 1.

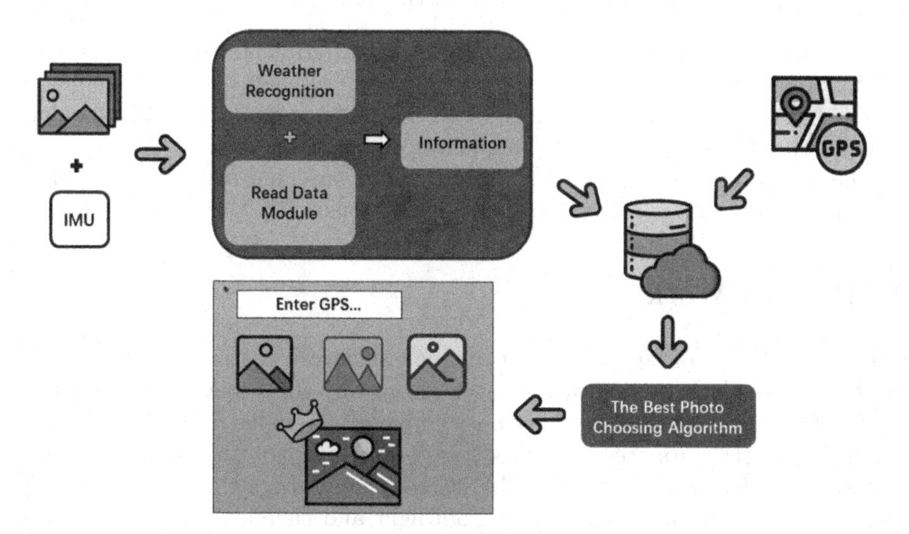

Fig. 1. The schematic diagram of our proposed system. It consists of an insertion module and a search module. The schematic diagram of our proposed system. It consists of an insertion module and a search module.

In this work, the proposed optimal image selection algorithm primarily employs (1) saliency detection, (2) color aesthetics, and (3) image quality assessment. Our objective is to recommend the most visually appealing images. Although visual aesthetics are subjective in general, the factors which influences the viewing experience can still be analyzed for the outdoor images. It is commonly believed that the main focus and color representation of images are the primary influencing factors, thus making our image selection technique based on these two factors [9]. When looking at a photo, different individuals might pay attention to different aspects. While some may notice the background, others could only focus on the people in the photo. Thus, it is possible to mimic the human visual system using saliency map detection, identifying the areas of the image that are more likely to attract a viewer's attention. An image's appeal is not only determined by the scene it depicts, but also by the brightness or colors that attract attention. Colors have a significant impact on people's moods and behaviors, and it is demonstrated in psychology that hue has quite an influence [3]. Finally, by approximating people's perceptions using image quality assessment, photos are scored and ranked, and recommendations are provided based on the evaluation. Through the optimal image selection algorithm, we will be able to obtain images that are similar to human aesthetic preferences.

This paper presents an approach that utilizes the location information, cloud data, and an optimal image selection algorithm based on deep learning for image retrieval. Our system recommends photos with the highest evaluation scores. The proposed optimal image selection method modifies the data preprocessing stage of NIMA [13] to improve the similarity between artificial intelligence and human aesthetic sense. The main contributions of this paper are as follows:

- Self-collected photos with sensor data in the experiments show the usability of system development.
- The NIMA technique based on the VGG16 model is used to modify the data before processing, improving the quality of photo selection.
- The photo retrieval system is realized by hardware, and operated in outdoor scenes, showing its practicability for applications.

2 Related Work

Conventional photography without a camera involves using photosensitive reaction technology to capture images. However, in recent years, artificial intelligence-based technologies have been employed to enable direct image acquisition. Yavuz [18] proposed a method for generating photo-like images using a Generative Adversarial Network (GAN) and the rhizome thinking model. By separating the photographic process from light and chemicals, GAN can learn to generate images that were not limited by the physical constraints of a camera sensor. The goal was to train the generator to produce more realistic images, using the rhizome thinking model as a framework for this process. They believed that the rhizome model reflected the characteristics of short-term memory with no fixed structure or linear progression. Hence, Yavuz's GAN model, which had no predetermined learning path and generated images through a continuous process of bifurcation in the middle, can be seen as a manifestation of this associative and non-hierarchical thinking. The resulting images were coherent and connected in a rhizomatic way, consistent with the rhizome thinking model proposed by Deleuze and Guattari [15].

Daily life is closely associated with weather phenomena, and the weather conditions can be inferred from the images taken during outdoor activities. Therefore, weather information can be regarded as a form of photo data. In the field of image recognition, deep learning has played a crucial role in developing artificial intelligence and computer vision. Xia *et al.* [6] used several CNN network architectures AlexNet, VGGNet, GoogLeNet, ResNet50 to test and compare weather data. Among them, GoogLeNet had the highest recognition accuracy at 86.67%. Kang *et al.* [7] proposed a weather classification system based on GoogLeNet. It was compared with MKL [19] and AlexNet respectively, and the result showed that GoogLeNet performed significantly better than MKL and AlexNet.

The human visual system receives a tremendous amount of sensor data from the environment. To reduce the information complexity, saliency detection algorithms prioritize the most visually prominent information within a given scene, reflecting the way humans tend to focus on the most attention-grabbing element. Zhao *et al.* introduced the Pyramid Feature Attention (PFA) network in an early work for detecting image saliency [20]. In the PFA architecture, a context-aware pyramid feature extraction module was employed to extract high-level features at multiple scales. The channel attention module then selected the appropriate scale to generate salient regions. Finally, the spatial attention module filtered noise to refine the boundaries of the salient regions. Kroner *et al.* introduced a convolutional neural network (CNN) for large-scale image classification, utilizing a pre-training technique. The proposed network architecture is based on VGG16 and consists of an encoder-decoder structure with multiple parallel convolutional layer modules. It enables the extraction of multi-scale features to predict salient objects in images [8].

Psychologists believe that vision is the primary sense of human beings, and color is considered to have the greatest influence [12]. It is also commonly believed that color has a significant impact on people's emotions and feelings. The impact of colors on people's psychological state is attributed to the origin of colors from nature and the feelings associated with natural scenes. This is a fundamental influence on human psychology. Adams *et al.* point out that there is no notable difference in color perception between men and women, and individuals from diverse cultural backgrounds exhibit similar attitudes towards the emotional responses evoked by colors [1]. The Luscher color test is a personality assessment tool [11], which is used to evaluate the subject's personality characteristics with eight color cards. Ram *et al.* conducted a study in which 944 respondents were asked to provide 20 emotional associations for 12 different colors [16]. The color-emotion associations were then quantified using neural networks, support vector machine (SVM), and 10-fold cross-validation with the variation in both age and gender.

3 Method

The schematic diagram of the proposed system for image acquisition using location information and cloud data is illustrated in Fig. 1. The system is a combination of the insert and search modules. In the insert module, users can select the photos to be stored in the cloud database. The image information, including EXIF data and weather recognition results, will be extracted. In the search module, users can retrieve photos by providing the location information and other characteristics provided by EXIF. Then, the system searches for similar images in the cloud database, evaluates their quality, and ranks them by score. Finally, the system recommends the best photos and provides feedback to the user.

3.1 System and Database

To gather the photo information, including location and three-axis attitude angles, we create a dataset with daytime images captured from several scenic spots. The information is collected using mobile phones and a mobile application called Sensor Logging APP. It records the three-dimensional data of the accelerometer, gyroscope, and magnetometer. The collected data can be converted into a CSV file for offline use. We store the photos and related information in the cloud database using the insert module. This information will be then used to test the system.

The system's GUI is developed using PyQt5 graphical interface library. The management system window is created using QMainWindow, and the insert and search windows are built using QDialog. To select and connect to the system, the management system utilizes QPushButton. The image acquisition module reads the photo information and three-axis attitude angles. It also employs weather recognition to obtain the weather conditions in the photos. Once the information is obtained, the system saves both the photos and related information into the cloud data created using MySQL.

3.2 Insert Module

In the acquisition system, two methods are employed to obtain the image information. The first method involves reading image content and calculating the information from external sources. The second method involves using graph identification to obtain information.

Image Information. Upon entering the insert module and selecting an image, the system reads its information using EXIF. It includes the image shooting time, exposure time, and GPS latitude and longitude information. The Sensor Logger APP is used to obtain a CSV file containing 3D accelerometer data, which is then used to calculate the three-axis attitude angle of the image, specifically the roll, pitch, and yaw values. This is achieved by converting the three-dimensional data of the gyroscope and magnetometer. The following formula is used to calculate the attitude angle value:

$$R = \frac{180}{\pi} \tan^{-1} \frac{accelY}{\sqrt{accelX \cdot accelX + accelZ \cdot accelZ}}$$
$$P = \frac{180}{\pi} \tan^{-1} \frac{accelX}{\sqrt{accelY \cdot accelY + accelZ \cdot accelZ}}$$
$$Y = \frac{180}{\pi} \tan^{-1} \frac{-mag_y}{mag_x}$$

The roll and pitch are determined based on the three-dimensional data obtained from the accelerometer, while the yaw is computed using the three-dimensional data acquired from the magnetometer.

$$mag_x = magX \cos P + magY \sin R \sin P + magZ \cos R \sin P$$
$$mag_y = magY \cos R - magZ \sin R$$

We collect the nine-axis sensor data at the frequency of 1 Hz, which results in approximately 5 or 6 data points per photo. To determine the three-axis attitude angle, we calculate the time difference between the data point and the photo. If it is within one second, the nine-axis sensor data is then used to calculate the three-axis attitude angle.

Weather Recognition. Due to the influence of weather conditions, the quality of captured images might be affected. To address this issue, we incorporate the weather recognition into our system to determine the weather condition in the image. The network for weather recognition is trained using a scale image weather dataset, Image2Weather [5]. It consists of over 180,000 photos of five weather types (sunny, foggy, snowy, cloudy, rainy) and 28 attribute information such as temperature and humidity. In order to better reflect local climate, we select about 20,000 photos of three common weather types (sunny, cloudy, rainy) taken in Taiwan as part of our data collection, as well as some images collected by ourselves. To improve the recognition accuracy, GoogLeNet is currently adopted for training and testing.

3.3 Search Module

After providing the location of the scene for photo capture, the search module will look for similar information in the cloud database, and use an algorithm to calculate and recommend the best images. There are two different ways to enter the location information: manually typing the GPS coordinates of the place of interest, or selecting an image taken at that location and using the EXIF module to extract the GPS latitude and longitude information from it.

To ensure that the results generated by the best image selection algorithm are compliant with the human aesthetic preferences, we use the AVA dataset for training [13]. The dataset contains around 255,000 images with scores, semantic labels and style labels assigned by 200 amateur and professional photographers based on their aesthetic sense. It utilizes the rating scale from 1 to 10, with 10 representing the best quality, and the average rating score for the AVA dataset is approximately 5.5. This work mainly focuses on the outdoor image acquisition, which is more complex than the indoor scenes in general. Thus, we use a saliency detection technique to derive the regions of attention in the images. The network is an encoder-decoder structure based on the VGG16 architecture [8]. It is trained for 100 epochs, with the image size of 240×320, batch size of 1, and the learning rate of 10^{-5}.

In photo aesthetics, we incorporate the concept of color aesthetics since the color representation will greatly affect the viewing experience, in addition to the main objects in the image. For training, we the AVA dataset is adopted. We use Pseudo Color to convert the number of colors in the images to 12, as shown in Fig. 2. There are 11 basic colors selected based on the work of Kay *et al.* [4], and an additional turquoise is suggested by Mylonas *et al.*'s color naming experiment conducted on the Internet [14]. Colors are grouped based on similar emotional associations proposed by Ram *et al.* [16]. The colors with higher quantification values signify common emotions within that color group as illustrated in Fig. 3. It is believed that colors in an image evoke common emotions, whether they are joyful or sad, are pure and straightforward. If there are too many colors in an image, it can lead to complex emotions and diminish the overall viewing experience.

Fig. 2. The 12 colors used for pseudo color conversion.

Actual\ Predicted	black gray purple	blue green turquoise	brown	orange yellow	pink red	white
black gray purple	0.632	0.209	0.023	0.102	0.032	0.003
blue green turquoise	0.084	0.699	0.029	0.116	0.053	0.018
brown	0.310	0.320	0.243	0.111	0.010	0.005
orange yellow	0.123	0.356	0.052	0.398	0.067	0.005
pink red	0.065	0.184	0.014	0.168	0.550	0.019
white	0.087	0.627	0.004	0.057	0.128	0.096

Fig. 3. The confusion matrix for color grouping.

We propose two methods to convert images of 16 million colors into 12 colors. The first one is a direct conversion from all colors to 12 colors, and the second one involves reducing the number of colors in the image using color quantization before the conversion to 12 colors. In the first method, we calculate the Euclidean distance between the original pixel value and the value of the 12 colors, where the smallest distance indicates that the original pixel value is most similar to one of the colors. As for the second method, we use the K-Means clustering color quantization algorithm to reduce the number of colors in the image to 32. The Euclidean distance between the quantized pixel value and the value of 12 colors is then calculated. Similarly, the smallest distance indicates that the pixel value is closest to one of the colors. We also employ the CIEDE2000 color difference formula to compute the variation in color between the original image and the one transformed into 12 colors. ΔE (total color difference) is based on the three color values, $\Delta L^{\star}$, $\Delta a^{\star}$, $\Delta b^{\star}$, in the rectangular coordinate system. After converting the pixel values of both images to the Lab color space, we apply the CIEDE2000 color difference formula to calculate the color difference for each pixel. They are averaged to derive the color difference value of the image. Once the saliency region is obtained from the saliency map detection, the image is converted to the one with 12 colors. We calculate the pixel value of the saliency region and check if it falls within the same color category, and then adjust the score accordingly. If the image contains only one color group, it implies that the image expresses a single emotion. If the image contains various color groups, it implies that there are different emotions conveyed in the image. In either case, we multiply the color ratio of the image by the quantized value and add the resulting value to the original score of the image.

To recommend the best photos from the cloud image database to users, the quality evaluation, score calculation and ranking are necessary. We use the NIMA method [17] for image quality assessment, which predicts human opinion scores using convolutional neural networks, not just a high or low quality classification. We use the AVA dataset with adjusted scores for network training and prediction. The backbone network we use is VGG16 for the NIMA method, and the last layer of the VGG16 network is replaced by a fully connected layer that outputs quality scores of 10 categories. The architecture diagram is illustrated in Fig. 4. We train the network using the AVA dataset after saliency detection and pseudo color processing, with 100 epochs, 256×256 image input cropped to 224×224, the batch size of 16 and learning rate of 5×10^{-4}.

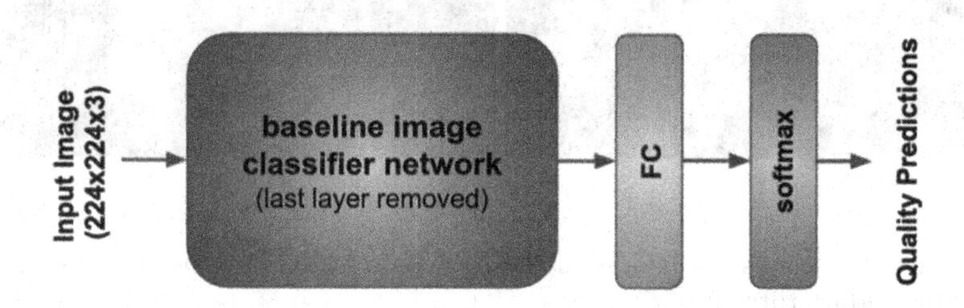

Fig. 4. The modified NIMA network architecture used in this work.

4 Experiments

In this section, we evaluate the effectiveness of our optimal image selection algorithm and image retrieval system. We first compare our best image selection algorithm with the original method, followed by the performance evaluation of the best image selection algorithm through human aesthetics. Finally the developed system is presented. We adopt the modified and original AVA datasets and use the VGG16 model for comparison. For testing, we use self-collected images to demonstrate the results. To test if our optimal image selection algorithm is compliant with people's aesthetics, we conducted user surveys using images captured from the same location but at different angles and weather conditions. It shows that the images with the same color tone have slightly higher scores. Moreover, most images with bright colors have higher scores on the modified dataset (Fig. 5).

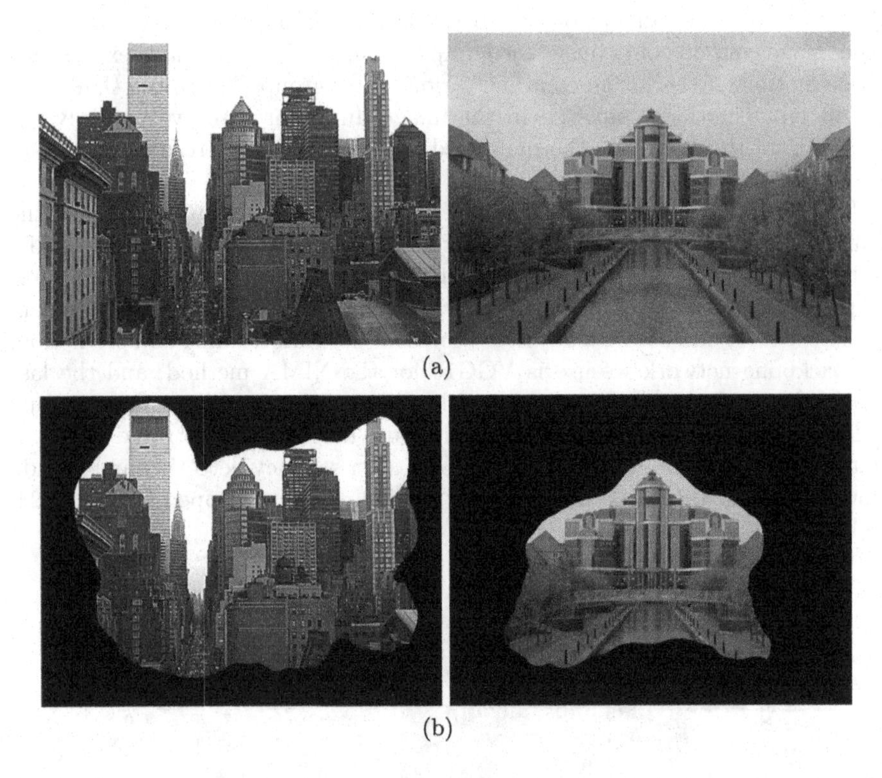

Fig. 5. The top contains the images in the AVA dataset, and the bottom presents the results after adding BitwiseAND.

We utilize a Raspberry Pi board to implement a solid data acquisition system. It is used to capture images and location information for storage and retrieval in the cloud database. However, due to hardware limitations, we also employed computer assistance. Figure 6 depicts the hardware flow chart of the system development. Raspberry Pi is adopted as our system implementation tool because of its compact size and ease of use for the outdoor scenes. On the Raspberry Pi board, we incorporate a camera module, a GPS module, and a Sense HAT expansion board to collect image time, location information, and three-axis attitude angle information. MariaDB is adopted as the cloud database for our Raspberry Pi system. To connect to the MariaDB database, we use the MySQL Connector module implemented with Python. The information stored in the database

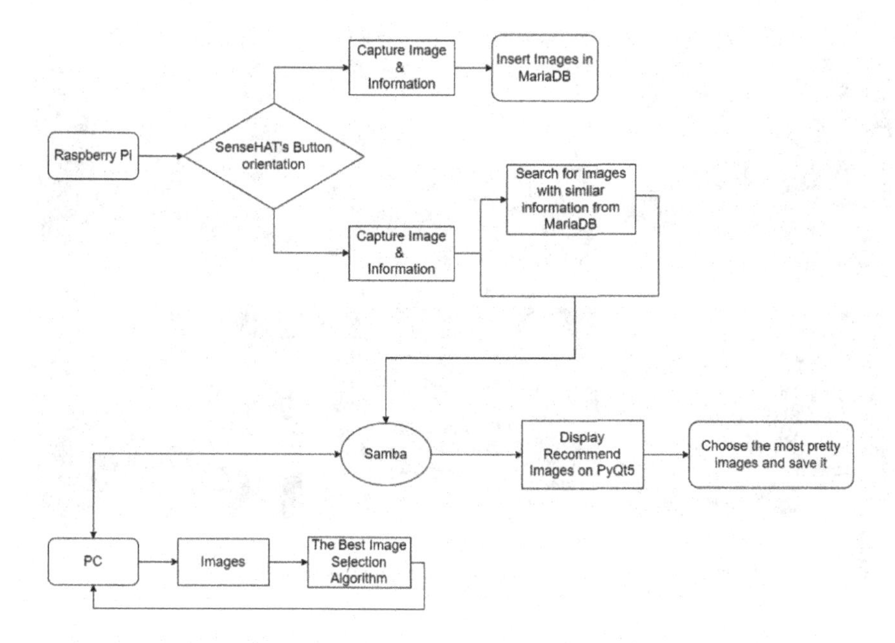

Fig. 6. The hardware system flow of the proposed technique.

for each photo includes its name, acquisition time, geographic coordinates, and three-axis attitude angle.

To select the image capture or retrieval modules on Raspberry Pi, we use multi-control buttons on the Sense HAT expansion board, with button directions acting as trigger keys. Although the three-axis attitude angle information can be obtained, the GPS signal transmission involves receiving data at a 9600 baud rate, with about 0.5 s. We set the received data to be GNGLL positioning information, which returns time and GPS latitude and longitude. However, GNGLL data may not be received every time postback information is triggered, necessitating multiple operations to obtain the data smoothly. The image capture system is activated by setting the upward direction of the multi-control buttons on the Sense HAT expansion board. Once the GPS module starts receiving data in the GNGLL format, the images are captured by the camera and the acquisition time is recorded by the Sense HAT expansion board. The location coordinates collected by the GPS module are then stored in the cloud database.

If we set the direction of the multi-control buttons on the Sense HAT expansion board to be down, it activates the retrieval module. In this mode, we also set the GPS to receive data in the GNGLL format. The images captured by the camera can be saved to a local folder. To overcome the limitations of the display connected to Raspberry Pi, the optimal image selection algorithm is executed on the computer instead. We utilize the Samba server running on Linux to establish a network between the Raspberry Pi board and the computer. After the computer reads the images, the optimal image selection algorithm calculates

Fig. 7. The results of the best image selection algorithm.

their aesthetic scores and identifies the top three images. These images are then sent back to the Raspberry Pi. Next, the GUI interface built with PyQt5 will be displayed on Raspberry Pi. This interface presents three images with the highest scores, and users can subjectively choose the images based on the preferences. Figure 7 shows several best image selection results with the proposed algorithm.

5 Conclusion

This paper develops an image acquisition system using location information and cloud data. It consists of two main modules: an image insertion module and an image retrieval module. The system enables sensor data recording for image information acquisition, storage of desired images, and efficient search for the images. Furthermore, the system utilizes an aesthetic score-based algorithm to recommend the best-looking images to the users. As the current publicly available datasets do not meet our requirements, we conduct our own data collection of photos and nine-axis sensor data. The best image selection algorithm and the entire system on weather recognition are then evaluated. To ensure diversity in our dataset, we capture photos under three different weather conditions, which meet our needs for variability in both weather and color presentation. The future work will focus on the identification of weather condition for the improvement of best-photo selection.

References

1. Adams, F.M., Osgood, C.E.: A cross-cultural study of the affective meanings of color. J. Cross Cult. Psychol. **4**(2), 135–156 (1973)
2. AlZayer, H., Lin, H., Bala, K.: AutoPhoto: aesthetic photo capture using reinforcement learning. In: 2021 IEEE/RSJ International Conference on Intelligent Robots and Systems (IROS), pp. 944–951. IEEE (2021)
3. Asarkar, S.V., Phatak, M.V.: Effects of color on visual aesthetics sense. In: Bhalla, S., Kwan, P., Bedekar, M., Phalnikar, R., Sirsikar, S. (eds.) Proceeding of International Conference on Computational Science and Applications. AIS, pp. 181–194. Springer, Singapore (2020). https://doi.org/10.1007/978-981-15-0790-8_19
4. Berlin, B., Kay, P.: Basic Color Terms: Their Universality and Evolution. University of California Press (1991)
5. Chu, W.T., Zheng, X.Y., Ding, D.S.: Image2weather: a large-scale image dataset for weather property estimation. In: 2016 IEEE Second International Conference on Multimedia Big Data (BigMM), pp. 137–144. IEEE (2016)
6. Xia, J., Xuan, D., Tan, L., Xing, X.: ResNet15: weather recognition on traffic road with deep convolutional neural network (2020)
7. Kang, L.W., Chou, K.L., Fu, R.H.: Deep learning-based weather image recognition. In: 2018 International Symposium on Computer, Consumer and Control (IS3C), pp. 384–387. IEEE (2018)
8. Kroner, A., Senden, M., Driessens, K., Goebel, R.: Contextual encoder-decoder network for visual saliency prediction. Neural Netw. **129**, 261–270 (2020)
9. Lin, H.Y., Chang, C.C., Chou, X.H.: No-reference objective image quality assessment using defocus blur estimation. J. Chin. Inst. Eng. **40**(4), 341–346 (2017)
10. Lin, H.Y., Wu, Z.Y.: Development of automatic gear shifting for bicycle riding based on physiological information and environment sensing. IEEE Sens. J. **21**(21), 24591–24600 (2021)
11. Lüscher, M.: The Luscher Color Test. Simon and Schuster (1971)
12. Mahnke, F.H.: Color, Environment, and Human Response: An Interdisciplinary Understanding of Color and Its Use as a Beneficial Element in the Design of the Architectural Environment. Wiley, Hoboken (1996)
13. Murray, N., Marchesotti, L., Perronnin, F.: AVA: a large-scale database for aesthetic visual analysis. In: 2012 IEEE Conference on Computer Vision and Pattern Recognition, pp. 2408–2415. IEEE (2012)
14. Mylonas, D., MacDonald, L.: Augmenting basic colour terms in English. Color. Res. Appl. **41**(1), 32–42 (2016)
15. Purcell, M.: A new land: Deleuze and Guattari and planning. Plann. Theory Pract. **14**(1), 20–38 (2013)
16. Ram, V., et al.: Extrapolating continuous color emotions through deep learning. Phys. Rev. Res. **2**(3), 033350 (2020)
17. Talebi, H., Milanfar, P.: NIMA: neural image assessment. IEEE Trans. Image Process. **27**(8), 3998–4011 (2018)
18. Yavuz, O.: Novel paradigm of cameraless photography: methodology of AI-generated photographs. Proc. EVA Lond. **2021**, 207–213 (2021)
19. Zhang, Z., Ma, H.: Multi-class weather classification on single images. In: 2015 IEEE International Conference on Image Processing (ICIP), pp. 4396–4400. IEEE (2015)
20. Zhao, T., Wu, X.: Pyramid feature attention network for saliency detection. In: Proceedings of the IEEE/CVF Conference on Computer Vision and Pattern Recognition, pp. 3085–3094 (2019)

Improved Obstructed Facial Feature Reconstruction for Emotion Recognition with Minimal Change CycleGANs

Tim Büchner[1(✉)], Orlando Guntinas-Lichius[2], and Joachim Denzler[1]

[1] Computer Vision Group, Friedrich Schiller University Jena, 07745 Jena, Germany
{tim.buechner,Joachim.Denzler}@uni-jena.de

[2] Department of Otorhinolaryngology, Jena University Hospital, 07747 Jena, Germany
Orlando.Guntinas@med.uni-jena.de

Abstract. Comprehending facial expressions is essential for human interaction and closely linked to facial muscle understanding. Typically, muscle activation measurement involves electromyography (EMG) surface electrodes on the face. Consequently, facial regions are obscured by electrodes, posing challenges for computer vision algorithms to assess facial expressions. Conventional methods are unable to assess facial expressions with occluded features due to lack of training on such data. We demonstrate that a CycleGAN-based approach can restore occluded facial features without fine-tuning models and algorithms. By introducing the minimal change regularization term to the optimization problem for CycleGANs, we enhanced existing methods, reducing hallucinated facial features. We reached a correct emotion classification rate up to 90% for individual subjects. Furthermore, we overcome individual model limitations by training a single model for multiple individuals. This allows for the integration of EMG-based expression recognition with existing computer vision algorithms, enriching facial understanding and potentially improving the connection between muscle activity and expressions.

Keywords: Image Restoration · Facial Features · CycleGAN · Minimal Change

1 Introduction

Facial expression recognition [12] is crucial in various research areas such as psychology, medicine, and computer vision. In computer vision, occlusion of facial features presents a challenge for existing algorithms, as most assume a fully visible face and lack consideration for occluded features in available datasets.

We focused on detecting six basic emotions (anger, disgust, fear, happiness, sadness, and surprise) defined by Ekman and Friesen [4], which are characterized by specific facial muscle activation. Accurate measurement of muscle activation

J. Blanc-Talon et al. (Eds.): ACIVS 2023, LNCS 14124, pp. 262–274, 2023.
https://doi.org/10.1007/978-3-031-45382-3_22

typically requires recordings detected via surface electrodes (sEMG). These electrodes and their cables cover parts of the faces. Combining EMG-based facial expression recognition with computer vision algorithms can enhance understanding of the underlying facial muscle activity. Our study involved 36 healthy subjects performing the six basic emotions four times, with recordings taken both with and without attached sEMG surface electrodes. This allowed for comparison between mimicked and targeted expressions. We employed the ResMaskNet [20] architecture for emotion detection but found that it struggled to handle occluded facial features. This resulted in only prediction of `anger` and `surprised`. Mimicked expressions, like `disgusted`, were recognized only 2 out of 528 times.

Büchner et al. [3] attempted to restore facial features by interpreting face coverage as a learnable style between uncovered and covered faces. Despite promising results, the method has drawbacks: it requires separate models for each individual, and it memorizes uncovered faces resulting in inadequate hallucination of occluded features, see Fig. 3. We extended Büchner et al.'s work by introducing new regularization terms to the optimization problem, enabling a single model to be trained for multiple individuals. We increased individual accuracy from 33.8% (random guessing) up to 90% and demonstrated emotion detection on individuals not part of the training set, providing generalizability to unseen individuals. Fine-tuning the general model to specific individuals with minimal data further improved results.

We conducted ablation studies to enhance the backbone network, significantly reducing the number of parameters and computational cost while maintaining comparable results. We evaluated emotion classification accuracy and the following qualitative metrics: Frenchet Inception Distance (FID) [17], Structural Similarity Index (SSIM) [25], and Learned Perceptual Image Patch Similarity (LPIPS) [27]. These advancements are crucial for EMG-based facial expression recognition applications. The reduced parameters, improved generalization, and enhanced visual quality enable a more efficient, robust implementation. Additionally, these improvements could benefit live or therapeutic applications requiring visual feedback.

2 Related Work

Restoring occluded facial features is challenging due to their invisibility in the input images. Generative approaches, such as Generative Adversarial Networks (GANs) [6], can learn anatomically correct facial features from non-occluded faces, making them suitable for this task.

GANs have demonstrated strong results in image generation, particularly in medical applications [26]. However, they are typically trained on specific datasets and struggle to generalize to unseen data. Facial generation research [10,11,16] employs GANs to create realistic faces from specific datasets, indistinguishable from real faces [24]. These works focus on non-occluded faces, thus generating only non-occluded facial images.

Restoring hidden facial features is more challenging since GANs must learn facial features from non-occluded faces. Li et al. [15] used GANs to restore artificially altered faces; however, the unrealistic changes limit its applicability to real-world data. Moreover, the restored facial expressions do not match the original ones. Alternative methods attempt to use GANs for transferring facial attributes [18] to modify faces.

Some approaches attempted to generalize GANs to unseen data. For instance, Zhu et al. [28] demonstrated that CycleGANs can translate images between domains without paired data, allowing each generator to learn a specific domain style. CycleGANs can thus translate covered faces to uncovered faces in medical applications, treating sEMG coverage as a domain style.

Abraiam and Eklund [1] used CycleGANs to restore obfuscated faces in MRI images, focusing on patient privacy and side views with no facial expression differences. However, they did not evaluate the quality of restored faces. In contrast, our work focused on facial feature restoration and quality evaluation.

We build upon Büchner et al. [3], using the CycleGAN architecture for facial feature restoration. While their architecture remained unchanged, every individual required a separately trained model which could hallucinate facial features. We introduce a new regularization term to the optimization problem, enforcing minimal changes and also enabling a single model to be trained for multiple individuals.

3 Method

Restoring facial features can be viewed as a style transfer problem, as surface electrodes are applied consistently across individuals. Following Büchner et al. [3], we use the CycleGAN architecture for facial feature restoration, with covered faces as the source domain and uncovered faces as the target domain. The two generators and discriminators are trained adversarially, using GAN loss [6], cycle consistency loss [28], and identity loss [22]. However, the base model (Fig. 3) hallucinates facial features or changes uncovered areas, affecting the original facial expression. This issue arises due to the absence of this constraint in the original training objective.

3.1 CycleGAN Architecture for Facial Feature Restoration

The CycleGAN architecture by Zhu et al. [28], consists of two generators and two discriminators, shown in Fig. 1. The generators, trained adversarially [6], translate images between domains while maintaining cycle consistency [28], allowing the translated images to be reverted to the original domain. Discriminators distinguish between real and fake generated faces. To improve visual quality, generators are trained using an identity loss [22], preserving color composition between input and output. We hypothesize that the identity loss encourages generators to learn input faces' facial features, enabling better feature restoration.

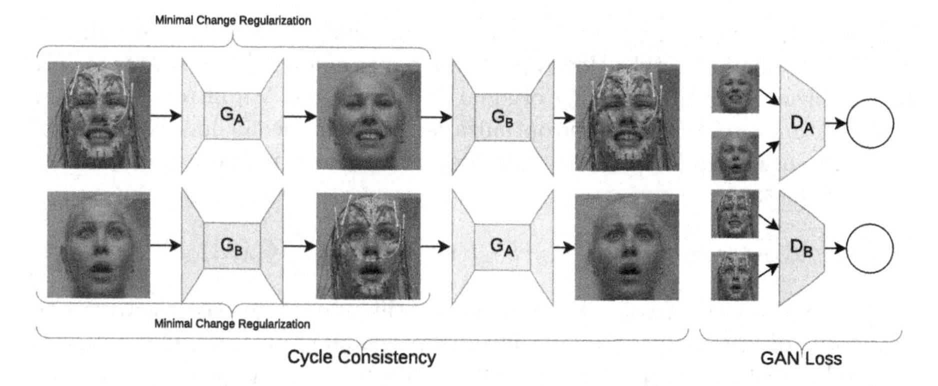

Fig. 1. The CycleGAN architecture for facial feature restoration: The generators G_A and G_B translate images between domains A and B. The minimal change regularization ensures that the generators do not modify uncovered areas.

Our generator's backbone network is a residual neural network [7]. The generator comprises two down-sampling blocks, n residual blocks, and two up-sampling blocks. In ablation studies, we explored the impact of varying residual block numbers in the generator. We found that $n = 5$ provided a balance between visual quality and computational cost and inference time. To minimize artifacts from deconvolution layers, we replace them with nearest-neighbor up-sampling layers followed by convolution layers. We use instance normalization as our normalization layer, because they are more suitable for style transfer [23]. This required a batch size of 1 during training to effectively learn the task. We maintained the discriminator architecture from [3,28], using the PatchGAN architecture [9]. We fixed the number of discriminator layers to 3, as this is sufficient for solving the task.

3.2 Optimization Problem with Minimal Change Regularization

The GAN loss (Eq. 1) describes the adversarial interaction between the generator and discriminator. The generator attempts to fool the discriminator by generating uncovered faces. Cycle consistency loss (Eq. 2) ensures accurate reattachment of the surfaces electrodes. The identity loss (Eq. 3) enables the generator to learn the covered facial features:

$$\mathcal{L}_{GAN} = D_A(A, G_A(B)) + D_B(B, G_B(A)), \tag{1}$$

$$\mathcal{L}_{cycle} = \lambda_A \cdot \mathcal{L}_{L1}(G_B(G_A(A)), A) + \lambda_B \cdot \mathcal{L}_{L1}(G_A(G_B(B)), B), \tag{2}$$

$$\mathcal{L}_{idt} = \lambda_{idt}(\cdot \mathcal{L}_{L1}(G_B(A), A) + \cdot \mathcal{L}_{L1}(G_A(B), B)). \tag{3}$$

While the identity loss helps generators to learn facial features, it is insufficient for correct facial feature restoration. As seen in Fig. 3, the generator hallucinated facial features not present in the original face, contrary to our goal. Limitations shown in [3] result from changes in uncovered face areas.

In style transfer tasks, every pixel of the input image might change according to the target domain's style. However, we wanted to enforce the generator to only change covered areas, preserving original facial features. We introduced a new regularization term (Eq. 4) to the optimization problem, penalizing the generator for significant changes and reducing the reconstruction error between input and output images:

$$\mathcal{L}_{MC} = \lambda_{MC_A} \cdot \mathcal{L}_{L1}(G_B(B), B) + \lambda_{MC_B} \cdot \mathcal{L}_{L1}(G_A(A), A). \tag{4}$$

Due to the GAN loss, generators must still remove surface electrodes to deceive the discriminator. We introduced hyperparameters λ_{MC_A} and λ_{MC_B} to weight the regularization term, enabling control over each domain's regularization amount. Notably, this regularization allows for simultaneous training on multiple individuals. Since surface electrodes are positioned consistently across individuals, the model learns to preserve uncovered face areas. Consequently, the model is not limited to a single individual and can be applied to unseen individuals. The final optimization problem (Eq. 5) comprises the GAN loss, cycle consistency loss, identity loss, and minimal change loss:

$$\mathcal{L} = \mathcal{L}_{GAN} + \mathcal{L}_{cycle} + \mathcal{L}_{idt} + \mathcal{L}_{MC}. \tag{5}$$

The optimization problem's individual components can be weighted using the corresponding λ hyperparameters. Setting λ_{MC_A} and λ_{MC_B} to 0 reverts to the original CycleGAN optimization problem.

4 Dataset

We evaluated our method on 36 test subjects [21] without medical conditions affecting facial expressions, such as facial paralysis. Using frontal cameras at 1280×720 pixels resolution and 30 fps, we recorded subjects mimicking the six basic emotions [4] four times each. The instruction order was randomized, and we conducted two recording sessions per subject, two weeks apart, obtaining alternating facial expressions videos with neutral expressions in between.

Each subject had two recordings without surface electrodes (baseline) and four recordings with surface electrodes to ensure accurate sEMG measurements. We used Fridlund and Cacioppo's [5] and Kuramoto et al.'s [13] schemes, applying 62 surface electrodes in total. The dataset comprised 71 recordings with electrodes and 138 without. We had 1704 emotions as ground truth for uncovered expressions and 3312 emotions requiring accurate reconstruction for electrode-covered faces. An overview of the six basic emotions with and without surface electrodes is shown in Fig. 3.

Table 1. Evaluation of emotion classification without (groundtruth), with, and removed surface electrodes and visual quality metrics: We achieved a higher classification rate on restored faces than on covered faces.

	Emotion Classification Accuracy							Visual Metrics		
sEMG	Anger	Disgusted	Fearful	Happy	Sad	Surprised	Overall	SSIM (↑)	LPIPS (↓)	FID (↓)
Without	61.6%	72.5%	28.1%	90.8%	47.1%	85.2%	64±10%	0.63±0.08	0.10±0.04	0.50±0.74
Attached	88.0%	0.3%	20.2%	21.9%	3.8%	68.8%	34±10%	0.38±0.05	0.25±0.02	10.46±2.10
Removed	66.2%	49.0%	11.9%	72.1%	42.5%	60.8%	**54±16%**	**0.66±0.09**	**0.08±0.04**	**0.35±0.48**

5 Experiments and Results

Our experiments focused on accurate facial feature reconstruction, validated with emotion classification accuracy on restored faces and their visual quality. To assess visual quality, we used the Fréchet Inception Distance (FID) [8], the Structural Similarity Index (SSIM) [25], and the Learned Perceptual Image Patch Similarity (LPIPS) [27].

We employed the Residual Masking Network (ResMaskNet) [20], a ResNet-18 [7] architecture with a mask branch, for emotion classification. We did not fine-tune the model on uncovered or covered faces, ensuring unbiased performance analysis and demonstrating correct restoration. On the uncovered faces we have a mean accuracy of 64.1% as baseline. However, if test subjects fail to accurately replicate emotions, the estimation prediction may deviate from the ground truth. For unobstructed faces, one individual has a correct classification accuracy of 37.5%, while another achieves 91.5%. Thus, we expect a high variance in the emotion classification accuracy. We classified the 1704 uncovered face expressions and 3312 covered face expressions as a baseline for emotion classification accuracy, see Table 1. The emotion with the highest activation was selected as the predicted emotion, excluding the neutral expression.

The accuracy for covered faces is 33.8%. The model predominantly predicts the emotions `angry` and `surprised`, as shown in Fig. 2, indicating performance no better than random guessing between two classes. Emotions like `disgusted` and `sad` are predicted correctly only 23 times out of the expected 1104 times. Our goal was to improve emotion classification accuracy using our method without altering the ResMaskNet architecture. We aimed to restore faces through a data-centric approach, enhancing emotion classification performance.

Backbone Network Size. We conducted experiments to determine an optimal backbone generator network size for achieving satisfactory results. The CycleGAN architecture was trained on various subsets for different depth and feature size combinations of the training data. Quantitative results are displayed in Table 2. Our findings indicate that emotion classification accuracy does not significantly improve with more training data. We measured a five percent point difference to the baseline accuracy and the ResMaskNet does not random guess between two classes anymore. However, the visual quality of restored faces

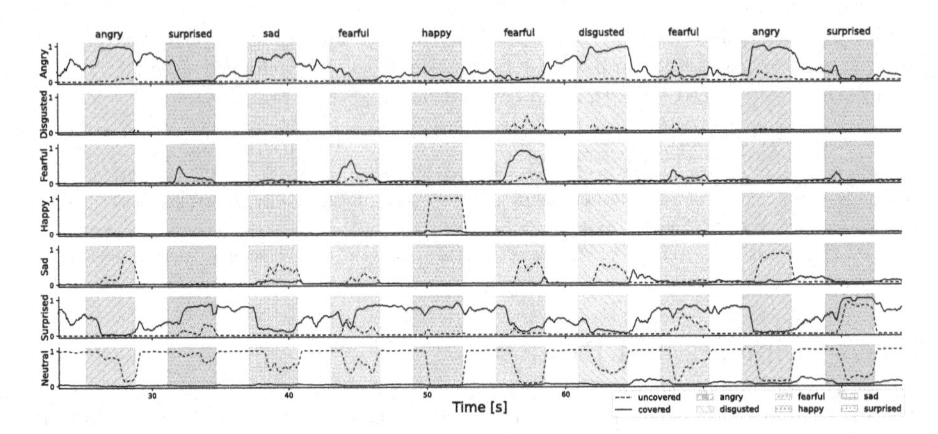

Fig. 2. Time series excerpt of emotion activations by ResMaskNet [20]: Dashed lines represent predicted emotions for uncovered faces, while solid lines represent predictions for covered faces. ResMaskNet struggles to correctly classify emotions for covered faces and predominantly activates for *angry* or *sad*. A neutral expression is never predicted for covered faces. (Best viewed digitally.)

increased with both larger amounts of training data and larger backbone networks. Fine details, highlighted in Fig. 3, are better preserved with larger backbone networks and might not be measurable with these metrics. Considering the tradeoff between visual quality and training time, we opted for a generators with three residual blocks and a feature dimension of 64. All further experiments utilize these model configurations.

Table 2. Hyperparameter ablation study for the generators: The hyperparameter impact is minimal but the qualitative image generation improves.

Features	Depth	Emo. Max	Emo. Acc. (↑)	SSIM (↑)	LPIPS (↓)	FID (↓)
32	3	0.88	0.56±0.17	0.70±0.06	0.07±0.02	0.17±0.07
	5	0.83	0.56±0.15	0.70±0.07	0.07±0.02	0.25±0.16
	7	0.88	0.58±0.15	0.69±0.06	0.07±0.02	0.34±0.21
64	3	0.92	0.56±0.15	0.70±0.08	0.06±0.02	0.17±0.12
	5	0.88	0.58±0.15	0.64±0.08	0.09±0.04	0.37±0.59
	7	0.88	0.59±0.15	0.69±0.07	0.07±0.02	0.23±0.18

5.1 Minimal Change Regularization

The proper restoration of faces was already shown by Büchner et al. [3] but their model might hallucinate features, see Fig. 3. Thus, we investigated on the one side if the minimal change regularization improves the results and on the other side if it reduces the hallucination of features. We trained the CycleGAN architecture

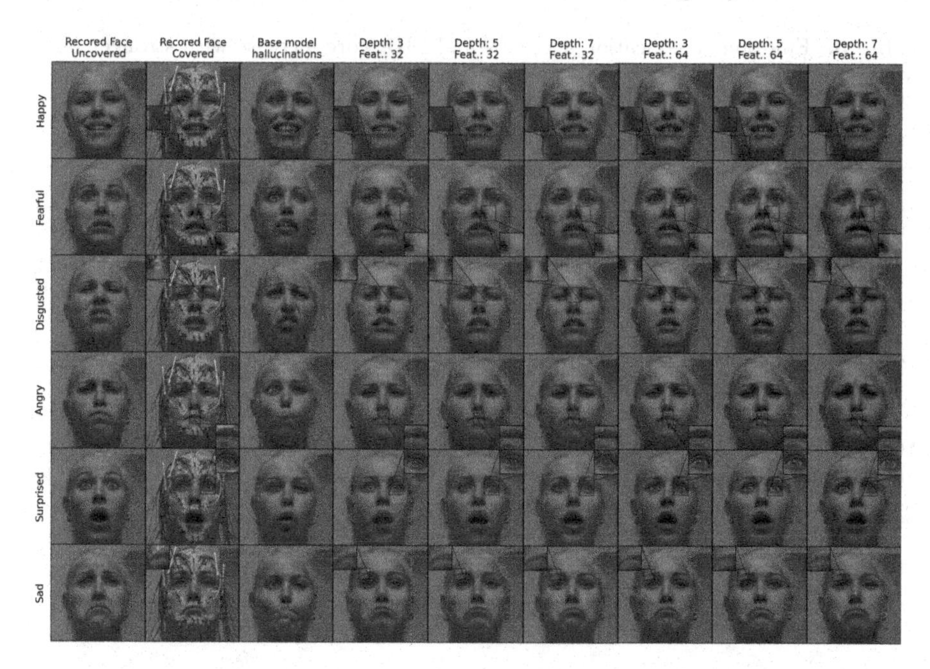

Fig. 3. Quality difference for backbone network sizes: The third column shows the base model without minimal change regularization, leading to hallucinated features. In contrast, the remaining columns incorporate minimal change regularization and tuned hyperparameters, showcasing improved restoration. (Best viewed digitally.)

with different weightings of minimal change regularization for each individual test subject. Setting λ_{MC} to 0 is equivalent to the baseline model without regularization. The results in Table 3 show that the emotion classification accuracy increases with stronger minimal change regularization. This indicates that the additional loss term is no hindrance for the model to learn the task of removing the surface electrodes. We observed that increasing the λ_{MC} value increases overall performance. The restoration cannot correct an initial wrong mimicking of a facial expression. Thus, lowering the mean accuracy and impacting the measurable performance. This effect does not impact the visual metrics and they improve with higher λ values. We observed that hallucinations, such as in Fig. 3, are not present anymore in the results, see Fig. 4. However, an increased λ_{MC} value leads to a more blurry result but more accurate head orientation. Thus, small details might be lost in the restoration process. Thus, depending on the application, one can choose a suitable λ value to either preserve the facial features or to preserve the facial orientation.

5.2 Generalization Capabilities

The CycleGAN architecture has the problem of not being able to generalize to unseen individuals as shown by Büchner et al. [3]. Thus, we investigate the gen-

Table 3. Emotion classification accuracy and FID score increase with stronger regularization: SSIM and LPIPS did not change significantly.

λ_{MC}	Emo. Max	Emo. Acc.(↑)	SSIM (↑)	LPIPS (↓)	FID (↓)
0.0	0.79	0.46±0.16	0.64±0.09	0.09±0.04	0.41±0.66
0.1	0.83	0.48±0.16	0.64±0.09	0.09±0.04	0.37±0.67
0.2	0.88	0.47±0.15	0.64±0.08	0.09±0.03	0.45±0.64
0.3	0.83	0.47±0.15	0.64±0.09	0.09±0.04	0.45±0.71
0.4	0.83	0.48±0.15	0.64±0.08	0.09±0.03	0.35±0.57
0.5	0.92	0.54±0.16	0.66±0.09	0.08±0.04	0.35±0.48

Table 4. Generalization capabilities of the CycleGAN architecture with minimal change regularization: The model is able to generalize to unseen individuals.

Trained On	Emo. Max	Emo. Acc. (↑)	SSIM (↑)	LPIPS (↓)	FID (↓)
False	0.83	0.53±0.15	0.60±0.06	0.12±0.03	0.53±0.84
True	0.88	0.55±0.18	0.61±0.09	0.10±0.04	0.33±0.29

eralization capabilities of the CycleGAN architecture with the minimal change regularization. We did a 6-fold cross validation on the 36 individuals, yielding 30 individuals for training and 6 individuals for testing. We set λ_{MC} to 0.5 for all experiments to enforce the minimal change regularization. Table 4 shows the results of the generalization experiments. The generalized model achieves a better performance on individual test subjects inside and outside the training set. Thus, we assume that the model learns to accurately remove the surface electrodes. This, in turn, leads to a better emotion classification accuracy on unseen individuals and we removed existing limitations of the work by Büchner et al. [3]. The qualitative results are shown in Fig. 4 and we see that the model is able to remove the surface electrodes for unseen individuals.

Additionally, we investigated the impact of fine-tuning the model on the unseen individuals. We analyzed which combination of available training data and training epochs yields the best results. Our results in Fig. 5 indicates that fine-tuning improved only the visual quality of the restored faces but not the emotion classification accuracy. We assume that important facial features are already learned by the model. Fine-tuning does not significantly enhance emotion classification accuracy; however, it improved visual quality compared to the baseline. Moreover, utilizing a smaller portion of training data and a higher number of epochs produces results similar to using larger training data fractions and fewer epochs.

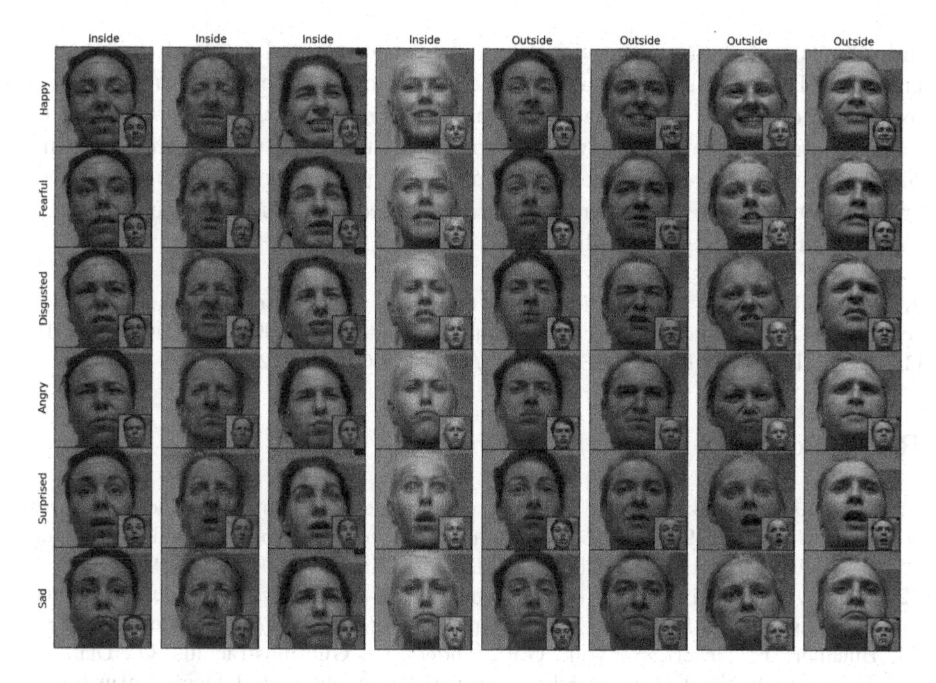

Fig. 4. Qualitative results of the generalization capabilities using the minimal change regularization: The model removes surface electrodes for individuals outside the training set. Important facial features are unchanged but the head shape is altered slightly. Uncovered expressions are shown in the lower right corner.

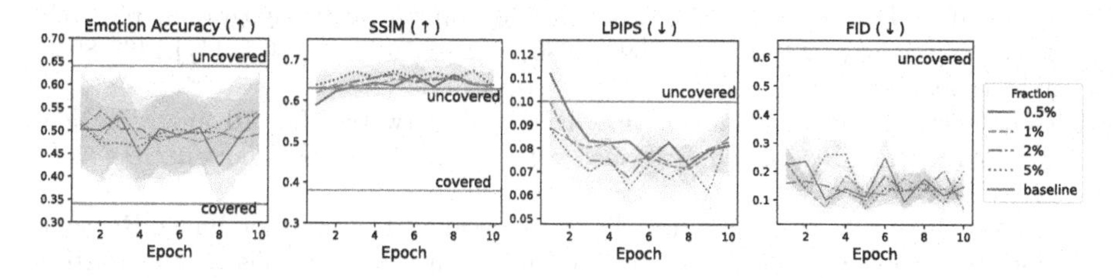

Fig. 5. Fine-tuning of the model on unseen individuals: The emotion classification accuracy does not improve but the visual quality of the restored faces does independent of the amount of training data and epochs.

6 Conclusion

Our study demonstrated that CycleGAN with minimal change regularization effectively restores individuals' faces with EMG surface electrodes, improving emotion classification accuracy. This regularization does not impede the model's ability to remove electrodes, and eliminates CycleGAN limitations like hallucinations, making the restored faces suitable for emotion classification tasks. Our approach enables direct utilization of existing methods without fine-tuning on our data. Additionally, we demonstrated its ability to generalize to unseen individuals and improve the visual quality of the restored faces without hallucinations. Here we showed that now sEMG-based measurements can be jointly used with computer vision-based techniques to enhance the comprehension of facial anatomy. The data-driven approach seamlessly integrates into existing pipelines, enabling real-time restoration of individuals' faces with surface electrodes. Thus, it can be used in applications where electrodes are used for facial muscle stimulation, such as physical therapy [2,14,19].

Acknowledgments. This work has been funded by the Deutsche Forschungsgemeinschaft (DFG - German Research Foundation) project 427899908 BRIDGING THE GAP: MIMICS AND MUSCLES (DE 735/15-1 and GU 463/12-1).

References

1. Abramian, D., Eklund, A.: Refacing: reconstructing anonymized facial features using GANS. 2019 IEEE 16th International Symposium on Biomedical Imaging (ISBI 2019), pp. 1104–1108 (2019). https://doi.org/10.1109/ISBI.2019.8759515
2. Arnold, D.: Selective surface electrostimulation of the denervated zygomaticus muscle. Diagnostics **11**(2), 188 (2021). https://doi.org/10.3390/diagnostics11020188
3. Büchner, T., Sickert, S., Volk, G.F., Anders, C., Guntinas-Lichius, O., Denzler, J.: Let's get the FACS straight - reconstructing obstructed facial features. In: International Conference on Computer Vision Theory and Applications (VISAPP), pp. 727–736. SciTePress (2023). https://doi.org/10.5220/0011619900003417
4. Ekman, P., Friesen, W.V.: Facial Action Coding System. Consulting Psychologists Press (1978)
5. Fridlund, A.J., Cacioppo, J.T.: Guidelines for human electromyographic research. Psychophysiology **23**(5), 567–589 (1986). https://doi.org/10.1111/j.1469-8986.1986.tb00676.x
6. Goodfellow, I.J.,et al.: Generative Adversarial Networks. In: Advances in Neural Information Processing Systems, vol. 27 (2014). https://doi.org/10.48550/arXiv.1406.2661
7. He, K., Zhang, X., Ren, S., Sun, J.: Deep Residual Learning for Image Recognition. In: Proceedings of the IEEE Conference on Computer Vision and Pattern Recognition, pp. 770–778 (2016). https://doi.org/10.48550/arXiv.1512.03385
8. Heusel, M., Ramsauer, H., Unterthiner, T., Nessler, B., Hochreiter, S.: GANs Trained by a two time-scale update rule converge to a local nash equilibrium. In: Advances in Neural Information Processing Systems, vol. 30 (2017). https://doi.org/10.48550/arXiv.1706.08500
9. Isola, P., Zhu, J.Y., Zhou, T., Efros, A.A.: Image-to-image translation with conditional adversarial networks. In: Proceedings of the IEEE Conference on Computer Vision and Pattern Recognition, pp. 1125–1134 (2017). https://doi.org/10.48550/arXiv.1611.07004
10. Kammoun, A., Slama, R., Tabia, H., Ouni, T., Abid, M.: Generative adversarial networks for face generation: a survey. ACM Comput. Surv. **55**(5), 1–37 (2023)
11. Karras, T., Laine, S., Aittala, M., Hellsten, J., Lehtinen, J., Aila, T.: Analyzing and improving the image quality of StyleGAN. In: 2020 IEEE/CVF Conference on Computer Vision and Pattern Recognition (CVPR), pp. 8107–8116 (2020). https://doi.org/10.1109/CVPR42600.2020.00813

12. Klingner, C.M., Guntinas-Lichius, O.: Mimik und Emotion. Laryngo-Rhino-Otologie **102**(S 01), S115–S125 (2023). https://doi.org/10.1055/a-2003-5687
13. Kuramoto, E., Yoshinaga, S., Nakao, H., Nemoto, S., Ishida, Y.: Characteristics of facial muscle activity during voluntary facial expressions: imaging analysis of facial expressions based on myogenic potential data. Neuropsychopharmacol. Rep. **39**(3), 183–193 (2019). https://doi.org/10.1002/npr2.12059
14. Kurz, A., Volk, G.F., Arnold, D., Schneider-Stickler, B., Mayr, W., Guntinas-Lichius, O.: Selective electrical surface stimulation to support functional recovery in the early phase after unilateral acute facial nerve or vocal fold paralysis. Frontiers Neurol. **13**, 869900 (2022)
15. Li, Y., Liu, S., Yang, J., Yang, M.H.: Generative face completion. In: Proceedings of the IEEE Conference on Computer Vision and Pattern Recognition, pp. 3911–3919 (2017)
16. Liu, M., Li, Q., Qin, Z., Zhang, G., Wan, P., Zheng, W.: BlendGAN: implicitly GAN blending for arbitrary stylized face generation. In: Neural Information Processing Systems (2021)
17. Liu, S., Wei, Y., Lu, J., Zhou, J.: An improved evaluation framework for generative adversarial networks. CoRR (2018). https://doi.org/10.48550/arXiv.1803.07474
18. Liu, Y., Li, Q., Sun, Z., Tan, T.: A ^{3} GAN: an attribute-aware attentive generative adversarial network for face aging. IEEE Trans. Inf. Forensics Secur. **16**, 2776–2790 (2021). https://doi.org/10.1109/TIFS.2021.3065499
19. Loyo, M., McReynold, M., Mace, J.C., Cameron, M.: Protocol for randomized controlled trial of electric stimulation with high-volt twin peak versus placebo for facial functional recovery from acute Bell's palsy in patients with poor prognostic factors. J. Rehabil. Assistive Technol. Eng. **7**, 2055668320964142 (2020). https://doi.org/10.1177/2055668320964142
20. Luan, P., Huynh, V., Tuan Anh, T.: Facial expression recognition using residual masking network. In: IEEE 25th International Conference on Pattern Recognition, pp. 4513–4519 (2020)
21. Mueller, N., Trentzsch, V., Grassme, R., Guntinas-Lichius, O., Volk, G.F., Anders, C.: High-resolution surface electromyographic activities of facial muscles during mimic movements in healthy adults: a prospective observational study. Frontiers Hum. Neurosci. **16**, 1029415 (2022)
22. Taigman, Y., Polyak, A., Wolf, L.: Unsupervised cross-domain image generation. CoRR (2016). https://doi.org/10.48550/arXiv.1611.02200
23. Ulyanov, D., Vedaldi, A., Lempitsky, V.: Instance normalization: the missing ingredient for fast stylization. CoRR (2017). https://doi.org/10.48550/arXiv.1607.08022
24. Wang, X., Guo, H., Hu, S., Chang, M.C., Lyu, S.: GAN-generated Faces Detection: a survey and new perspectives. CoRR (2022)
25. Wang, Z., Bovik, A., Sheikh, H., Simoncelli, E.: Image quality assessment: from error visibility to structural similarity. IEEE Trans. Image Process. **13**(4), 600–612 (2004). https://doi.org/10.1109/TIP.2003.819861
26. Yi, X., Walia, E., Babyn, P.: Generative adversarial network in medical imaging: a review. Med. Image Anal. **58**, 101552 (2019). https://doi.org/10.1016/j.media.2019.101552
27. Zhang, R., Isola, P., Efros, A.A., Shechtman, E., Wang, O.: The unreasonable effectiveness of deep features as a perceptual metric. In: Proceedings of the IEEE Conference on Computer Vision and Pattern Recognition, pp. 586–595 (2018). https://doi.org/10.48550/arXiv.1801.03924

28. Zhu, J.Y., Park, T., Isola, P., Efros, A.A.: Unpaired image-to-image translation using cycle-consistent adversarial networks. In: Proceedings of the IEEE Conference on Computer Vision and Pattern Recognition, pp. 2223–2232 (2017). https://doi.org/10.48550/arXiv.1703.10593

Quality Assessment for High Dynamic Range Stereoscopic Omnidirectional Image System

Liuyan Cao[1], Hao Jiang[1], Zhidi Jiang[2], Jihao You[1], Mei Yu[1], and Gangyi Jiang[1(✉)]

[1] Faculty of Information Science and Engineering, Ningbo University, Ningbo 315211, China
jianggangyi@126.com

[2] College of Science and Technology, Ningbo University, Ningbo 315300, China

Abstract. This paper focuses on visual experience of high dynamic range (HDR) stereoscopic omnidirectional image (HSOI) system, which includes such as HSOI generation, encoding/decoding, tone mapping (TM) and terminal visualization. From the perspective of quantifying coding distortion and TM distortion in HSOI system, a "no-reference (NR) plus reduced-reference (RR)" HSOI quality assessment method is proposed by combining Retinex theory and two-layer distortion simulation of HSOI system. The NR module quantizes coding distortion for HDR images only with coding distortion. The RR module mainly measures the effect of TM operator based on the HDR image only with coding distortion and the mixed distorted image after TM. Experimental results show that the objective prediction of the proposed method is better compared some representative method and more consistent with users' visual perception.

Keywords: Stereoscopic Omnidirectional Image · HDR · Quality Assessment · No-Reference · Reduced-Reference · Retinex Theory

1 Introduction

Omnidirectional images can provide users with a sense of immersion and interaction within 360° × 180° field of view (FOV) [1–3]. Due to the large FOV, the luminance of scene may be very inconsistent. High dynamic range (HDR) stereoscopic omnidirectional image (HSOI) system can record real scene information [2], and it consists of HSOI generation, coding/decoding, tone mapping (TM), and visualization with head-mounted display (HMD), as shown in Fig. 1. The generated HSOI is compressed with the JPEG XT standard and transmitted to the client. Then, TM operator (TMO) is used to compress the dynamic range of the decoded HSOI so as to adapt to HMD to display the image. Since these processes will introduce distortion and result in a decrease in HSOI quality, how to establish a reliable objective prediction model to monitor the quality of HSOIs is an important issue to be studied.

General image quality assessment (IQA) methods can be divided into the 2D-IQA and 3D-IQA. For 2D-IQA, some representative 2D-IQA metrics were proposed, such as IL-NIQE [4], GWH-GLBP [5], OG [6], BRISQUE [7], SISBLIM [8], dipIQ [9], BMPRI [10], and so on. For 3D-IQA, Liu et al. [11] designed a S3D INtegrated Quality (SINQ)

J. Blanc-Talon et al. (Eds.): ACIVS 2023, LNCS 14124, pp. 275–286, 2023.
https://doi.org/10.1007/978-3-031-45382-3_23

Predictor, Shen et al. [12] used deep learning to explore high-level features to characterize complex binocular effects. For stereoscopic omnidirectional IQA (SOIQA), Zhou et al. [3] combined invariant features, visual salience and position priori of projection, and extended the framework to other projection formats. Qi et al. [13] presented a viewport perception based blind SOIQA method (VP-BSOIQA). For TM-IQA, Gu et al. [14] designed the blind tone mapping quality index (BTMQI) by adopting the methods of naturalness interference and structure preservation. Jiang et al. [15] designed a blind tone mapping image IQA (BTMIQA) method with the help of global and aesthetic features.

In HSOI system, an HSOI at the server is coded by JPEG XT and transmitted to the client, where it is decoded and processed with TMO into a distorted HSOI for visualization on the standard dynamic range HMD, so that the coding distortion and TM distortion will be produced [2]. The process of TM may enhance or mask the coding distortion, hence the actual visual content presented to users in subjective assessment is the result of the joint effect of these two types of distortion. Retinex theory points out that the color of an object observed by human eye is determined by the color of the object itself and the surrounding lighting environment [16]. The color of the light source illuminated on the surface of the object is the final perceived visual signal, and these two factors can be separated.

In this paper, combing Retinex theory and the processing flow of HSOI system, a quality assessment method for HSOI system (denoted as QA-HSOI) is proposed to quantify the distortions in the processing process of the system. No-reference (NR) and reduced-reference (RR) modules are established to measure coding distortion and TM distortion, respectively. Additionally, the idea of intrinsic image decomposition based on Retinex theory is also adopted in preprocessing of the NR module.

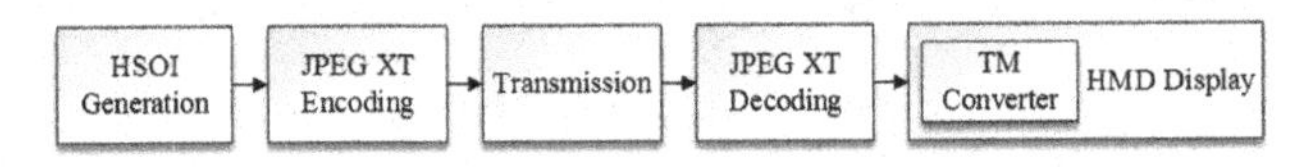

Fig. 1. The HSOI system.

2 Proposed QA-HSOI Method

From the perspective of quantifying distortion in HSOI system, a QA-HSOI method with the design of NR module (denoted as NR_{hdr}) and RR module (denoted as $RR_{hdr,ldr}$) is proposed, as shown in Fig. 2. For NR_{hdr}, considering that HSOI contains rich scene information, the HSOI only with coding distortion is decomposed into two parts: illumination and reflectance components, and the blocking artifact and naturalness are utilized as characteristic attributes. $RR_{hdr,ldr}$ takes the HSOI only with coding distortion as the reference to evaluate the HSOI after TM. The color fidelity and perceptual hashing differences are measured in the single-channel module, while the double-channel module includes the fusion channel and the competition channel. Finally, the random forest (RF) model is trained and then used to predict the objective quality of HSOI by taking all the perceptual features.

Let $\boldsymbol{I}$ denote an HSOI with ERP format, $\boldsymbol{I} = \{\boldsymbol{I}_E, \boldsymbol{I}_B\}$, where $\boldsymbol{I}_E$ and $\boldsymbol{I}_B$ correspond to the HSOI's equator and bipolar regions, respectively. Let $\boldsymbol{V}$ be the viewport sequence of the HSOI, $\boldsymbol{S}$ be the plane saliency map and $\boldsymbol{S}^{LR}$ be the binocular product saliency map. The features extracted by NR_{hdr} are expressed as blocking artifact feature $\boldsymbol{F}_1 = \{f_i^L, f_i^R, f_r^L, f_r^R\}$ and naturalness feature $\boldsymbol{F}_2 = \{f_n^L, f_n^R\}$. The features extracted by $\text{RR}_{\text{hdr,ldr}}$ are expressed as: color fidelity feature $\boldsymbol{F}_3 = \{f_c^L, f_c^R\}$, perceptual hashing feature $\boldsymbol{F}_4 = \{f_p^L, f_p^R\}$, fusion feature $\boldsymbol{F}_5 = \{f_{fc}^g, f_{fc}^s\}$ and competition feature $\boldsymbol{F}_6 = \{f_{cc}^g, f_{cc}^{ed}\}$.

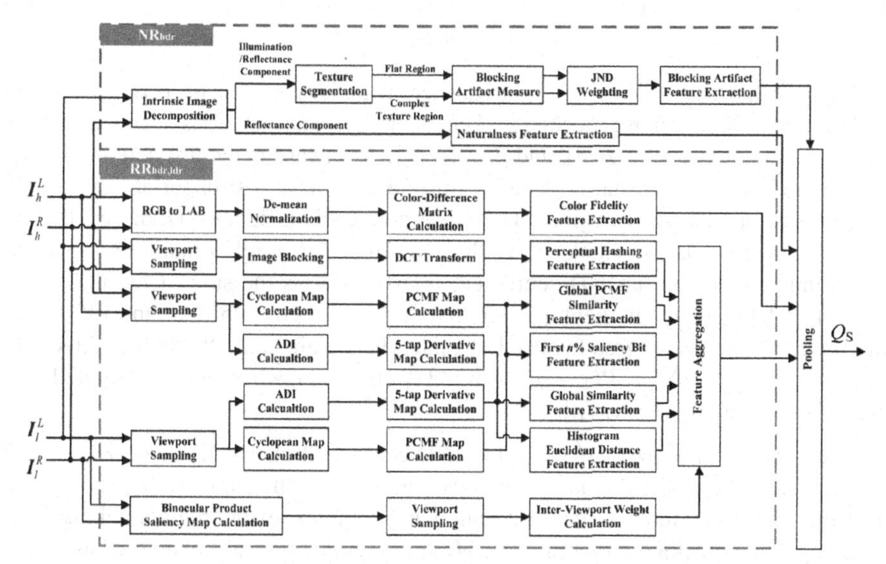

Fig. 2. The framework of the proposed QA-HSOI method.

2.1 The NR Module (Denoted as NR_{hdr})

HSOI has a wide dynamic range and rich illumination information. The illumination and reflectance components can be obtained with intrinsic image decomposition, the former represents the light/dark information after the interaction of light and the object itself, while the latter can reflect the color of the object. In feature extraction stage, texture segmentation is carried out for both of the two components, and Tchebichef moment is used to measure the blocking artifact. But naturalness is measured only against the reflectance component.

1) Intrinsic Image Decomposition: Let $\boldsymbol{I}_h{}^L/\boldsymbol{I}_h{}^R$ denote the left/right images of HDR images with coding distortion only where the subscript h indicates HDR image, and the intrinsic image decomposition is performed on them. Here we only take the left image of HDR as the example, whose obtained luminance eigenmap and reflectivity eigenmap are denoted as $\boldsymbol{\zeta}_i$ and $\boldsymbol{\zeta}_r$ respectively, and the processing of the right image is similar to that of the left image.

2) Feature Extraction: To measure coding distortion, texture segmentation in [17] is performed on $\boldsymbol{\zeta}_i$ and $\boldsymbol{\zeta}_r$ to obtain the flat region $\{\boldsymbol{\zeta}_{i,f},\boldsymbol{\zeta}_{r,f}\}$ and the complex texture region $\{\boldsymbol{\zeta}_{i,c},\boldsymbol{\zeta}_{r,c}\}$. For the flat and complex texture regions, the Tchebichef moment [18] is used to measure the blocking artifact resulted from coding. For 8 × 8 blocks of an image, the Tchebichef moments are denoted as $\boldsymbol{T} = \{t_{i,j}; 0 \leq i \leq 7, 0 \leq j \leq 7\}$. Let τ be a very small constant, the image blocking artifact score in horizontal direction is expressed as follows

$$Q_T = \frac{\sum_{i=0}^{7} |t_{i,7}| + \tau}{(\sum_{i=0}^{7} \sum_{j=0}^{7} |t_{i,j}|) - |t_{0,0}| + \tau} \tag{1}$$

Suppose that the size of $\boldsymbol{I}_h{}^L$ are $W \times H$, where W and H represent the height and width respectively. After the above processing, Tchebichef score matrix $\boldsymbol{M}_c$ is obtained with the size of $floor(W/8) \times floor(H/8)$, and $floor(\cdot)$ is a downward integral function. Texture masking effect indicates that complex texture regions have a stronger ability to hide distortion compared to flat regions. Here, just noticeable difference (JND) coefficient of local image blocks is calculated by the JND model in pixel domain. For an input 8 × 8 image block $\boldsymbol{P}$, its JND coefficient is expressed as the sum of the maximum value of each pixel after gradient filtering in four directions (0°, 45°, 90° and 135°). The filtering coefficient is referred to [19]. Let $g_o(x,y)$ be the gradient filter in four direction, $G_o(x,y) = \frac{1}{16}\sum_{i=1}^{5}\sum_{j=1}^{5}\boldsymbol{P}(x-3+i, y-3+j)g_o(x,y)$, where $o = 1, 2, 3, 4$. Then, $JND = sum(\max_{o=1,2,3,4}\{|G_o(x,y)|\})$.

The Tchebichef score matrix $\boldsymbol{M}_c$ and JND coefficient matrix $\boldsymbol{M}_j$ of $\{\boldsymbol{\zeta}_{i,f},\boldsymbol{\zeta}_{r,f}\}$ and $\{\boldsymbol{\zeta}_{i,c},\boldsymbol{\zeta}_{r,c}\}$ are obtained after the above steps. The JND coefficient is further used as the distortion sensitivity index to fuse flat region and complex texture region. Taking $\boldsymbol{\zeta}_i$ as an example, the blocking artifact matrix $\boldsymbol{M}_i$ after regional fusion is expressed as

$$\boldsymbol{M}_i = \sum_{k=f,c} \boldsymbol{M}_{c_{\zeta_{i,k}}} \cdot \exp\left(-\left(\boldsymbol{M}_{j_{\zeta_{i,k}}}\right)^2 \Big/ \left(2 \times 0.3^2\right)\right) \tag{2}$$

Similarly, the blocking artifact matrix of $\boldsymbol{\zeta}_r$ is expressed as $\boldsymbol{M}_r$. For $\boldsymbol{M}_i/\boldsymbol{M}_r$, the matrix values are arranged in descending order and the mean of the first $n\%$ ($n =$ 10,30,50,70,90) is taken as the final blocking artifact feature $\boldsymbol{f}_i/\boldsymbol{f}_r$. $\boldsymbol{I}_h{}^L$ and $\boldsymbol{I}_h{}^R$ are treated in the same way, and the blocking artifact feature is expressed as $\boldsymbol{F}_1 = \{\boldsymbol{f}_i{}^L, \boldsymbol{f}_i{}^R, \boldsymbol{f}_r{}^L, \boldsymbol{f}_r{}^R\}$.

Then, the naturalness is measured according to $\boldsymbol{\zeta}_r$. Firstly, $\boldsymbol{\zeta}_r$ is converted from RGB space to antagonistic space $\boldsymbol{\kappa}_1 \to rg$, $\boldsymbol{\kappa}_2 \to by$, where rg and by represent the red-green and yellow-blue opsin pairs in the retina; $\boldsymbol{\kappa}_1 = R - G$ and $\boldsymbol{\kappa}_2 = 0.5(R+G) - B$. For each antagonistic channel, its mean and standard deviation are expressed as μ and σ, respectively. μ and σ can be fitted by Gaussian function and *Beta* function respectively, referring to Eq. (3) and Eq. (4)

$$P_m(m) = (1/\sqrt{2\pi})\exp(-(m-\mu_m)/(2\sigma_m^2)) \tag{3}$$

$$P_d(d) = \frac{(1-d)^{\beta_d - 1} d^{\alpha_d - 1}}{Beta(\alpha_d, \beta_d)} \tag{4}$$

The joint probability distribution is defined as $N = (1/L)P_m P_d$, L is the normalized factor. $\mu_m, \sigma_m, \alpha_d$ and β_d are all constants and can be referred to [14]. The fitting results and N of antagonistic channels κ_1-κ_2 are taken as the naturalness feature $\boldsymbol{f}_n$. Finally, the naturalness feature of the left and right views are expressed as $\boldsymbol{F}_2 = \{\boldsymbol{f}_n{}^L, \boldsymbol{f}_n{}^R\}$.

2.2 The RR Module (Denoted as $\mathbf{RR_{hdr,ldr}}$)

$\mathrm{RR_{hdr,ldr}}$ is designed to quantify the distortion effect of TMO. For single-channel, the color distortion caused by TMO is firstly measured by the color-difference matrix. Then, considering that TMO can mask or enhance coding distortion, a similarity measurement method combining discrete cosine transform (DCT) domain and perceptual hashing is designed to measure the visual change of coding distortion. For double-channel, the phase congruency based Monogenic filter is used to measure the cyclopean map, and the absolute difference image is measured by the 5-tap derivative map. The sum proportion of saliency values of each viewport is used as the feature weighting coefficient to integrate the features of all viewports.

1) Viewport Sampling: The used viewport sampling method relies heavily on the fact that the user pays more attention to the equator region. Suppose the viewport's field of view is φ, the equator region $\boldsymbol{I}_E$ refers to the latitude range of $[-\varphi/2, \varphi/2]$, and the bipolar regions $\boldsymbol{I}_B$ are the latitude range of $(\varphi/2, 90°]$ and $[-90°, -\varphi/2)$. In $\boldsymbol{I}_E$, M viewports are generated evenly at equal angle intervals. For $\boldsymbol{I}_B$, the point with the largest pixel value in the binocular product saliency map $\boldsymbol{S}^{LR}$ is used as the center point of the viewport. For $\boldsymbol{I}_B$, there are two viewports. The total number of viewports is $M + 2$. Let $\boldsymbol{S}^L$ and $\boldsymbol{S}^R$ be the saliency maps of the left/right views $\boldsymbol{I}_l{}^L/\boldsymbol{I}_l{}^R$ where the subscript l indicates low dynamic range (LDR) image after TM, then, $\boldsymbol{S}^{LR}$ is calculated by $\boldsymbol{S}^{LR}(i, j) = \boldsymbol{S}^L(i, j) \cdot \boldsymbol{S}^R((i, j) + d_{i,j})$, and $d_{i,j}$ is the disparity at the pixel of (i,j).

2) Single-Channel Module: For the color fidelity features, the most intuitive global change is color distortion when HSOI is mapped from HDR to LDR domain. Some TMOs may saturate the image tone and make it look false. Some TMOs may darken the image altogether. The color-difference matrix $\boldsymbol{\Delta E}_{00}$ is used to describe the color change caused by TM. Take $\boldsymbol{I}_h{}^L/\boldsymbol{I}_l{}^L$ for example, they are converted from RGB space to Lab space. De-mean normalization is used to eliminate the interference of image content to color distortion. Let $\boldsymbol{\Delta E}_{00}$ be the matrix composed of the color difference values of $\boldsymbol{I}_h{}^L$ and $\boldsymbol{I}_l{}^L$ at the corresponding pixels, it can be calculated as

$$\boldsymbol{\Delta E}_{00} = \sqrt{\left(\frac{\Delta L'}{k_L S_L}\right)^2 + \left(\frac{\Delta C'}{k_C S_C}\right)^2 + \left(\frac{\Delta H'}{k_H S_H}\right)^2 + R_T\left(\frac{\Delta C'}{k_C S_C}\right)\left(\frac{\Delta H'}{k_H S_H}\right)} \quad (5)$$

where $\Delta L'$, $\Delta C'$ and $\Delta H'$ represent the brightness, chroma and hue differences of the corresponding pixel positions of $\boldsymbol{I}_h{}^L$ and $\boldsymbol{I}_l{}^L$,. S_L, S_C, S_H and R_T represent brightness, chroma, tone compensation and tone rotation, respectively. k_L, k_C, and k_H default to 1.

For $\boldsymbol{\Delta E}_{00}$, the second order moment variance, the third order moment skewness and the fourth order moment kurtosis are further calculated as the color fidelity feature $\boldsymbol{f}_c$. Finally, the color fidelity features of the left and right views are $\boldsymbol{F}_3 = \{\boldsymbol{f}_c{}^L, \boldsymbol{f}_c{}^R\}$.

In addition to color distortion, single-channel module should also pay attention to the visual change of coding distortion. Perceptual hashing encodes each image as a string of fingerprint characters. Here, perceptual hashing is used to measure structural degradation due to coding distortion. Specifically, for ${\boldsymbol{I}_h}^L$ and ${\boldsymbol{I}_l}^L$, viewport sampling is firstly carried out to obtain viewport sequences ${\boldsymbol{V}_h}^L$ and ${\boldsymbol{V}_l}^L$. DCT is performed for each viewport image, and the image block size is set as 5 × 5. The DC component is removed, and the low/medium/high frequency bands in the AC components are retained and expressed as ***LF***, ***MF*** and ***HF***, respectively. Then, for each frequency band component, the mean value of the frequency band ε_F is calculated first, and each coefficient of the frequency band is compared with ε_F successively. Let ${\boldsymbol{B}_l}^L$ denote one of the image blocks of ${\boldsymbol{V}_l}^L$, its ***LF*** binary fingerprint code ${\boldsymbol{H}_l}^{L,LF}$ is defined as

$$\boldsymbol{H}_l^{L,LF} = \begin{cases} 1, & if\ \boldsymbol{B}_l^{L,LF}(i,j) \geq \varepsilon_F^{LF} \\ 0, & otherwise \end{cases} \tag{6}$$

Similarly, the ***LF*** binary fingerprint code of ${\boldsymbol{B}_h}^L$ is represented as ${\boldsymbol{H}_h}^{L,LF}$. The perceptual hashing difference is defined as $Hamming({\boldsymbol{H}_h}^{L,LF},{\boldsymbol{H}_l}^{L,LF})$, and $Hamming(\cdot)$ is a Hamming distance function. The Hamming distances are calculated in ***LF***, ***MF*** and ***HF***, respectively, and the sum of Hamming distances of all image blocks in one frequency band is taken as the feature of an image in such frequency band.

3) Double-Channel Module: Binocular perception is one of the important characteristics of HSOI system. Binocular effect is particularly obvious when the human eye is viewing stereoscopic images, especially when the distortion levels of the left and right images are not consistent, that is, asymmetric distortion. The initial performance is fusion, that is, the corresponding relationship between two views is established first, and binocular competition will occur in regions that cannot correspond.

Considering that the images viewed are processed with TMO, the binocular characteristics of HSOI are measured in $RR_{hdr,ldr}$. The proposed double-channel module includes fusion channel and competition channel. Cyclopean map is commonly used to represent binocular fusion and many theoretical models have been proposed to calculate the cyclopean map, such as eye-weighted, vector sum and gain control models. Here we use gain control theory because it well explains Fechner's paradox and cyclopean perception. Taking $\boldsymbol{V}_{l,m}^L$ and $\boldsymbol{V}_{l,m}^R$ as an example, the subscript m is used to represent the m-th viewport, their cyclopean map is denoted as $\boldsymbol{C}_l$.

For the cyclopean maps $\boldsymbol{C}_h/\boldsymbol{C}_l$ of the viewport images $\boldsymbol{V}_{h,m}/\boldsymbol{V}_{l,m}$, phase congruency based Monogenic filter (PCMF) map [20] is adopted to estimate the structural degradation degree of the cyclopean map.

The monogenic signal of an image is analyzed according to three dimensions of row x, column y and scale s, $f_M(x,y,s) = f_P(x,y,s) + f_x(x,y,s) + f_y(x,y,s)$. The PCMF map is calculated by $P_{CMF}(x,y,s) = \omega\left[1 - a\cos(\frac{E_P}{A_\sum+\varepsilon}) - \frac{T_P}{A_\sum+\varepsilon}\right]$, where E_P is local energy, $A_\sum$ is the local amplitude, T_P is the noise threshold and ω is a weight coefficient.

$$E_P = \sqrt{\left(\sum_{s=1}^{n} f_P(x,y,s)\right)^2 + \left(\sum_{s=1}^{n} f_x(x,y,s)\right)^2 + \left(\sum_{s=1}^{n} f_y(x,y,s)\right)^2} \tag{7}$$

$$A_\sum = \sum\nolimits_{s=1}^{n} \sqrt{f_P^2(x,y,s) + f_x^2(x,y,s) + f_y^2(x,y,s)} \tag{8}$$

Let $\boldsymbol{P}_{ch}/\boldsymbol{P}_{cl}$ denote the PCMF map of $\boldsymbol{C}_h/\boldsymbol{C}_l$, the global PMCF similarity $\boldsymbol{f}_g$ is defined as the sum of similarity matrix $\boldsymbol{M}_{fc}$ obtained as follows

$$\boldsymbol{M}_{fc} = \frac{2 \times \boldsymbol{P}_{ch} \times \boldsymbol{P}_{cl} + \tau}{\boldsymbol{P}_{ch}^2 + \boldsymbol{P}_{cl}^2 + \tau} \tag{9}$$

We also define a first $n\%$ saliency bit feature $\boldsymbol{f}_s$ as an auxiliary feature of $\boldsymbol{f}_g$. Specifically, the pixel values of the saliency map of the corresponding viewport image are arranged in descending order, and all pixel coordinates $\{a,b\}$ corresponding to the first $n\%$ ($n = 10,30,50,70,90$) saliency bits are filtered respectively. The mean of all values at $\{a,b\}$ are screened out in $\boldsymbol{M}_{fc}$ as the feature of the first $n\%$ saliency bits. $\boldsymbol{f}_g$ and $\boldsymbol{f}_s$ are calculated on four scales. The final fusion channel features are denoted as $\boldsymbol{F}_5 = \{\boldsymbol{f}_{fc}{}^g, \boldsymbol{f}_{fc}{}^s\}$.

For competition channel, absolute difference image (ADI) is used here to characterize the differences between the two views. Taking $\boldsymbol{V}_{l,m}^L$ and $\boldsymbol{V}_{l,m}^R$ as an example, its ADI is defined as $\boldsymbol{A}_{l,m} = \left|\boldsymbol{V}_{l,m}^L - \boldsymbol{V}_{l,m}^R\right|$. Similarly, the HDR version of ADI is represented as $\boldsymbol{A}_{h,m}$. ADI reflects two kinds of information. One is approximate contour information. Usually, more contour stimulation will lead the competition. The other is content differences between the two views. Here, the 5-tap derivative (T_5) maps is used to describe this special structure, where 5-tap refers to the 5 parameters set when performing derivative filtering. Different derivative maps have different parameters [21]. Here, let x and y be the horizontal and vertical directions, and $\boldsymbol{\psi}_x$, $\boldsymbol{\psi}_y$, $\boldsymbol{\psi}_{xx}$, $\boldsymbol{\psi}_{yy}$ and $\boldsymbol{\psi}_{xy}$ be five derivative maps. The former two represent first-order derivatives, and the latter three are second-order derivatives. The five derivative maps can complement each other. Let $\boldsymbol{\psi}_{\upsilon,h}/\boldsymbol{\psi}_{\upsilon,l}$ ($\upsilon = x,y,xx,yy,xy$) denote the T_5 map of $\boldsymbol{A}_{h,m}/\boldsymbol{A}_{l,m}$. Then the global similarity $\boldsymbol{f}_{cc}{}^g$ is defined as the sum of the similarity matrix $\boldsymbol{M}_{cc}$ obtained as follows

$$\boldsymbol{M}_{cc} = \frac{2 \times \psi_{\upsilon,h} \times \psi_{\upsilon,l} + \tau}{\psi_{\upsilon,h}^2 + \psi_{\upsilon,l}^2 + \tau}, \quad (\upsilon = x, y, xx, yy, xy) \tag{10}$$

The histogram Euclidean distance $\boldsymbol{f}_{cc}{}^{ed}$ is defined as an auxiliary feature of $\boldsymbol{f}_{cc}{}^g$. $\boldsymbol{\psi}_{\upsilon,h}/\boldsymbol{\psi}_{\upsilon,l}$ are quantized to 10-bins, and their Euclidean distances are calculated.

The features of the final competition channels are expressed as $\boldsymbol{F}_6 = \{\boldsymbol{f}_{cc}{}^g, \boldsymbol{f}_{cc}{}^{ed}\}$.

2.3 Quality Prediction

The proposed method combines NR_{hdr} and $RR_{hdr,ldr}$ for feature extraction. Finally, 90-dimensional features are obtained to form a feature vector, denoted as $\boldsymbol{F} = \{\boldsymbol{F}_1, \boldsymbol{F}_2, \boldsymbol{F}_3, \boldsymbol{F}_4, \boldsymbol{F}_5, \boldsymbol{F}_6\}$. Then, a RF-based mapping model is trained to complete the mapping task from feature space. Let f_{RF} denote the trained model of RF, the predicted quality score Q_S can be expressed as $Q_S = f_{\mathrm{RF}}(\mathrm{F})$.

3 Experimental Results and Discussions

The proposed method is tested and compared with some representative IQA methods on the SHOID dataset [2], whose source sequences (left images) are shown in Fig. 3. In the experiments, the K-fold cross-validation is used to divide the test set and training set.

Three classical indicators are used to measure the accuracy of regression tasks: Pearson linear correlation coefficient (PLCC), Spearman rank order correlation coefficient (SROCC) and root mean squared error (RMSE).

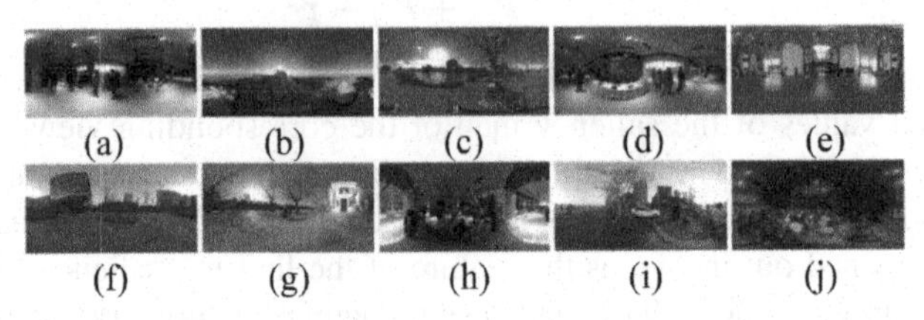

Fig. 3. The SHOID source sequences (left images). (a) - (j) Scene 1 - Scene 10.

3.1 Overall Performance Comparison

To test the performance of the proposed method, it is compared with the representative NR-IQA methods, including seven 2D-IQA methods (IL-NIQE [4], GWH-GLBP [5], OG [6], BRISQUE [7], SISBLIM [8], dipIQ [9], BMPRI [10]), one 3D-IQA method (SINQ [11]), two TM-IQA methods (BTMQI [14] and BTMIQA [15]), and one SOIQA method (VP-BSOIQA [13]). All methods based on supervised learning use K-fold cross-validation for model training. To ensure the fairness and reliability of data, all methods are tested according to the programs published by the original authors. Table 1 lists the values of PLCC, SROCC and RMSE for all methods on the SHOID dataset, with the best performance metrics highlighted in bold. From Table 1, some observations are gotten as follows. (1) For the seven 2D-IQA methods, the top two performing methods are OG and BRISQUE. OG extracts gradient domain features, while the latter extracts spatial features. SISBLIM and GWH-GLBP are specifically designed for multi-distortion images, their objective performance is mediocre. This may be because they are not suitable for the distortion type of SHOID dataset. The HSOI system includes coding distortion as well as TM distortion, which requires targeted feature extraction method to improve the effectiveness. (2) SINQ considers stereoscopic properties, while TM-IQA methods take into account TM distortion, each corresponding to a property in the HSOI system. TM-IQA methods have better performance than SINQ. This shows that for the dataset, human eyes may be more sensitive to TM distortion. (3) Theoretically, SOIQA method should be superior to 3D-IQA and TM-IQA, followed by 2D-IQA. Because SOIQA correspond to two characteristics of the HSOI system, but 3D-IQA and TM-IQA only correspond to one characteristic each. Indeed, these three classes of methods are generally superior to 2D-IQA methods, and the performance differences of 2D-IQA methods can be attributed to feature suitability. (4) The proposed QA-HSOI method comprehensively considers omnidirectional, stereoscopic and HDR characteristics, thus, its performance indexes are better than the other eleven methods. It implies that the designed objective evaluation idea for the distortion type of HSOI system is effective, and the proposed method can be regarded as an effective quality metric.

Table 1. Objective Assessment Results of Different Methods.

	Method	PLCC	SROCC	RMSE
2D-IQA	IL-NIQE [4]	0.5452	0.5424	1.5354
	GWH-GLBP [5]	0.6677	0.6652	1.3634
	OG [6]	0.7741	0.7649	1.1595
	BRISQUE [7]	0.7273	0.7230	1.2571
	SISBLIM [8]	0.5928	0.5725	1.4750
	dipIQ [9]	0.6290	0.6193	1.4238
	BMPRI [10]	0.4837	0.4495	1.6030
3D-IQA	SINQ [11]	0.6804	0.6732	1.3422
TM-IQA	BTMQI [14]	0.7720	0.7690	1.1629
	BTMIQA [15]	0.7067	0.7037	1.2959
SOIQA	VP-BSOIQA [13]	0.7614	0.7532	1.1873
Proposed	QA-HSOI	0.9144	0.9047	0.7415

3.2 Analysis on Different Feature Sets

The proposed method contains six perceptual feature sets $\boldsymbol{F} = \{\boldsymbol{F}_1,\boldsymbol{F}_2,\boldsymbol{F}_3,\boldsymbol{F}_4,\boldsymbol{F}_5,\boldsymbol{F}_6\}$, which are as follows: blocking artifact feature $\boldsymbol{F}_1 = \{f_i{}^L,f_i{}^R,f_r{}^L,f_r{}^R\}$, naturalness feature $\boldsymbol{F}_2 = \{f_n{}^L,f_n{}^R\}$, color fidelity feature $\boldsymbol{F}_3 = \{f_c{}^L,f_c{}^R\}$, perceptual hashing feature $\boldsymbol{F}_4 = \{f_p{}^L,f_p{}^R\}$, fusion feature $\boldsymbol{F}_5 = \{f_{fc}{}^g,f_{fc}{}^s\}$ and competition feature $\boldsymbol{F}_6 = \{f_{cc}{}^g,f_{cc}{}^{ed}\}$. The RF model is used to train each feature set individually and then reports its performance. Table 2 shows the results of ablation experiment, and some observations can be obtained as follows. (1) For the six single feature set, $\boldsymbol{F}_1$, $\boldsymbol{F}_2$, $\boldsymbol{F}_3$ and $\boldsymbol{F}_4$ perform relatively well. Among them, $\boldsymbol{F}_1, \boldsymbol{F}_2, \boldsymbol{F}_3$ are evaluated in ERP format, while $\boldsymbol{F}_4$ is evaluated in viewport. Compared with $\boldsymbol{F}_5$ and $\boldsymbol{F}_6$, they can be regarded as global information, which indicates that ensuring the integrity of global information plays a great role in improving performance. Secondly, the performance of perceptual hashing feature $\boldsymbol{F}_4$ is the best, which explains the effectiveness of multi-channel and multi-band feature extraction to a certain extent. (2) For $\mathrm{NR_{hdr}}$, the performance of $\boldsymbol{F}_1 + \boldsymbol{F}_2$ is relatively stable, although it is not significantly improved compared to their performance alone, but we still retain them. Because $\boldsymbol{F}_1$ and $\boldsymbol{F}_2$ consider characteristics from different perspectives, it is beneficial to improve the overall performance. (3) For $\mathrm{RR_{hdr,ldr}}$, $\boldsymbol{F}_3 + \boldsymbol{F}_4$ is extracted by single-channel module, and $\boldsymbol{F}_5 + \boldsymbol{F}_6$ is extracted with double-channel module. Although the performance of the former is better, the double-channel features improve PLCC by 0.0261 on the basis of the single-channel, which is a considerable gain, indicating that it is necessary and effective to consider the stereoscopic effect. (4) For PLCC, $\mathrm{NR_{hdr}}$ reaches 0.8037, $\mathrm{RR_{hdr,ldr}}$ reaches 0.8654, and the overall modules reach 0.9144. HDR images and LDR images can be comprehensively utilized and assisted each other to further improve the accuracy of the objective model.

Table 2. The Performance of Different Feature Sets.

Models	Features	PLCC	SROCC	RMSE
Single Feature Set	$\boldsymbol{F}_1$	0.8036	0.8023	1.0901
	$\boldsymbol{F}_2$	0.8034	0.8028	1.0905
	$\boldsymbol{F}_3$	0.7019	0.6991	1.3045
	$\boldsymbol{F}_4$	0.8128	0.8066	1.0670
	$\boldsymbol{F}_5$	0.3292	0.3028	1.7295
	$\boldsymbol{F}_6$	0.4815	0.4781	1.6052
$\mathrm{NR}_{\mathrm{hdr}}$	$\boldsymbol{F}_1$+$\boldsymbol{F}_2$	0.8037	0.8033	1.0898
$\mathrm{RR}_{\mathrm{hdr,ldr}}$	$\boldsymbol{F}_3$+$\boldsymbol{F}_4$	0.8393	0.8305	0.9958
	$\boldsymbol{F}_5$+$\boldsymbol{F}_6$	0.5344	0.5193	1.5481
	$\boldsymbol{F}_3$+$\boldsymbol{F}_4$+$\boldsymbol{F}_5$+$\boldsymbol{F}_6$	0.8654	0.8552	0.9177
Total	$\boldsymbol{F}$	0.9144	0.9047	0.7415

Table 3. The Objective Assessment Results of Symmetric/Asymmetric Distortion.

Types	Methods	Symmetric distortion		Asymmetric distortion		Total	
		PLCC	SROCC	PLCC	SROCC	PLCC	SROCC
3D-IQA	SINQ	0.7802	0.7760	0.5841	0.5671	0.6804	0.6732
TM-IQA	BTMQI	0.8092	0.8061	0.7310	0.7225	0.7720	0.7690
	BTMIQA	0.7686	0.7696	0.6383	0.6267	0.7067	0.7037
	HIGRADE-2	0.8940	0.8834	0.7299	0.7204	0.8475	0.8410
SOIQA	VP-BSOIQA	0.8161	0.8006	0.6877	0.6936	0.7614	0.7532
Proposed	QA-HSOI	0.9435	0.9285	0.8862	0.8800	0.9144	0.9047

3.3 Analysis on Symmetric/Asymmetric Distortion

To verify the consideration of binocular characteristics in this paper, Table 3 shows some comparison results including 3D-IQA, TM-IQA, SOIQA and the proposed method. The SHOID dataset is divided into the symmetric and asymmetric distortions to train the RF models, respectively. Table 3 lists the performance indexes of these methods. It can be found that the performance of all methods for symmetric distortion is better than that of asymmetric distortion, and the overall performance is between asymmetric distortion and symmetric distortion. This shows that asymmetric distortion should be considered in HSOI system, and the better the performance of asymmetric distortion is, the better the overall performance is.

Finally, some discussions are as follows. The HSOI system includes various perceptual characteristics, such as omnidirectional, stereoscopic and HDR. This paper proposes a new quality assessment method combining NR_{hdr} and $RR_{hdr,ldr}$. The former quantifies coding distortion and the latter measures the effect of TMO. The SHOID dataset is used to verify the performance of the proposed method. However, in some challenging situations, the work of this paper may leave some room for improvement. The first is for omnidirectional characteristics. The viewport sampling method in this paper is based on the viewing habits of users. However, such a method is difficult to accurately simulate the user's behavior in the actual viewing process. In the future, it is necessary to further combine HMD interactive sensor to develop a more effective viewport sampling method. The second is for binocular perception. How to simulate binocular effect is always the focus of SIQA research. In general, the research on HSOI system is still in the exploratory stage. HSOI system contains a variety of perceptual characteristics, which pose challenges to future research work.

4 Conclusions

High dynamic range stereoscopic omnidirectional image (HSOI) system contains coding distortion and tone mapping (TM) distortion. How to monitor visual experience of the HSOI system is an important issue. From the perspective of quantifying coding distortion and TM distortion in HSOI generation process, a quality assessment method of "no-reference (NR) plus reduced-reference (RR)" has been proposed. The NR module decomposes HSOI only with coding distortion. The corresponding features are extracted according to the distortion characteristics of the decomposed image. The NR module quantizes coding distortion, while the RR module measures TM distortion introduced by TM operator. The RR module takes HSOI only with coding distortion as the reference to evaluate the HSOI after TM, and considering the stereoscopic characteristics of HSOI, the features are extracted with single-channel and double-channel modules. Finally, the feasibility of the proposed method is verified. Experimental results show that the proposed method is superior to the existing state-of-the-art methods and can be used as an effective evaluator for HSOI systems.

Acknowledgements. This work was supported in part by the National Natural Science Foundation of China under Grant Nos. 61871247, 62071266 and 61931022, and Science and Technology Innovation 2025 Major Project of Ningbo (2022Z076).

References

1. Liu, Y., Yin, X., Wang, Y., Yin, Z., Zheng, Z.: HVS-based perception-driven no-reference omnidirectional image quality assessment. IEEE Trans. Instrum. Measur. **72**, art no. 5003111 (2023)
2. Cao, L., You, J., Song, Y., Xu, H., Jiang, Z., Jiang, G.: Client-oriented blind quality metric for high dynamic range stereoscopic omnidirectional vision systems, Sensors **22**, art no.8513 (2022)
3. Zhou, X., Zhang, Y., Li, N., Wang, X., Zhou, Y., Ho, Y.-S.: Projection invariant feature and visual saliency-based stereoscopic omnidirectional image quality assessment. IEEE Trans. Broadcast. **67**(2), 512–523 (2021)

4. Zhang, L., Zhang, L., Bovik, A.C.: A feature-enriched completely blind image quality evaluator. IEEE Trans. Image Process. **24**(8), 2579–2591 (2015)
5. Li, Q., Lin, W., Fang, Y.: No-reference quality assessment for multiply-distorted images in gradient domain. IEEE Sig. Process. Lett. **23**(4), 541–545 (2016)
6. Liu, L., Hua, Y., Zhao, Q., Huang, H., Bovik, A.C.: Blind image quality assessment by relative gradient statistics and adaboosting neural network. Sig. Process. Image Commun. **40**, 1–15 (2016)
7. Mittal, A., Moorthy, A.K., Bovik, A.C.: No-reference image quality assessment in the spatial domain. IEEE Trans. Image Process. **21**(12), 4695–4708 (2012)
8. Gu, K., Zhai, G., Yang, X., Zhang, W.: Hybrid no-reference quality metric for singly and multiply distorted images. IEEE Trans. Broadcast. **60**(3), 555–567 (2014)
9. Ma, K., Liu, W., Liu, T., Wang, Z., Tao, D.: DipIQ: blind image quality assessment by learning-to-rank discriminable image pairs. IEEE Trans. Image Process. **26**(8), 3951–3964 (2017)
10. Min, X., Zhai, G., Gu, K., Liu, Y., Yang, X.: Blind image quality estimation via distortion aggravation. IEEE Trans. Broadcast. **64**(2), 508–517 (2018)
11. Liu, L., Liu, B., Su, C., Huang, H., Bovik, A.C.: Binocular spatial activity and reverse saliency driven no-reference stereopair quality assessment. Sig. Process. Image Commun. **58**, 287–299 (2017)
12. Shen, L., Chen, X., Pan, Z., Fan, K., Li, F., Lei, J.: No-reference stereoscopic image quality assessment based on global and local content characteristics. Neurocomputing **424**, 132–142 (2021)
13. Qi, Y., Jiang, G., Yu, M., Zhang, Y., Ho, Y.-S.: Viewport perception based blind stereoscopic omnidirectional image quality assessment. IEEE Trans. Circ. Syst. Video Technol. **31**(10), 3926–3941 (2021)
14. Gu, K., et al.: Blind quality assessment of tone-mapped images via analysis of information, naturalness, and structure. IEEE Trans. Multimedia **18**(3), 432–443 (2016)
15. Jiang, G., Song, H., Yu, M., Song, Y., Peng, Z.: Blind tone-mapped image quality assessment based on brightest/darkest regions, naturalness and aesthetics. IEEE Access **6**, 2231–2240 (2018)
16. Xu, J., et al.: STAR: A structure and texture aware Retinex model. IEEE Trans. Image Process. **29**, 5022–5037 (2020)
17. Chi, B., Yu, M., Jiang, G., He, Z., Peng, Z., Chen, F.: Blind tone mapped image quality assessment with image segmentation and visual perception, J. Vis. Commun. Image Represent. **67**, art. no. 102752 (2020)
18. Li, L., Zhu, H., Yang, G., Qian, J.: Referenceless measure of blocking artifacts by Tchebichef kernel analysis. IEEE Sig. Process. Lett. **21**, 122–125 (2014)
19. Yang, X., Ling, W., Lu, Z., Ong, E., Yao, S.: Just noticeable distortion model and its applications in video coding. Sig. Process. Image Commun. **20**(7), 662–680 (2005)
20. Wang, L., Zhang, C., Liu, Z., Sun, B.: Image feature detection based on phase congruency by Monogenic filters. In: Proceedings of theChinese Control and Decision Conference, pp. 2033–2038 (2014)
21. Farid, H., Simoncelli, E.P.: Differentiation of discrete multidimensional signals. IEEE Trans. Image Process. **13**(4), 496–508 (2004)

Genetic Programming with Convolutional Operators for Albatross Nest Detection from Satellite Imaging

Mitchell Rogers[1](✉), Igor Debski[2], Johannes Fischer[2], Peter McComb[3], Peter Frost[5], Bing Xue[4], Mengjie Zhang[4], and Patrice Delmas[1]

[1] University of Auckland, Auckland CBD, Auckland 1010, New Zealand
{mitchell.rogers,p.delmas}@auckland.ac.nz
[2] Biodiversity Systems and Aquatic Unit, Department of Conservation, Wellington, New Zealand
[3] Oceanum Ltd., New Plymouth, New Zealand
[4] Victoria University of Wellington, Kelburn, Wellington 6012, New Zealand
{bing.xue,mengjie.zhang}@ecs.vuw.ac.nz
[5] Science Support Service, 87 Ikitara Road, Whanganui 4500, New Zealand

Abstract. Conservation efforts in remote areas, such as population monitoring, are expensive and laborious. Recent advances in satellite resolution have made it possible to achieve sub-40 cm resolution and see small objects. This study aimed to count potential southern royal albatross nests based on the remote Campbell Island, 700 km south of New Zealand. The southern royal albatross population is declining, and due to its remoteness, there is an urgent need to develop new remote sensing methods for assessing the population. This paper proposes a new tree-based genetic programming (GP) approach for binary image segmentation by extracting shallow convolutional features. An ensemble of these GP segmentation models and individual GP models were compared with a state-of-the-art nnU-Net segmentation model trained on manually labelled images. The ensemble of shallow GP segmentation trees provided significantly more interpretable models, using <1% the number of convolutions while achieving performance similar to that of the nnU-Net models. Overall, the GP ensemble achieved a per-pixel F1 score of 75.44% and 123 out of 166 correctly identified nest-like points in the test set compared with the nnU-Net methods, which achieved a per-pixel F1 score of 74.49% and 129 out of 166 correctly identified nest-like points. This approach improves the practicality of machine learning and remote sensing for monitoring endangered species in hard-to-reach regions.

Keywords: Remote sensing · Genetic programming · Semantic segmentation · Southern royal albatross

1 Introduction

Advances in remote sensing have had a tremendous impact on conservation [4]. Through very high-resolution (VHR) satellite imaging, it is possible to obtain

J. Blanc-Talon et al. (Eds.): ACIVS 2023, LNCS 14124, pp. 287–298, 2023.
https://doi.org/10.1007/978-3-031-45382-3_24

daily data for remote regions, thereby allowing year-round monitoring of inaccessible areas [9]. Previous remote sensing research has led to many impressive applications in machine learning and computer vision, such as detecting boats [11] and counting elephant herds [5]. As the spatial resolution of satellite imaging has improved to sub-40 cm resolution, the range of possible applications has expanded to enable monitoring of birds in remote regions, such as the albatross [3,6,8].

This study aims to detect and count southern royal albatross nests on a ridge of remote Campbell Island, south of New Zealand. The sub-Antarctic Campbell Island group lies 700 km south of New Zealand's South Island. The subject of this study, the southern royal albatross, has the largest wingspan of any bird, but the population is declining at an alarming rate [15]. Currently, manual species counting methods are laborious, infrequent, and expensive. Traversing the entire island on foot requires a full-day hike, and manual counting of nests requires weeks to months of work. An alternative is to assess the area via a helicopter, which disturbs wildlife and is expensive, with various logistical and health and safety hurdles. Owing to Campbell Island's remoteness, transporting conservationists and helicopters takes weeks, meaning it is not currently possible to regularly count the population.

Over the last 15 years, several advances in image classification and segmentation have been based on convolutional neural networks (CNNs) [14], which apply a series of local convolutions to images to extract increasingly higher-level features. These features range from simple features, such as edges or corners, to more complex features, such as cars or faces. Over this period, researchers have created increasingly complex models by adding new layer types and tweaking loss functions to incrementally increase the performance [10]. State-of-the-art deep learning methods often require large datasets and extensive expert knowledge to implement and tweak for maximum performance in new applications. These models tend to extract redundant or overly complex features [13] and are difficult to interpret. As such, they are often referred to as black boxes. Genetic Programming (GP) offers an explainable, simple, and self-configuring alternative for designing these architectures [19] and can learn generalisable models from a small number of training samples [2].

Simple, explainable approaches that do not require extensive AI expert knowledge, large-scale datasets, or expensive GPUs are crucial for wildlife conservation efforts. Our paper improves on previous albatross-counting deep learning methods [3] to create an interpretable computer vision model to count potential southern royal albatross (*Diomedea epomophora*) nests year-round from satellite images.

This paper aims to develop a new GP-based approach with convolutional filters using an explainable and flexible tree-based program structure for binary image segmentation. The contributions of this study can be summarised as follows:

1. A new flexible GP-program structure utilising convolutional filters, various activation functions, and logical operators to evolve binary image segmentation models.
2. An evaluation of the proposed GP approach along with an ensemble of GP individuals compared to the state-of-the-art segmentation method from the biomedical image segmentation domain: nnU-Net [10].
3. An explainable model for year-round counting of possible southern royal albatross nests on Campbell Island.

2 Proposed Approach

This section describes the proposed convolutional-GP approach for image segmentation, including the algorithm, program structure, fitness function, and function set.

2.1 Algorithm Overview

The framework of this convolutional-GP approach follows that of a typical evolutionary algorithm. First, an initial population of individuals (trees) are randomly initialised using the *ramped half-and-half* population generation method. The *ramped half-and-half* method ensures that trees are generated with a high depth variety, with a 50% chance that all input leaves lie at the predefined maximal depth or are randomly grown to different depths. The internal nodes of the trees are selected from a function set, and the leaf nodes are selected from the terminal set and other available inputs, such as random constants, which could play a role of coefficients for each feature/terminal or some specific parameter values (to be specified later). In this implementation, the function set consists of various logical and morphological operators, convolutional operators, and activation functions. The terminal set defines the values that the convolutional filters and thresholds can take. Each individual takes RGB images as input and, by applying an evolved series of functions, produces binary masks classifying each pixel as foreground and background.

Each individual's performance/fitness is evaluated by comparing the output and ground truth masks. For each generation, a new population of individuals is created to replace the previous population by selecting the best-performing individuals and three genetic operators: *subtree crossover*, *subtree mutation*, and *elitism*. *subtree crossover* randomly selects a node from two trees and swaps the subtree at that node, *subtree mutation* generates a new subtree at a randomly selected node, and *elitism* copies the best individuals unchanged to the next generation. New individuals are repeatedly created and evaluated until the maximum number of generations is reached, and the best-performing individual is returned. The steps involved in this process are illustrated in Fig. 1. This final program is expected to extract low-level discriminatory features to distinguish between foreground and background classes.

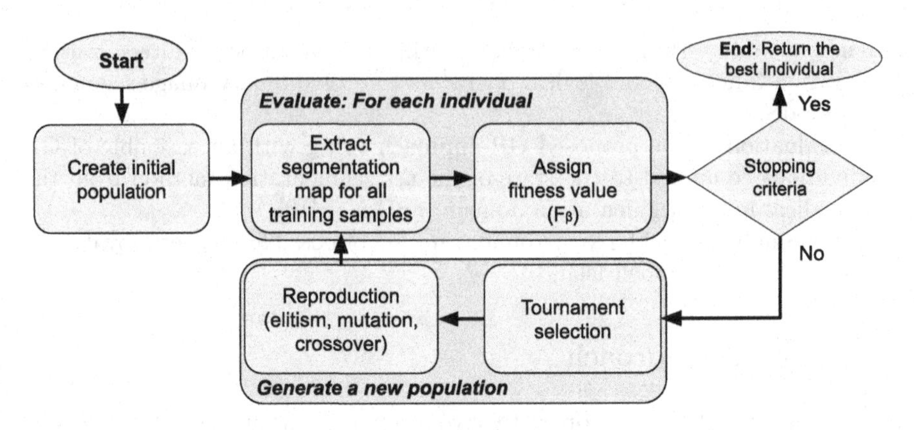

Fig. 1. The evolutionary process of GP.

2.2 Program Structure

In this approach, a flexible program structure was developed to integrate convolutional functions and thresholding operations into a single tree. The program structure uses four types of operations: convolutions, normalisation functions, activation functions, and logical operators. Based on the type constraints of strongly typed GP [16], each function expects specific input types corresponding to the outputs of other functions or terminal types, and each function provides a specific output type. The feature extraction branches of the tree connect convolutional filters with activation functions and thresholding operations, and the binary masks of each branch are combined with logical operators to form a single binary classification mask. Unlike the U-Net architectures [10,18] and previous convolutional-GP methods [1], the function set in this approach does not include downsampling or pooling functions to preserve small image details and the original image dimensions.

The number of consecutive convolution operators, size of the filtering windows, and number of feature extraction branches in the tree are flexible, allowing the GP method to evolve solutions with varying complexity through the depth and breadth of filters. This program structure enables GP to evolve trees focusing on simple (lower-level) features or deeper trees to find higher-level features. The functions and terminals used in the program structure are described in the following sections. Figure 2 shows an example tree.

2.3 Function Set

Based on the described program structure, four types of functions constitute the function set. The functions and their input types are listed in Table 1.

The **convolve** functions take either an input channel of the image or a previous convolution map along with an evolved filter and generate a new convolution map. The filter is applied to the image as a convolution.

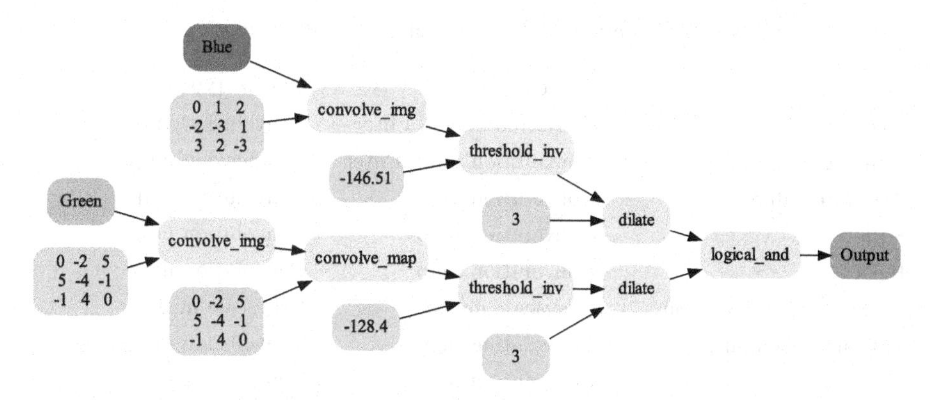

Fig. 2. An example of a GP individual. The channels of the image are processed by a series of functions and the root node (right) returns a binary segmentation mask. (Color figure online)

The activation functions transform the convolution map values. The **min-max normalisation** rescales the values between 0 and 255. This function can be applied to a convolution map or an input channel because normalisation is a common input preprocessing technique for deep learning. The other activation functions in this function set were **ReLU**, **sqrt**, and **absolute value**. The ReLU function transforms the image by applying a rectified linear unit function, returning the pixel value if it is positive; otherwise, it returns to zero. The square root function is protected to return zero if the value is negative.

The **threshold** and **threshold inverse** operators convert a convolved image into a binary image using a given threshold value (τ). Logical operators **OR**, **AND**, **XOR**, and **NOT** are applied to the binary images and can be chained to combine multiple thresholded feature maps. A binary image may also be **eroded** or **dilated** to decrease or increase the size of nest components masked by the image.

The convolution and threshold functions are required by each individual, with all others being optional. These are the only two functions that change the input type of an image into a different output type. All other functions are optional, and evolutionary processes automatically determine their suitability.

2.4 Terminal Set

The values of the convolutional filters, thresholds, and kernel sizes are selected randomly from a distribution of values. The convolutional filters have sizes of 3×3, 5×5, or 7×7 with values randomly selected from -5 to 5. The threshold values are randomly sampled from values between -150 and 150, as the input feature maps may be positive or negative floating-point values. Finally, the kernel sizes were either 3×3 or 5×5 with a square structuring element.

Table 1. Function set for this convolutional-GP method

Function	Input types	Output types
Convolve	Grayscale image, Filter	Convolution map
Convolve	Convolution map, Filter	Convolution map
Absolute value	Convolution map	Convolution map
Sqrt	Convolution map	Convolution map
ReLU	Convolution map	Convolution map
Min-max normalisation	Grayscale image	Grayscale image
Min-max normalisation	Convolution map	Convolution map
Threshold	Convolution map, τ	Binary image
Threshold Inverse	Convolution map, τ	Binary image
Logical OR	2 Binary images	Binary image
Logical XOR	2 Binary images	Binary image
Logical AND	2 Binary images	Binary image
Logical NOT	Binary image	Binary image
Erosion	Binary image, k	Binary image
Dilation	Binary image, k	Binary image

2.5 Fitness Function

Due to the significant imbalance between foreground and background pixels in this experiment, F_β score was chosen as the fitness function (Eq. 1). By changing the value of β, the equation penalises false positives at a higher or lower rate than that of false negatives. A beta value of 0.2 was chosen to apply a higher weight to penalise false positives more.

$$F_\beta = \frac{(1+\beta^2)TP}{(1+\beta^2)TP + (\beta^2)FN + FP} \tag{1}$$

3 Experimental Design

3.1 Data Collection

In this study, we used satellite data from the Faye Bump, north of Mount Faye, on Campbell Island, south of New Zealand. This island is home to the majority (>99%) of the southern royal albatross population [17]. An RGB satellite image ($6,668 \times 3,335$ pixels) of the region of interest was taken on the 4th of June, 2021 from the Maxar WorldView-3 satellite (DigitalGlobe, Inc., USA) was used for training in this study. Each pixel has a spatial resolution of approximately 31 cm^2. The annotated locations are shown in Fig. 3b.

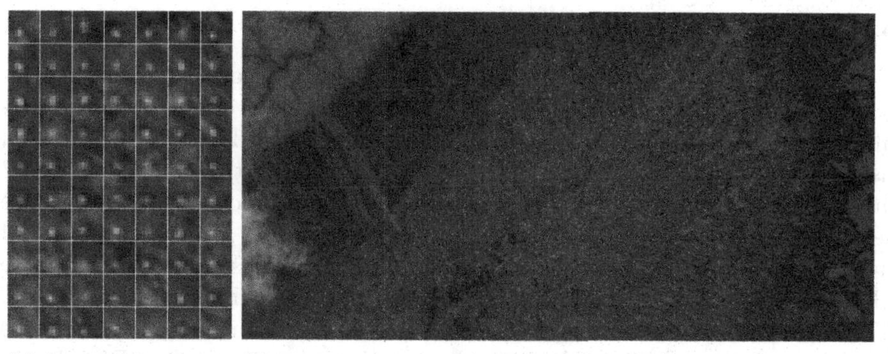

(a) Zoomed in examples of possible nests. (b) Locations of annotated points (magenta dots). WorldView-3 © 2021 Maxar Technologies.

Fig. 3. Visual examples of the possible nest points and their locations along the Faye Bump, north of Mount Faye, Campbell Island.

Experts from the New Zealand Department of Conservation (DoC) provided annotated possible nest points. These points may correspond to occupied nests, abandoned nests covered in guano, or other nest-like points. Both sexes of albatross share incubation time and typically sit in nests for 78–80 days, followed by a further 35–42 days for brooding and guarding the chick [12]. The satellite images were collected outside the typical incubation time, so real nests are characterised by the colour of the chick that sits in the nest after the incubation period and the surrounding guano [20]. Due to the remoteness of this region, the labelled points were not validated against any GPS points collected on the ground.

These locations were manually refined into a binary segmentation mask. A pixel mask was chosen as opposed to bounding boxes [5] because the objects were typically only a few pixels wide. The binary mask represents the manually annotated nest locations as ones and zeros for the background pixels. For evaluation, pixels surrounding the ground truth nest pixels were ignored because it was impossible to confidently label these pixels as part of the nest or background, and any misclassification would not be detrimental to performance. The evaluation mask was defined as the inverse of the external gradient using a 5-by-5 square structuring element (the original mask was subtracted from the dilated mask). The image contained 651 potential nest locations. Because of the cost of data acquisition, only a single image was used. This image was subdivided into a 256-by-256-pixel grid, creating 338 tiles/images. The tiles were then split into training and test sets with 253 and 85 images, containing 485 and 166 potential nest points, respectively.

3.2 Parameter Settings

The key parameters of the GP-based method were set according to the settings commonly used in the GP community [1]. Individuals evolved over 50 generations, with a population size of 1000. The initial population was generated using the Ramped half-and-half method, with an initial tree depth between 3 and 8 and a maximum depth of 9. After fitness evaluations, individuals were selected for combination with tournament selection using a tournament size of five. As with [1], a larger total mutation rate of 0.49 was utilised to promote filter changes and a crossover rate of 0.5, with the remaining 0.01 for elitism. The mutation rate was broken down into 0.3 for terminal mutation and 0.19 for subtree mutation. The GP method was implemented using the Distributed Evolutionary Algorithms in Python (DEAP) package [7]. The experiments were run for ten independent runs. The best individual from each run was used to create the binary segmentation masks for each test sample. An ensemble of the best individual from each of these ten runs was also constructed by labelling each pixel as a nest if at least one individual labelled that pixel as a nest.

3.3 Benchmark Methods

The new GP approach was compared with a nnU-Net segmentation model [10] trained on the albatross nest images. The nnU-Net model was trained using five-fold cross-validation on the training set. Finally, nnU-Net automatically determines which configuration or ensemble of two configurations to use for inference based on the Dice coefficient from cross-validation [10].

The predicted masks of the five nnU-Net models from five-fold cross-validation, the automatically-configured nnU-Net ensemble model, the best GP individuals, and the ensemble of GP individuals were compared to the ground-truth labels using per-pixel and per-nest metrics. The per-pixel metrics were precision, recall, and F1 Score, based on the binary classification of each pixel. The per-nest metrics considered whether the predicted masks overlapped with labelled nests. If there was some overlap, it was labelled as a correct nest, a prediction that did not overlap with a labelled nest was considered a false nest, and a labelled nest that did not overlap with a prediction was considered a missed nest.

4 Experimental Results

In this section, the experimental results of the GP-based method are compared with those of the nnU-Net model, and the evolved operators and structures of GP individuals are examined.

4.1 Overall Test Performance

Table 2 lists the average and ensemble per-pixel and per-nest results of the GP method and nnU-Net models evaluated on the test set. Among the individual

Table 2. Overall test set performance of the GP-based method compared with the average of the individual nnU-Net models and the ensemble of nnU-Net models.

	Per-pixel metrics			Per-nest metrics		
	Precision	Recall	F1 Score	Correct nests	Missed nests	False nests
GP-method (average)	93.85%	41.69%	57.28%	74.8	91.2	2.2
GP-method (ensemble)	79.04%	72.15%	75.44%	123	43	12
nnU-Net (average)	92.35%	62.42%	74.49%	128.4	37.6	10.6
nnU-Net (ensemble)	90.78%	60.9%	72.9%	129	37	12

models, the nnU-Net models achieved the highest performance, correctly identifying an average of 128.4 nest-like points out of 166. There was minimal variation among the five models from cross-validation, and the configured ensemble of the two best models performed slightly worse in terms of per-pixel metrics on the test set.

The best GP individuals had extremely low false-positive rates, with some having per-pixel precision of 1.0. However, recall rates were significantly lower than those of the nnU-Net models, and they could only detect 45% of the total number of nests on average. Given the shallow nature of the GP-based method, a single function tree with a constrained depth cannot come close to the deep feature extraction ability of nnU-Net.

The ensemble of the best GP trees, where each pixel was labelled as a nest if at least one individual classified it as a nest, achieved a detection rate comparable to the nnU-Net models. The combination of the ten best GP individuals predicted a similar number of false nests as the nnU-Net model and correctly identified 123. These results indicate that a shallow model that extracts a broad range of features may be better suited for segmenting small objects than deep features from large CNN models.

The nnU-Net models had a lower per-pixel recall than the GP method ensemble but a higher number of correct nests because the GP method's fitness function did not penalise false positives surrounding correct nest pixels as the nnU-Net loss function did. Many evolved function trees would predict abnormally large nest objects by using functions such as dilation. Further investigation found the GP-ensemble method labelled 42.58% of the pixels surrounding the labelled ground truth values as nests, compared with an average for each individual of 11.94% and only 0.27% of pixels from the nnU-Net ensemble.

4.2 GP Individuals

This subsection analyses the top individual obtained in each GP run. Figure 4 presents an example of one of the best individuals. The most commonly selected image channel across the best individuals was blue (18 times), followed by red (ten times), and green (three times), being the least likely. The green channel is likely the least informative because of the large amount of green vegetation on the island. Only three out of ten individual trees contained function branches applying more than one convolutional operation, indicating that the extracted features were typically responses from a single convolution filter. The nnU-Net architecture, on the other hand, applied 26 convolutional blocks, each with between 32 and 512 convolutional filters, meaning that the GP individuals used <1% the number of convolutional filters of the nnU-Net model. Typically, the filters were 3×3. These occurred 28 times in the convolutions compared to 5×5 and 7×7 filters used four and five times, respectively. As the nest objects were typically only a few pixels wide, the evolved filters did not need to be very large. Only two out of the 31 feature extraction branches utilised the min-max normalisation function. The min-max normalisation function may yield inconsistent results for small images, where the results of the convolution operations may have varying pixel intensity distributions between images.

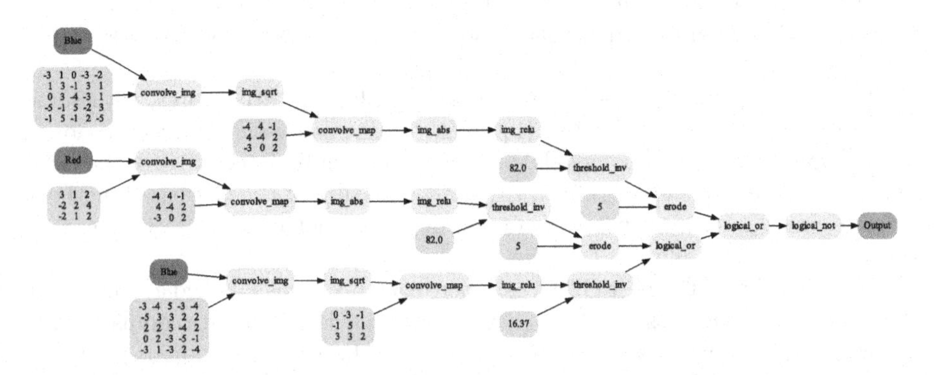

Fig. 4. An example of one of the best individuals from the ten individual runs. (Color figure online)

Eight individuals included dilation operations, increasing the size of the nest components, and two had erosion operations. Due to these operations, the GP individuals often labelled pixels directly surrounding the labelled nest as nests. Some GP individuals contain structurally ineffective code sequences, such as the absolute value followed by the ReLU function in the above example. These are commonly referred to as introns and can be manually pruned from the resulting tree; otherwise, they have no effect on model performance. Natural selection during the evolutionary process removes functionally detrimental code sequences.

5 Conclusion

This study developed a new GP-based image segmentation approach to count possible Southern Royal Albatross nests on remote Campbell Island. This goal has been successfully achieved by applying a new flexible program structure using convolutional operations evolved to detect human-annotated points. The experimental results indicated that an ensemble of evolved GP function trees can achieve results that are competitive with state-of-the-art deep learning segmentation models, such as nnU-Net. The ensemble of models also requires less inference computation, typically applying less than ten convolutions or morphological filters, and other logical operations. Despite being in an ensemble configuration, the GP individuals are also easily interpretable, unlike deep learning approaches. Future work will extend these models to various seasons and a larger section of Campbell Island.

This study presents a proof of concept for simple convolutional-based machine learning methods detecting small objects using remote sensing data. Future models will be trained on multiple images across the breeding season to measure the year-round model performance more accurately. Changes in vegetation colour across the year may impact the precision and recall of the GP-based and deep learning-based models. Previous studies on albatross nest counting from satellites used images from multiple islands, including various colony densities, to build more robust models [3,8]. Their models considered the variation between six observers and chose to train their U-Net on the majority vote of labelling in the absence of verified GPS points. Our models achieved similar F1 scores to their models, which varied between the islands. We aim to extend this work to different times of the albatross breeding cycle and verify the performance with known GPS coordinates to classify nests as occupied or unoccupied.

Acknowledgments. We thank Kāi Tahu and Kaitiaki Rōpū ki Murihiku for allowing us to work on their taonga. We also thank DOC Murihiku for their logistical support.

References

1. Bi, Y., Xue, B., Zhang, M.: An evolutionary deep learning approach using genetic programming with convolution operators for image classification. In: 2019 IEEE Congress on Evolutionary Computation (CEC), pp. 3197–3204 (2019)
2. Bi, Y., Xue, B., Zhang, M.: Genetic programming with image-related operators and a flexible program structure for feature learning in image classification. IEEE Trans. Evol. Comput. **25**(1), 87–101 (2021)
3. Bowler, E., Fretwell, P.T., French, G., Mackiewicz, M.: Using deep learning to count albatrosses from space: assessing results in light of ground truth uncertainty. Remote Sens. **12**(12), 1–18 (2020)
4. Corbane, C., et al.: Remote sensing for mapping natural habitats and their conservation status - new opportunities and challenges. Int. J. Appl. Earth Obs. Geoinf. **37**, 7–16 (2015). Special issue on earth observation for habitat mapping and biodiversity monitoring
5. Duporge, I., Isupova, O., Reece, S., Macdonald, D.W., Wang, T.: Using very-high-resolution satellite imagery and deep learning to detect and count African elephants in heterogeneous landscapes. Remote Sens. Ecol. Conserv. **7**(3), 369–381 (2021)
6. Ferreira, A.C., et al.: Deep learning-based methods for individual recognition in small birds. Methods Ecol. Evol. **11**(9), 1072–1085 (2020)

7. Fortin, F.A., Rainville, F.M.D., Gardner, M.A., Parizeau, M., Gagné, C.: DEAP: evolutionary algorithms made easy. J. Mach. Learn. Res. **13**(70), 2171–2175 (2012)
8. Fretwell, P.T., Scofield, P., Phillips, R.A.: Using super-high resolution satellite imagery to census threatened albatrosses. Ibis **159**(3), 481–490 (2017)
9. Hoffman-Hall, A., Loboda, T.V., Hall, J.V., Carroll, M.L., Chen, D.: Mapping remote rural settlements at 30 m spatial resolution using geospatial data-fusion. Remote Sens. Environ. **233**, 1–19 (2019)
10. Isensee, F., Jaeger, P.F., Kohl, S.A.A., Petersen, J., Maier-Hein, K.H.: nnU-Net: a self-configuring method for deep learning-based biomedical image segmentation. Nat. Methods **18**(2), 203–211 (2021)
11. Kanjir, U., Greidanus, H., Oštir, K.: Vessel detection and classification from spaceborne optical images: a literature survey. Remote Sens. Environ. **207**, 1–26 (2018)
12. Marchant, S., Higgins, P., Ambrose, S.: Handbook of Australian, New Zealand & Antarctic Birds: Volume I, Ratites to Ducks, vol. 1. Oxford University Press, Oxford (1990)
13. Menghani, G.: Efficient deep learning: a survey on making deep learning models smaller, faster, and better. ACM Comput. Surv. **55**(12), 1–37 (2023)
14. Minaee, S., Boykov, Y., Porikli, F., Plaza, A., Kehtarnavaz, N., Terzopoulos, D.: Image segmentation using deep learning: a survey. IEEE Trans. Pattern Anal. Mach. Intell. **44**(7), 3523–3542 (2022)
15. Mischler, C., Wickes, C.: Campbell Island/Motu Ihupuku Seabird Research & Operation Endurance February 2023. POP2022-11 final report prepared for Conservation Services Programme. Department of Conservation (2023)
16. Montana, D.J.: Strongly typed genetic programming. Evol. Comput. **3**, 199–230 (1995)
17. Moore, P.J.: Southern royal albatross on Campbell Island/Motu Ihupuku: solving a band injury problem and population survey, 2004–08. DOC research and development series, 333, Publishing Team, Department of Conservation, Wellington, New Zealand (2012)
18. Ronneberger, O., Fischer, P., Brox, T.: U-Net: convolutional networks for biomedical image segmentation. In: Navab, N., Hornegger, J., Wells, W.M., Frangi, A.F. (eds.) MICCAI 2015. LNCS, vol. 9351, pp. 234–241. Springer, Cham (2015). https://doi.org/10.1007/978-3-319-24574-4_28
19. Suganuma, M., Shirakawa, S., Nagao, T.: A genetic programming approach to designing convolutional neural network architectures (2017)
20. Westerskov, K.: The nesting habitat of the royal albatross on Campbell Island, vol. 6, pp. 16–20. New Zealand Ecological Society (1958)

Reinforcement Learning for Truck Eco-Driving: A Serious Game as Driving Assistance System

Mohamed Fassih[1,2], Anne-Sophie Capelle-Laizé[1(✉)], Philippe Carré[1], and Pierre-Yves Boisbunon[2]

[1] XLIM UMR CNRS 7252, Université de Poitiers, 11 Bd Marie et Pierre Curie, 86360 Futuroscope Chasseneuil, France
{mohamed.fassih,anne.sophie.capelle,philippe.carre}@univ-poitiers.fr
[2] Strada, 10 rue Jean Mermoz, 79300 Bressuire, France

Abstract. Making fuel-economy for vehicles is an important and current challenge in particular for professionals of transportation. In this article, we address the challenge of providing a driving serious game based on artificial intelligence in order to significantly reduce the fuel consumption for trucks. Our proposition is based on a machine learning process consisting of a Self-Organizing Network (SOM) for clustering and subsequent reinforcement learning to deliver precise recommendations for eco-driving. Driving experts provide us knowledge in order to model the actions-rewards process. Experiments conducted on simulated data demonstrate that the recommendations are coherent and enable drivers to adopt eco-driving behavior.

Keywords: Eco-driving · Self-organizing Network · Reinforcement Learning · Serious game

1 Introduction

Eco-driving has emerged as a crucial aspect within the field of road freight transport. The requirements for adapting an eco-driving behavior are becoming stronger: transport companies aims to reduce their fuel consumption and CO2 emissions for economic and environmental issues. Logistic and transportation enterprises benefit of research in numerical data. Today, most of the trucks are equipped with numerous sensors that provide real-time information. When coupled with driving indicators, they can effectively contribute to punctually optimizing driving efficiency. For example, gear shift indicator can be coupled to engine speed.

In this work, we aim to overtake this simple and punctual driving-assistance by a complete eco-driving system which assist the driver in the long term journey. By the analysis of vehicle technical data such as vehicle speed, real-time consumption or braking state our system should be able to evaluate driving

J. Blanc-Talon et al. (Eds.): ACIVS 2023, LNCS 14124, pp. 299–310, 2023.
https://doi.org/10.1007/978-3-031-45382-3_25

quality and propose actions that globally optimize fuel consumption and CO2 emissions. Our solution can be viewed as a serious game where driver has to improve their driving score. This work is jointly developed with the company *Strada*[1] specialized in Transport Management System (TMS) for fleet management optimization. *Strada* possesses a vast amount of transport data due to its 4800 customers and approximately 5800 connected vehicles. The company's primary goal is to offer drivers a solution that enhances their eco-driving awareness and learning capabilities through a driving simulator.

The analysis of *Strada* requirements involve some assumptions and constraints. First, since it exists several driving modes and situations depending on the drivers, vehicles, journeys, roads, *etc*, the simulator must autonomously adapt to various contexts and provide precise driving recommendations tailored to each specific situation. Secondly, the solution should appears as a positive experience for the learner. Thus a gamification of the solution sounds an appropriate approach. As driving can be viewed as a continuous sequence of actions, our proposition is to develop a simulator that increase the driving performance of drivers in a whole sequence. With these hypotheses reinforcement learning algorithms sound well adapted to our problematic. Indeed, the reinforcement learning (RL) are another path of machine learning approach between unsupervised and supervised learning. The agent learns to behave in environment depending on future rewards, and has a goal to develop efficient policy to optimize a cumulus reward. From our point of view, RL is a adapted response to our concern: increase driving score in a gamification approach.

It exists large number of papers about Reinforcement Learning. We can cite the introduction to RL presented by Sutton in [1] and some surveys in [2–4]. Obviously, RL and deep RL are currently being employed in a wide range of practical applications with numerous and diverse use cases. In [5], authors present review of RL for Cybersecurity domain. In [6], healthy problems are treated. In [7], a survey of Deep RL for blockchain in industrial IoT is presented. Closer to our problematic, RL and Deep RL are also used in vehicle management but most of times it concerns autonomous driving context. One survey is proposed by Elallid et al. [8]. In particular context of eco-driving, some propositions exists but mainly concerns electric and personal vehicle [9–11] but trucks consumption assistance in transport of goods still little treated.

In this article, we propose a solution for truck eco-driving. As a chest game, our proposal consists in giving to a driver optimal recommendations (driving actions) that will increase a global performance considering fuel consumption using reinforcement learning algorithm. As RL is based on the principle of action-reward estimation, our solution should be able to identify current state of a vehicle and to provide local rewards. The originality of our proposition is to combine clustering approach for states estimation and RL. Rewards and states are defined within expert knowledge.

The remaining part of this paper is organized as follows. Section 3 dedicate to theoretical aspects of RL and the proposal description: RL basis are first intro-

[1] https://www.stradaworld.com/.

duced in Sect. 2 and general scheme of our proposal is described in Sect. 3.1. The use of experts' knowledge for actions-rewards modeling is presented in Sect. 3.2 and Sect. 3.3 describes how states are estimated. Section 4 presents our experiments and results. We conclude and propose some new perspectives in Sect. 5.

2 The Basis of Reinforcement Learning

Reinforcement Learning is a body of theory and algorithms for optimal decision making developed in the last twenty-five years. RL methods find useful approximate solutions to optimal-control problems that are too large or too ill-defined for classical methods such as dynamic programming. The main explanation and principle can be found in this reference [1] and we review here only the background concepts.

Reinforcement learning is a class of solutions for solving Markov Decision Systems defined such that:

- a set S of states,
- a set A of actions,
- a transition probability function $p : S \times A \rightarrow P_{sa}(.)$ that is the transition probabilities upon taking action a in state s,
- a reward function $R : S \times S \times A \rightarrow r$ which modelizes the reward $R(s_{t+1}, s_t, a_t)$, the expected rewards for state-action-next-state triples,
- a future discount factor γ, the discount rate determines the present value of future rewards.

Markov Decision Processes (MDPs) provide a framework for modeling decision making. The key feature of MDPs is that they follow the Markov Property; all future states are independent of the past given the present. In RL, the goal is to find a policy $\pi : S \rightarrow A$, that maximizes the "action-value function" of every state-action pair, defined as:

$$Q_\pi(s, a) = E_\pi \left[\sum_{t=0}^{T} \gamma^t R(s_{t+1}, s_t, a_t) | a_0 = a, s_0 = s \right] \quad (1)$$

In the previous equation, the expectation, noted E is over the state sequence $(s_0, s_1, ...)$ we pass through when we execute the policy π starting from s_0. In our setting, we use a finite time horizon T (it is a particular context, the end of the truck travel). $Q_\pi(s_0, a_0)$ is the expected cumulative reward received while starting from state s_0 with action a_0 and following policy π. The solution of an MDP is a policy π^* that for every initial state s_0 maximizes the expected cumulative reward.

Reinforcement Learning tests which actions are best for each state of an environment essentially by trial and error. The model sets a random policy to start, and each time one action is taken. This continues until the state being terminal (T in the previous equation).

To solve this problem a classical strategy is to use the Q-Learning algorithm. The Q-Learning algorithm was first introduced by [12], and is one of the most studied methods. Given an MDP, Q-Learning aims to calculate the corresponding optimal action value function Q^*, and thus we can choose any behavioural policy to gather experience from the environment.

The $Q(s, a)$ function is represented in tabular form, with each state-action pair (s, a) represented discretely [12]. The Q-Learning algorithm converges to an optimal policy by applying the following update rule at each step t:

$$Q(s_t, a_t) \leftarrow Q(s_t, a_t) + \alpha[R(s_t, a_t) + \gamma \max_a[Q(s_{t+1}, a)] - Q(s_t, a_t)] \quad (2)$$

$\alpha \in [0, 1]$ is the learning rate. With the ϵ-greedy strategy, the agent choose the optimal action with probability $(1 - \epsilon)$, and to choose a random action with probability ϵ. The value of the parameter can be varied over time, by decreasing it over the course of training.

3 A Reinforcement Learning Solution for Eco-Driving

In order to use the reinforcement learning for eco-driving it is necessary to be able to define precisely the state space S, the action space A and the reward function. Before detailing these different parameters in our application context, we provides in next section, an overview of the proposed solution for *Strada* eco-driving problematic.

3.1 General Proposition for Eco-Driving Simulator

The *Strada* driving experts have identified three distinct phases for heavy-duty vehicles driving: *acceleration* also called start-up phase, *rolling* phase and *braking* phase. These ones will be described in the next section. During each of these three phases, the engine characteristics and its performance are very different and the experts estimate that the types of actions a driver can do to improve its eco-driving are significantly different. Given thus point of view, we propose that couples $\{actions - rewards\}$ differ according to the driving phase.

Our global workflow for eco-driving recommendation is described Fig. 1. This proposition is divided into 5 stages. The first is the *data collection* which consists of measuring different driving and engine characteristics. The second is a *driving phase detection* in charge of identifying the driving phase of the sequence. The next one consists in *estimating* of the current *state* as an entry of the last stage where the *recommendation* is given. In the next sections, we describe how experts' knowledge can be used in our proposal.

3.2 Actions-Rewards Definition Using Expert Knowledge

As previously told, the *Strada* driving experts have identified three distinct situations in a time lapse sequence of driving truck: start-up phase, rolling phase and braking phase. The illustration can be seen in Fig. 2. For each driving sequence,

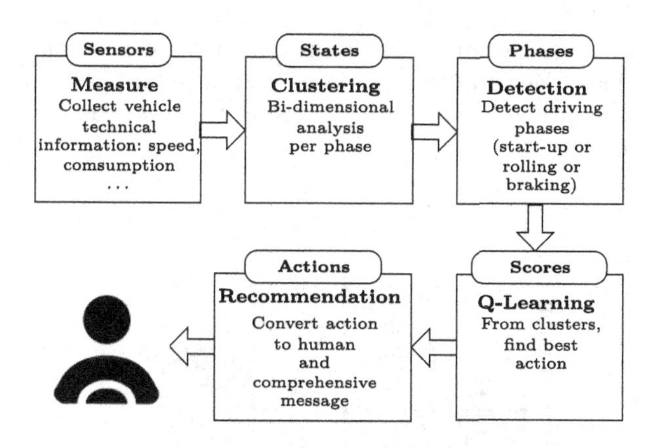

Fig. 1. Global workflow overview

the experts have identified significant technical parameters to monitor and evaluate the driver's adherence to the objective of minimizing fuel consumption.

The *Start-up phase* is characterized by a strong speed acceleration in a certain time slot: during this stage, the driver has to increase speed very quickly to reach the cruising speed of the vehicle. Factually, driver can act on the vehicle acceleration (more or less acceleration) and change gears. So, engine speed (in *rpm*) and acceleration (in %) can be used to characterize vehicle state during the *start-up phase*. This couple of measures defines the start-up phase feature space. Using RL, each driver's action can be perceived as a movement within the measurement space, necessitating the ability to assign a *state* and a localized *score* to each position in the feature space. *Strada*'s experts, based on their experience, propose to divide feature space into several sub-spaces and associate each sub-space to a local reward. Figure 3 describes their proposal. As we can see,

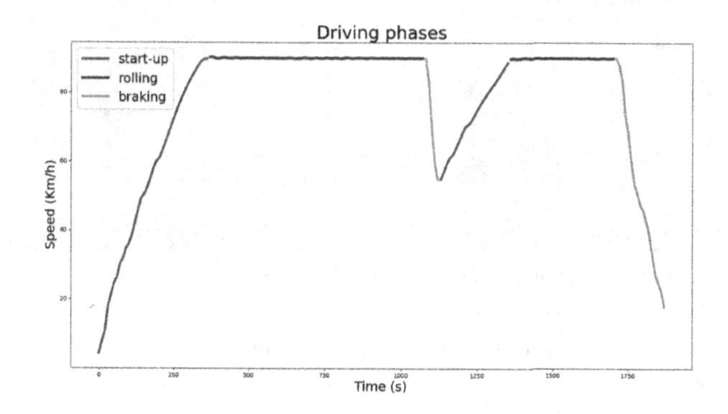

Fig. 2. The three driving phases

experts estimate a vehicle can be in 12 different states depending on speed and acceleration values. Each state is associated with a reward. The more higher is the reward, the more the engine's characteristics tends near in optimal position (green sub-spaces).

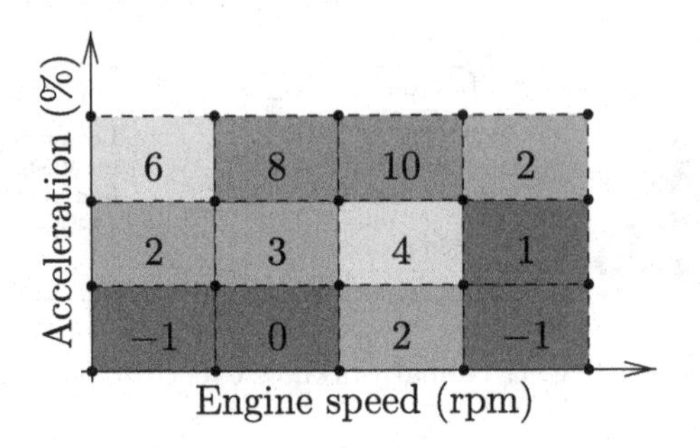

Fig. 3. Start-up phase feature space, states and *scoring* table (Color figure online)

On the same principle, scoring tables can be defined for the two others driving phases. Considering that *rolling phase* corresponds to a constant speed without acceleration neither deceleration and *braking phase* corresponds to a strong speed deceleration in a certain time slot, experts propose to retain the measures of variation speed and fuel consumption for the *rolling* phase and the measures of braking percentage and deceleration for the *braking* phase. Figure 4 presents the proposed scoring Tables for these phase and the Table 1 summarize measures and driver actions.

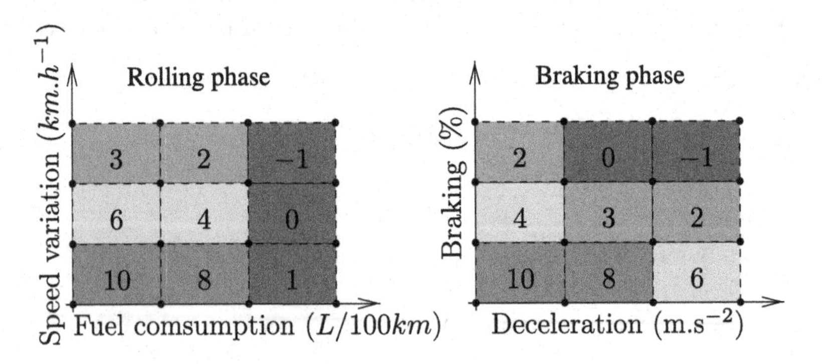

Fig. 4. Rolling and braking phase feature space, states and *scoring* tables (Color figure online)

Table 1. Driving expertise for RL application

Driving phase	Useful features/measures	Possible driver's actions
Start-up	engine speed, acceleration	increase acceleration, decrease acceleration, change gears
Rolling	speed variation, fuel consumption	use more inertia, use less inertia, use less regulator, use more regulator
Braking	braking percentage, deceleration	less deceleration, more deceleration, use more brake, more deceleration

3.3 Estimation of the *states*

Figures 3 and 4 define the possible *states* in our RL modeling. The choice by the experts of the number of states and of the values of rewards is based on their experience. It meets their main objective which is the determination of the optimal driving path in the least possible states.

However using *Q-Learning* method for RL implies to be able to identify at each iteration the current state (see Eq. 2) of the vehicle. The advantage of the *scoring* tables as defined by the experts is their simplicity. A simple couple of measures provide the state. However, at this point, experts are unable to precisely define the boundaries of each sub-space. Moreover, we have to keep in mind that the boundaries between each sub-space can not be universal. Indeed, many exterior parameters influence driving context (type of journey, cargo, vehicle...) and it seems obvious that the boundaries between states can not be static. They have to be estimated for each driving situation.

Some solutions exist to estimate states by combining the convolutional neural network with the Q-Learning algorithm as proposed in [13]. In this deep Q-Learning [13], a full-connected neural network is used to identify some states according to the input values during a learning state. But this process is very time consuming and requires a large data bases. This due to the fact that during the reinforcement learning algorithm it is necessary to learn the weights associated with the conventional neural network and the full connected part.

In our application, we propose an another strategy: a clustering process in order to identify the state of the vehicle and the sub-spaces boundaries. Our proposition is to use Self-Organizing Map (SOM) which is a well-known unsupervised learning tool. Some authors have already proposed to use SOM in RL [14,15]. In these articles, the SOM maps the input space in response to the real-valued state information, and each unit is then interpreted as a discrete state. This context closely aligns with our specific problem. So, we suggest utilizing this structure as it enables the application of a clustering process for state detection. Moreover, the huge advantage of SOM is its ability to preserves the topological properties of the input space. The number of neurons will correspond to the number of sub-spaces proposed by the experts.

The Fig. 5 illustrates SOM convergence. Applied to our process, the grid corresponds to the *score* space and blue cloud to real data measures.

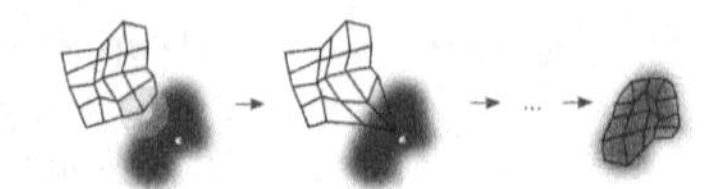

Fig. 5. *2D-SOM* (illustration from wikipedia) (Color figure online)

As SOM is a classical algorithm, so we just briefly resume its principles. Native works on SOM can be find in [16–18].

The SOM is a prominent unsupervised neural network, considered as a 2D mapping of the data group [18]. SOM net is made up of a number of nodes usually organized in a rectangular grid where an input vector x_i is link to a weigh vector. SOM algorithm is composed of 2 iterative stages: a competitive and a cooperative stage. The competitive stage aims to select the best neuron whereas the cooperative one adjust the weights. Topological aspects are driven during the update using neighborhood function also called kernel. So, in our proposition, the number of neurons equals the number of states. When SOM algorithm has converged, input vectors are labeled. The final position of the neurons are then used to estimate the limits of each subspace as defined by expert. Therefore, this learning stage facilitates the association of a state with each new pair of measurements for the Q-learning algorithm.

In that section, we have explained our proposed method based on a first clustering of the driving spaces followed by the RL stage. The next section details experimentation and results of the proposed method.

4 Results of the SOMQL Algorithm

In this part, we describe our experimentation, in particular the used data, the clustering stage and the driving recommendations.

4.1 Data Generation

Our recommendation system is based on Q-learning algorithm. The convergence of a such system suppose to own a large amount of data that covers the representation space: optimal policy is found by navigating through a stochastic maze in the feature space. Obviously, it is not realistic to obtain all the data using a truck inserted in a real traffic. First, some measures could be reached only if driver makes unsuitable actions which may cause damages to trunk. Moreover, some actions would have be dangerous in real traffic conditions. So, we decide to generate simulated data using *Euro Truck Simulator 2* [19] which is a software that proposed various type of truck and journeys (missions). Furthermore, by employing this simulator, we maintain a realistic serious game environment. Figure 6 illustrates a set of collected data for the three specific driving phase.

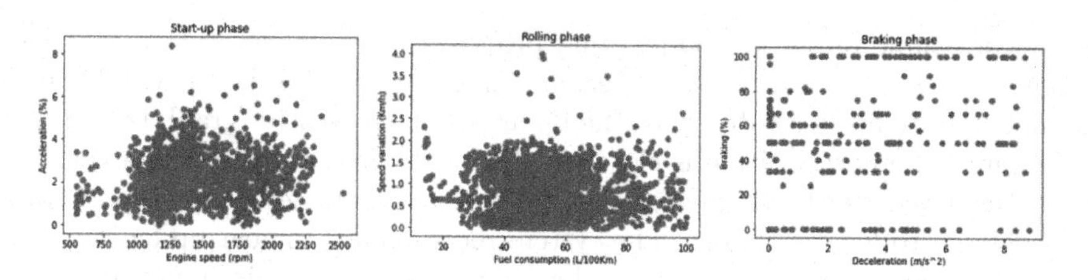

Fig. 6. Example of data corresponding to start-up, rolling and braking phases (from left to right)

4.2 States Estimation Using SOM

As described in Sect. 3.3, SOM is used in order to estimate the boundaries between the several possible states defined by experts. Indeed, the defined grids consist of multiple states/scores, but no numerical values have been specified. Based on the clusters estimated by SOM, it is possible to partition feature spaces into truck type-specific grids. Figures 7 illustrates the some estimated clustering and grid boundaries. Each subspace can then be associated with experts state and score in reference with the scoring tables presented in Figs. 3 and 4.

In the next section, we provide some result obtained by our recommendation system for optimal driving based on this clustering and the QL algorithm.

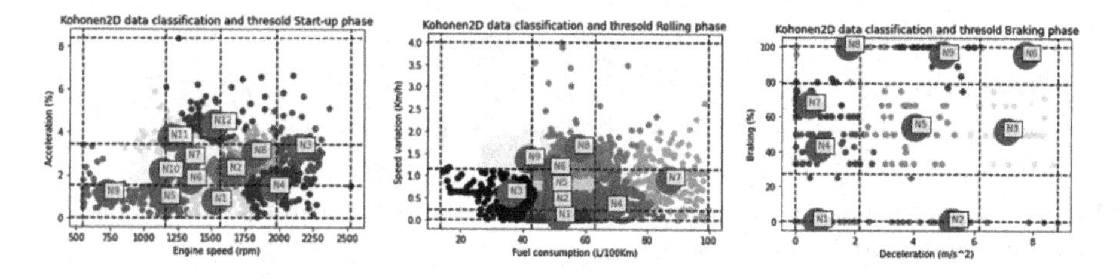

Fig. 7. Kohonen2D result associated with start-up, rolling and braking phases

4.3 Driving Recommendation Results and Discussions

In order to test the recommendation system, we simulate new data for the three specific driving phases and produce recommendations. Here, we describe and provide commentary on some specific cases. First, Table 2 provides some examples of measures obtained during *start-up* phase, their estimated score issued from clustering, and the final recommendation. A graph shows displacements in feature space if drivers apply driving recommendations.

For the *start-up* phase, in case 1 (first row of Table 2), the measures correspond to a low engine speed and a low acceleration. In that case, the thresholds learned with SOM indicate that the current state of the vehicle is associated to local score value of -1 and Q-learning proposes *more acceleration* recommendation. If the driver executes the recommended action, the point will move to the upper state as illustrated with the up arrow on the scatter graph (Fig. 8, left sub-figure); the new position is closer to the ideal position defined by experts.

In the case 2, with a high engine speed and a low acceleration, current state is associated to local -1 score. The system recommends to gear up (*next gear*). This action decreases the engine speed and induces a new position in the feature space by moving to the left. Finally, with case 3, we get a *previous gear* recommendation for a low engine speed and a high acceleration, going to the right of to current state.

Table 2. Recommendations for start-up phase (s: engine speed in rmp, a acceleration in m/s^2

Some data measures	Recommendations	Score
(1) $s = 600$, $a = 0.7$	*more acceleration*	-1
(2) $s = 2650$, $a = 1.1$	*next gear*	-1
(3) $s = 800$, $a = 8$	*previous gear*	6

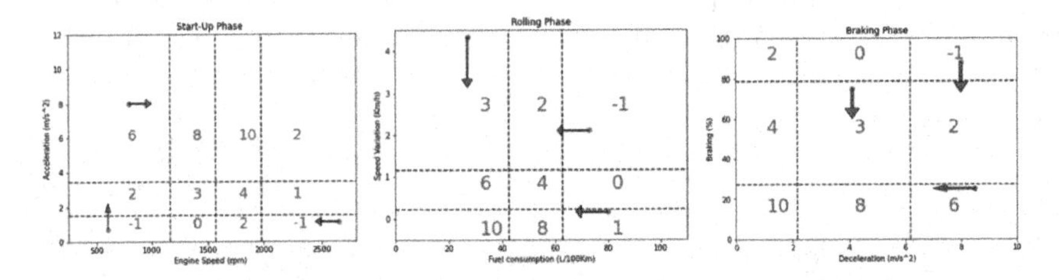

Fig. 8. Displacements in feature/state space - Scatter space (From left to right: for *start-up*, *rolling* and *braking* phases

The same analysis is performed for the rolling and braking phases. The Tables 3 and 4 sum up theses cases. We observe that the system is always trying to reach the high score zone which is the desired behavior for eco-driving optimization. The system seems to be efficient to provide accurate driving recommendations. The recommended actions successively move feature date toward "green" regions of the *score* tables as expected.

Theses results were presented to *Strada* experts to evaluate the global solution on a complete driving sequence. They estimate that the system offers consistent recommendations from an eco-driving behavior point of view. The recommendations offered by our system appear highly relevant. They perfectly meet the driving advises to give according to the simulated driving sequence. Right now, the solution can be viewed of as three parallel processes, one per driving phase. Individually, each process provides accurate recommendations for eco-driving. Obviously, driving a truck is a continuous process where *start-up*, *rolling*

Table 3. Recommendations for rolling phase (with fc: fuel consumption in $L/100$ Km, Δs: speed variation in Km/h)

Some data measures	Recommendations	Score
$fc = 73$, $\Delta s = 2.1$	*use more inertia*	-1
$fc = 80$, $\Delta s = 0.15$	*use more inertia*	$+1$
$fc = 27.1$, $\Delta s = 4.33$	*use more regulator*	$+3$

Table 4. Recommendations for braking phase (with d: deceleration in m/s^2, b: braking

Some data measures	Recommendations	Score
$d = 8$, $b = 88\%$	*use less brake*	-1
$d = 4.12$, $b = 75\%$	*use less brake*	$+3$
$d = 8.5$, $b = 25\%$	*less deceleration*	$+6$

and *braking* phases follow one another. Global solution includes automatic detection of the driving phase. Evaluation of global fuel consumption on a simulated travel has also to be done. Given these results, solution will be adapted to real truck. Others search should be engaged based on deep RL.

5 Conclusion

In the context of eco-driving, we presented a real-time recommendation system for a simulator-type driving training. The proposed system is based on the reinforcement learning. Since the states domain is continuous, we introduce an identification of discrete states by using a Self-organizing Map. All the parameters of our strategy are setting by qualified expert. Finally, we use our proposed system on simulated driving, and according to the expert, the recommendations are coherent and allows the driver to have an eco-driving behavior. Perspectives of this work are numerous. First, the modeling of states and actions has been deliberately simplified. However, improvement within each driving phase by refining the representation of states and actions could be made. Implementing, for example, more specific and precise actions for each of the phases is a very good start for improvement. Secondly, introducing Deep RL should simplify identifications of the current states for Q-learning. Despite the need for abundant data, acquiring real-world transportation data poses challenges for deep learning-based ML and RL systems. Thirdly, the proposal can be enriched by taking into account new information such as type of road during the travel (streets, roundabout, highway...), topology of missions or shipments.

Acknowledgments. This work was supported by the ANRT funding. Special thanks to the Strada team for their time, knowledge and expertise.

References

1. Sutton, R.S., Barto, A.G.: Introduction to Reinforcement Learning, 1st edn. MIT Press, Cambridge (1998)
2. Wiering, M., Van Otterlo, M.: Reinforcement Learning: State of the Art. Springer, Heidelberg (2012). https://doi.org/10.1007/978-3-642-27645-3
3. Garcıa, J., Fernández, F.: A comprehensive survey on safe reinforcement learning. J. Mach. Learn. Res. **16**(1), 1437–1480 (2015)
4. Kaelbling, L.P., Littman, M.L., Moore, A.W.: Reinforcement learning: a survey. J. Artif. Intell. Res. **4**, 237–285 (1996)
5. Adawadkar, A.M.K., Kulkarni, N.: Cyber-security and reinforcement learning – a brief survey. Eng. Appl. Artif. Intell. **114**, 105116 (2022). https://doi.org/10.1016/j.engappai.2022.105116
6. Coronato, A., Naeem, M., De Pietro, G., Paragliola, G.: Reinforcement learning for intelligent healthcare applications: a survey. Artif. Intell. Med. **109**, 101964 (2020). https://doi.org/10.1016/j.artmed.2020.101964
7. Frikha, M.S., Gammar, S.M., Lahmadi, A., Andrey, L.: Reinforcement and deep reinforcement learning for wireless Internet of Things: a survey. Comput. Commun. **178**, 98–113 (2021). https://doi.org/10.1016/j.comcom.2021.07.014
8. Elallid, B.B., Benamar, N., Hafid, A.S., Rachidi, T., Mrani, N.: A comprehensive survey on the application of deep and reinforcement learning approaches in autonomous driving. J. King Saud Univ. Comput. Inf. Sci. **34**(9), 7366–7390 (2022). https://doi.org/10.1016/j.jksuci.2022.03.013
9. Yeom, K.: Model predictive control and deep reinforcement learning based energy efficient eco-driving for battery electric vehicles. Energy Rep. **8**, 34–42 (2022). https://doi.org/10.1016/j.egyr.2022.10.040
10. Li, J., Wu, X., Xu, M., Liu, Y.: Deep reinforcement learning and reward shaping based eco-driving control for automated HEVs among signalized intersections. Energy **251**, 123924 (2022). https://doi.org/10.1016/j.energy.2022.123924
11. Du, G., Zou, Y., Zhang, X., Liu, T., Wu, J., He, D.: Deep reinforcement learning based energy management for a hybrid electric vehicle. Energy **201**, 117591 (2020). https://doi.org/10.1016/j.energy.2020.117591
12. Watkins, C.J., Dayan, P.: Q-learning. Mach. Learn. **8**(3–4), 279–292 (1992). https://doi.org/10.1007/BF00992698
13. Mnih, V., et al.: Playing atari with deep reinforcement learning, CoRR abs/1312.5602 (2013)
14. Osana, Y.: Reinforcement learning using Kohonen feature map probabilistic associative memory based on weights distribution. In: Advances in Reinforcement Learning. IntechOpen (2011)
15. Montazeri, H., Moradi, S., Safabakhsh, R.: Continuous state/action reinforcement learning: a growing self-organizing map approach. Neurocomputing **74**(7), 1069–1082 (2011)
16. Kohonen, T.: Self-organized formation of topologically correct feature maps. Biol. Cybern. **43**(1), 59–69 (1982). https://doi.org/10.1007/BF00337288
17. Kohonen, T.: Self-Organization and Associative Memory, vol. 8. Springer, Heidelberg (2012)
18. Kohonen, T.: Essentials of the self-organizing map. Neural Netw. **37**, 52–65 (2013). https://doi.org/10.1016/j.neunet.2012.09.018
19. Euro Truck Simulator 2. https://eurotrucksimulator2.com/

Underwater Mussel Segmentation Using Smoothed Shape Descriptors with Random Forest

David Arturo Soriano Valdez[1(✉)], Mihailo Azhar[2,4], Alfonso Gastelum Strozzi[3], Jen Hillman[2], Simon Thrush[2], and Patrice Delmas[4]

[1] Strong AI Lab, University of Auckland, Private Bag 92019, Auckland 1142, New Zealand
d.soriano.valdez@gmail.com
[2] George Mason Centre of the Natural Environment, University of Auckland, Private Bag 92019, Auckland 1142, New Zealand
mihailo.azhar@auckland.ac.nz
[3] Institute of Applied Scince and Technology, CDMX, Mexico City, Mexico
alfonso.gastelum@icat.unam.mx
[4] Intelligent Vision Systems Lab, University of Auckland, Private Bag 92019, Auckland 1142, New Zealand
p.delmas@auckland.ac.nz

Abstract. Segmentation of objects of interest is no longer a massive challenge with the adoption of machine learning and AI. However, feature selection and extraction are not trivial tasks in these approaches, and it is often necessary to introduce new methods for the creation of such features. Due to the lack of control over environmental conditions, for example turbidity and light scatter for underwater data, it is difficult to acquire color and texture features. However, it is still possible to obtain satisfactory shape features. This has led to the development of methods that can generate shape descriptors for use as features for data segmentation using machine learning methods such as random forest. In this work, we introduce a smoothed shape descriptor, which is the basis for a set of features used for the segmentation of underwater mussel structures with an accuracy of almost 90% based on manually labeled and measured mussel clusters by professional divers.

Keywords: Computer Vision · Machine Learning · SPH · Segmentation

1 Introduction

In mussel restoration ecology there is a need to rapidly quantify the spatial extent and three-dimensional (3D) geometry of restored mussel beds to assess the efficacy of restoration efforts. Traditional surveying methods are labour intensive utilising multiple divers and measurement tools such as weighted tapes and calipers. Limited both temporally and spatially due to diving logistics and safe diving practice, it is difficult to obtain high-resolution continuous data and large

J. Blanc-Talon et al. (Eds.): ACIVS 2023, LNCS 14124, pp. 311–321, 2023.
https://doi.org/10.1007/978-3-031-45382-3_26

coverage, on the growth of restored mussel beds over time. With the increased availability of open-source 3D reconstruction software, obtaining accurate geometrical data on underwater mussel structures only requires high-resolution photos or videos [3]. Utilising cameras allows marine ecologists to simplify underwater data collection procedures or introduce autonomous underwater vehicles (AUVs) into their monitoring programmes [8,16] however, the labour is shifted from in-situ measurements to isolating the benthic structures in the reconstructions using 3D software. The increasing volume of data gathered by modern sampling procedures will soon be untenable without reliable autonomous processing pipelines. Point cloud segmentation is one such approach suited for the autonomous processing of unstructured and irregular 3D data.

2 Related Work

Point cloud segmentation algorithms can be divided into two families of approaches, classical approaches, and deep learning approaches. Both families seek to group points based on a criterion of similarity using features extracted between the points in the point cloud. These features are typically based on geometrical and radiometric measures, e.g., position, normal and residuals when fitted to a shape.

Classical approaches focus on local features such as gradients, curvature, and distance as a discriminator. They include edge-based, region growing, model fitting and clustering approaches [9]. Methods like Histogram Oriented Gradients [31] (HOG) follow the approach of computing features to obtain a robust descriptor that can perform detection of objects [25], actions [29] and facial recognition [22] from images. This approach has been proven successful for face recognition on 3D meshes [10], but no underwater point cloud segmentation of mussels has been reported with this approach.

Deep learning approaches shift feature extraction and learning to large neural networks that learn the features that best discriminate and aggregate the points in the point cloud. Projection-based deep learning approaches leverage existing 2D image convolutional neural networks (CNN) approaches by projecting the 3D point clouds into multiple 2D images using virtual cameras. The images can encode shape, color and depth [6] but are prone to information loss due to the 3D-2D projection and occlusion. Other approaches discretise the point clouds into regular grids representations [30] on which 3D CNNs are applied, but due to the general unstructured nature of point clouds, some information can be lost, and artefacts may be introduced [1]. Point-based methods are well suited to irregularly shaped point clouds as features are learned per point [18] and have been extended to encompass neighbourhood information as well, thereby learning local and global features despite the added computational cost to search these neighbourhoods [12]. By and large, deep learning approaches are designed on small point clouds (1 m by 1 m with 4096 points) and focus on urban settings (scene and object). This is reflected in popular datasets like Shapenet [5] and benchmarking datasets such as Semantic KITTI [4] and Paris Lille 3D [4].

While the aforementioned 3D point cloud segmentation approaches have been tested for urban settings, there is little investigation and application in marine contexts. Work done by Runyan et al. has shown that using SparseConvNet, it is possible to obtain around 80% accuracy in segmenting coral reef point clouds generated from a custom-built frame containing two Nikon d7000 DSLR cameras [19]. Corals can have similar geometrical complexity to mussel reefs but tend to be much larger and continuous structures. Martin-Abadal et al. successfully used PointNet to segment underwater pipes [13], structures similar in shape and regularity to those commonly found in 3D point cloud datasets.

In the following sections, we introduce a straightforward method for segmenting underwater mussel structures. Our method uses a 3D point cloud generated by stereo reconstruction of the seafloor as input. This data was generated from footage of diving transects of different areas. The main characteristic of our approach is that it relies only on shape data from the 3D point clouds, as color and texture features are difficult to acquire in underwater real-world conditions. While the shape data may be obtained through different methods, we derived it from Smother Particle Hydrodynamics (SPH).

3 Methodology

The methodology section is divided into three sections: data preparation, shape descriptors computation and point cloud element classification.

3.1 Data Preparation

The data used in this work was acquired from locations in Mahurangi Harbour (36.4625°S 174.7225°E) and Kawau Bay (36.4192°S 174.7629°E). The underwater conditions of the first location present low visibility, while the second one is less muddy. However, both locations did present sediment, which requires to apply several image processing operations to improve data quality.

The data was acquired using a stereo-synchronized system consisting of dual GoPro Hero 3+ Black with a resolution of $1920 \times 1080@60FPS$. This system was stereo calibrated in order to be able to compute depth by calculating the parallax between the cameras. The data was later processed to obtain 3D structures using SfM from Meshroom[1].

The data acquisition process was challenging since the diver had to maintain a minimum distance of 0.5 m from the seafloor. The seafloor was composed of numerous sparse features, which resulted in the addition of artefacts to the 3D data.

Once the 3D data is generated, an expert needs to label the regions that contain mussels structures. This step is crucial and usually requires a considerable amount of time. This labeling is not precise due to the changes in turbidity near the locations of interest, which resulted in non-optimal color data and random texture quality. Even when this process is performed diligently, some boundary errors are unavoidable, even among experts. Figure 1 shows several examples of the 3D textured data along with their corresponding labeled data.

[1] https://github.com/alicevision/Meshroom.

Fig. 1. Different locations of 3D seafloor reconstruction with color and texture (top row). Corresponding labeled mussels structures colored in magenta (bottom row). (Color figure online)

3.2 Shape Descriptor

It is possible to use the labeled data to train a decision tree, which can segment the mussel structures. However, by looking at the color data, it is obvious that texture is not an ideal feature to perform any classification of structures. Results can be further degraded by turbidity variations around the locations of interest. In order to minimize the impact of the above on the segmentation process, a shape descriptor approach was selected for this work.

We propose the use of a shape descriptor based on the method of Smoothed Particle Hydrodynamics (SPH) [14]. A similar approach was used successfully for Archaeological studies [28]. This shape descriptor aims to provide the robustness of the local shape descriptors, such as splash shape descriptor [24], along with global operator descriptors, such as smoothing salient features [11].

The shape descriptor is defined as a field property A_s for each particle P (see Eq. 1), where the mass m and the density ρ are fixed in an arbitrary fashion. While the smoothing length h of the kernel W can be computed by considering the size of the features of the underwater structures, it was set manually across this work. In order to avoid changes in shape description and sensitivity, the smoothing kernel function W (see Eq. 2) was not modified from the one proposed in the original work, where R is the distance between particles and h the interaction radius of the kernel.

$$A_s(P_i) = \sum_{j=1}^{n} \frac{m_j}{\rho_j} A_s(P_j) W(r_i - r_j, h) \tag{1}$$

$$W(R,h) = \frac{315}{64\pi h^9} \begin{Bmatrix} (h^2 - R^2)^3 & 0 \leq R \leq h \\ 0 & R > h \end{Bmatrix} \quad (2)$$

Descriptors used in detection and classification are mainly based on region covariance and have proven to be quite successful [26,27]. Taking this into consideration, we added a region covariance computation to our shape descriptor. This was applied to the main feature used for our approach as well as the position covariance of 3D data. The main feature that we used for our mussel segmentation approach includes only the vertex normal.

Since the dimensions of the 3D point cloud of each transact exhibit a principal axis, along which the data is acquired, we decided to subdivide it into patches to provide regions with more regular dimensions. Following this approach we were able to use an octree with the same depth level for both the SPH and HOG method. The Depth of these octree was chosen by taking into consideration the 64 grid dimension proposed by Dalal [7]. Since we are aiming to obtain a cubic volume, we opted for an octree with a maximum depth level 6 for the HOG initial grid ($64 \times 64 \times 64$). This value is also used for the kernel size of the SPH shape descriptor, that way the gradient and SPH computation are equivalent region wise.

Using these conditions we tested both, the SPH shape descriptor, and HOG shape descriptor. To measure the performance we opted for several combinations of the SPH features and the complete set of HOG features, comparing four scores: precision, recall, F1 and accuracy.

3.3 Binary Classification

Object classification using decision trees along with texture and shape features has been proven successful on tasks that were challenging before machine learning adoption [2,20]. Our approach provides a set of features for the classification of objects. This novel shape descriptor, along with traditional ones, such as region covariance, deliver a set of multi-features, which are only based on 3D shape data. For performance comparison, we evaluated our approach outputs against the results obtained using solely HOG shape features.

We used the python implementation of random forest provided in the library Scikit-learn [15]. The random forest classifier was executed using one hundred estimators and the random state parameter was set at 42.

For our binary classification, we defined a single class, "Mussel", while everything else was considered as "NOT Mussel". The random forest classifier trained on data that was labeled as Mussel, as well as data labeled as "NOT Mussel", which typically includes seafloor, rocks, algae, and other underwater objects. While our focus is the detection of mussel structures, the approach can be modified to include more classes and detect other features.

4 Results

We tested our approach using a data set which contains nine 3D reconstructions from different underwater locations. Our data set contains over 12 million vertexes. Since the amount of transects is limited, we opted for a cross-validation approach, where we used each transect as training data, and then we tested our classifier on the remaining 8 transects.

The features selected are listed in Table 1. The same encoding is used in subsequent tables and figures in this paper. Those that are named "covariance" are computed using region covariance, while those with "SPH" in their name are computed using the smoothed shape descriptor.

Table 2 shows the average results for precision, recall, f1 and accuracy. Each row indicates which shape descriptor data was included for the random forest training. The most dominant feature is "NORMAL", while the computation of covariance has some of the best results. However, when combined with the SPH computation, all measures improve slightly. As such, the use of SPH can be considered as a fine-tuning tool. The most relevant results containing our SPH descriptor were obtained for experiment 9 (see 2), where using SPH features along with the SPH centroid features achieve a similar performance as the ones that used covariance or covariance and SPH.

Our SPH shape descriptor performed better than the shape descriptor based on HOG features in most cases, only experiments 1,2 and 5 got a lower score in precision, only on experiment 2 the score for precision, F1 and accuracy was worse than 20% compared to HOG. As expected position was never a reliable feature, but those which included features based on vertex normal, where the better performing, even better than HOG. In the case of experiments 3, 5, 6, 8 and 9, all of them performed better in all the scores we evaluated going from ranges of 2% to 9% in precision, 15% to 21% in Recall, 12% to 15% on F1, and 4% to 5% in Accuracy (see 2).

Table 1. Encoding of features for experiments.

Name	Code	Size
Position Covariance	A	9
Position SPH Covariance	B	9
Normal Covariance	C	9
Normal SPH Covariance	D	9
Normal SPH	E	3
Normal	F	3
Normal Centroid	G	3
Normal SPH Centroid	H	3
HOG Features	I	27

Table 2. Performance obtained using cross-validation in each experiment using different features.

Experiment ID	Precision	Recall	F1	Accuracy	Features								
1	58.38%	76.47%	65.66%	81.64%	✓								
2	32.41%	47.52%	37.72%	64.40%		✓							
3	77.57%	75.62%	75.72%	89.17%			✓						
4	72.03%	67.45%	68.53%	86.27%				✓					
5	76.17%	79.63%	77.07%	89.32%			✓		✓				
6	74.05%	81.21%	76.60%	88.72%	✓		✓		✓				
7	58.29%	77.89%	66.09%	81.60%	✓				✓				
8	74.95%	80.73%	76.92%	89.01%	✓		✓						
9	70.79%	78.66%	73.60%	87.24%					✓	✓	✓	✓	
10	68.72%	60.77%	61.50%	84.32%									✓
					A	B	C	D	E	F	G	H	I

Segmentation results are shown in Fig. 2. Figure 2 includes the 3D textured data (upper), mussel labeled data in magenta (middle), and segmentation result (bottom) in this section, true positive is colored green, true negative is blue, false positive is red and false negative is black. Figure 2a contains the best overall result, which achieved an accuracy of 95.9% with precision, recall and F1 scores of 0.866, 0.861, and0.863, respectively. Figure 2b shows the worst results in accuracy (86.3%). Finally, Fig. 2c presents the worst performance for recall and F1 (0.578, 0.694).

5 Discussion

As with 2D segmentation tasks, 3D segmentation benefits from the use of robust shape descriptors invariant to the orientation and scale differences present in 3D point clouds. Our method requires normalization of the orientation and scale differences between the point clouds to improve the robustness of the SPH descriptor, while HOG features readily extracted are scale and orientation invariant. Despite the difference in invariance, SPH shape descriptors outperformed HOG in accuracy and recall during segmentation. Our SPH shape descriptor requires further extension for scale and orientation invariance or should be used alongside those like HOG in 3D point cloud segmentation tasks.

For highly complex structures, segmentation pipelines should seek to adequately represent or leverage this geometrical complexity. Runyan et al. [19] used SparseConvNet to process point clouds similarly derived from SfM photogrammetry on complex coral structures. SparseConvNet operates on voxels requiring the creation of implicit surfaces from the point cloud using a fixed voxel resolution. The chosen voxel resolution however determines the minimum geometric complexity maintained (at the cost of computation). Comparatively,

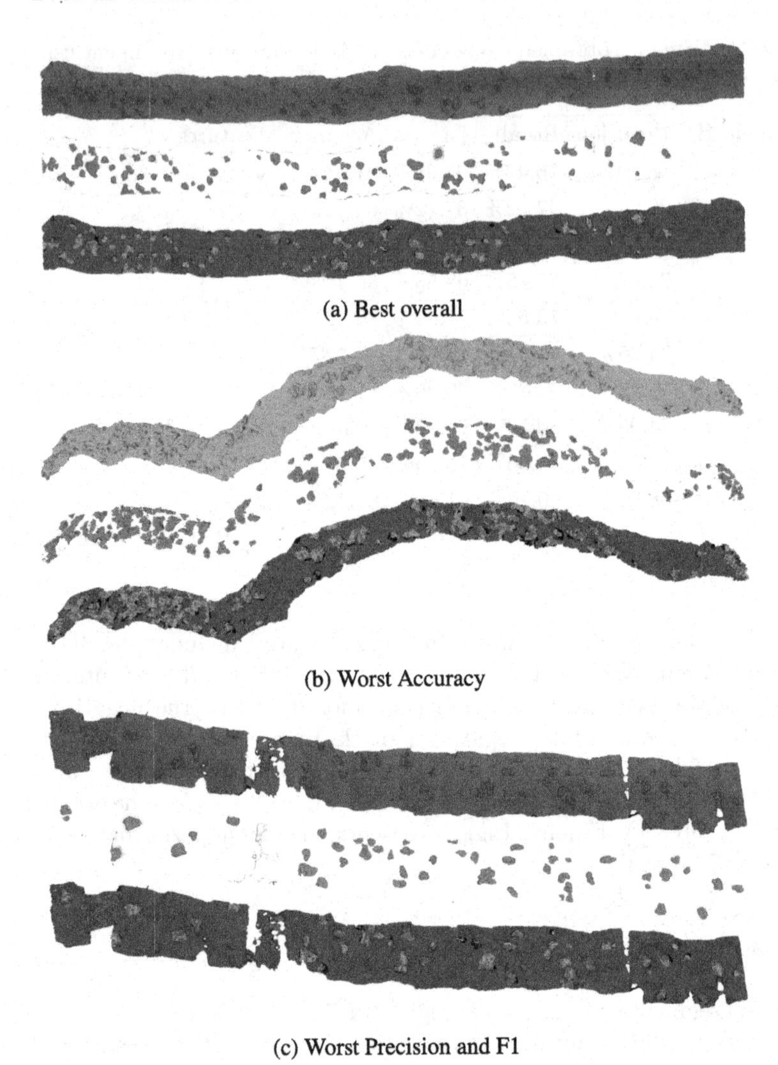

(a) Best overall

(b) Worst Accuracy

(c) Worst Precision and F1

Fig. 2. Successful segmentation of mussels structures using the features of experiment 6 (see Table 2.)

our approach works directly on the unstructured point cloud. SPH computes per vertex field properties that can be turned into individual features per vertex, where closer and more dense sets of elements significantly influence the features assigned to each vertex utilizing the geometric complexity of the full point cloud. Pointnet++ [18] similarly operates directly on the unstructured point cloud. While the features learned are orientation and scale-invariant, the network's ability to learn representative features for the naturally fractal mussel structures [23] should be investigated.

Mussels typically inhabit turbid waters where data acquisition constraints, such as visibility, are poor. While turbidity will also limit the quality of the 3D point cloud reconstruction, by utilizing only 3D shape as a feature in our approach, we mitigate the inconsistencies in the reconstructed texture that arise from backscatter and extend the applicability of our method to different colored water bodies without the need for further training. Furthermore, our method can leverage acoustic imaging approaches such as 3D sonar [21] and side-scan sonar [17] that circumvent optical issues with turbid water.

6 Future Work and Improvements

Our method proved effective for underwater mussel segmentation using only shape descriptors. While our shape descriptor has achieved similar performance to methods based on region covariance, it is more memory efficient and less computationally intensive since fewer floating-point operations are required.

The robustness of the shape descriptor is low since changes in scale, 3D artefacts and inconsistencies in the mesh data are the primary sources of segmentation errors. This can partially be tackled by filtering these artefacts as well as edge cleaning for better consistency. This step can be avoided if scaled and registered 3D point cloud data is provided.

Since our segmentation approach does not rely on texture data, it is ideal for underwater conditions with high turbidity and light variance. The segmentation results can be improved by a post-processing method for filtering smaller structures and filling holes in segmentation, for example, binary opening and closing; the only limitation is that it must be performed in 3D.

The addition of these methods to a real-time pipeline for reconstruction and segmentation is possible since most of the execution of the software can be accelerated by using Graphic Processing Units.

References

1. Akhtar, A., Gao, W., Li, L., Li, Z., Jia, W., Liu, S.: Video-based point cloud compression artifact removal. IEEE Trans. Multimed. **24**, 2866–2876 (2021)
2. Ali, J., Khan, R., Ahmad, N., Maqsood, I.: Random forests and decision trees. Int. J. Comput. Sci. Issues (IJCSI) **9**(5), 272 (2012)
3. Azhar, M., Hillman, J.R., Gee, T., Thrush, S., Delmas, P.: A low-cost stereo pipeline for semi-automated spatial mapping of mussel structures within mussel beds. Remote Sens. Environ. (Manuscript in review) (2023)
4. Behley, J., et al.: Semantickitti: a dataset for semantic scene understanding of lidar sequences. In: Proceedings of the IEEE/CVF International Conference on Computer Vision, pp. 9297–9307 (2019)
5. Chang, A.X., et al.: Shapenet: an information-rich 3d model repository. arXiv preprint arXiv:1512.03012 (2015)
6. Chang, Y.L., Fang, C.Y., Ding, L.F., Chen, S.Y., Chen, L.G.: Depth map generation for 2D-to-3D conversion by short-term motion assisted color segmentation. In: 2007 IEEE International Conference on Multimedia and Expo, pp. 1958–1961. IEEE (2007)

7. Dalal, N., Triggs, B.: Histograms of oriented gradients for human detection. In: 2005 IEEE Computer Society Conference on Computer Vision and Pattern Recognition (CVPR 2005), vol. 1, pp. 886–893. IEEE (2005)
8. Ferrari, R., et al.: 3D photogrammetry quantifies growth and external erosion of individual coral colonies and skeletons. Sci. Rep. **7**(1), 1–9 (2017)
9. Grilli, E., Poux, F., Remondino, F.: Unsupervised object-based clustering in support of supervised point-based 3D point cloud classification. Int. Arch. Photogrammetry Remote Sens. Spat. Inf. Sci. **43**, 471–478 (2021)
10. Li, H., Huang, D., Lemaire, P., Morvan, J.M., Chen, L.: Expression robust 3D face recognition via mesh-based histograms of multiple order surface differential quantities. In: 2011 18th IEEE International Conference on Image Processing, pp. 3053–3056 (2011). https://doi.org/10.1109/ICIP.2011.6116308
11. Li, X., Guskov, I.: Multiscale features for approximate alignment of point-based surfaces. In: Symposium on Geometry Processing, vol. 255, pp. 217–226 (2005)
12. Lu, B., Wang, Q., Li, A.: Massive point cloud space management method based on octree-like encoding. Arab. J. Sci. Eng. **44**, 9397–9411 (2019)
13. Martin-Abadal, M., PiÃČÂśar-Molina, M., Martorell-Torres, A., Oliver-Codina, G., Gonzalez-Cid, Y.: Underwater pipe and valve 3D recognition using deep learning segmentation. J. Mar. Sci. Eng. **9**(1), 5 (2020)
14. Monaghan, J.J.: Smoothed particle hydrodynamics. ARAA **30**, 543–574 (1992). https://doi.org/10.1146/annurev.aa.30.090192.002551
15. Pedregosa, F., et al.: Scikit-learn: machine learning in Python. J. Mach. Learn. Res. **12**, 2825–2830 (2011)
16. Pizarro, O., Eustice, R.M., Singh, H.: Large area 3-D reconstructions from underwater optical surveys. IEEE J. Oceanic Eng. **34**(2), 150–169 (2009)
17. Pulido, A., Qin, R., Diaz, A., Ortega, A., Ifju, P., Shin, J.J.: Time and cost-efficient bathymetric mapping system using sparse point cloud generation and automatic object detection. In: OCEANS 2022, Hampton Roads, pp. 1–8 (2022). https://doi.org/10.1109/OCEANS47191.2022.9977073
18. Qi, C.R., Yi, L., Su, H., Guibas, L.J.: Pointnet++: Deep hierarchical feature learning on point sets in a metric space. arXiv preprint arXiv:1706.02413 (2017)
19. Runyan, H., et al.: Automated 2D, 2.5 D, and 3D segmentation of coral reef pointclouds and orthoprojections. Front. Rob. AI **9** (2022)
20. Schroff, F., Criminisi, A., Zisserman, A.: Object class segmentation using random forests. In: BMVC, pp. 1–10 (2008)
21. SÃühnlein, G., Rush, S., Thompson, L.: Using manned submersibles to create 3d sonar scans of shipwrecks. In: OCEANS 2011 MTS/IEEE KONA, pp. 1–10 (2011). https://doi.org/10.23919/OCEANS.2011.6107130
22. Shu, C., Ding, X., Fang, C.: Histogram of the oriented gradient for face recognition. Tsinghua Sci. Technol. **16**(2), 216–224 (2011). https://doi.org/10.1016/S1007-0214(11)70032-3
23. Snover, M.L., Commito, J.A.: The fractal geometry of mytilus edulis l. spatial distribution in a soft-bottom system. J. Exp. Mar. Biol. Ecol. **223**(1), 53–64 (1998)
24. Stein, F., Medioni, G.: Structural indexing: efficient 2d object recognition. IEEE Trans. Pattern Anal. Mach. Intell. **14**(12), 1198–1204 (1992)
25. Surasak, T., Takahiro, I., Cheng, C.H., Wang, C.E., Sheng, P.Y.: Histogram of oriented gradients for human detection in video. In: 2018 5th International Conference on Business and Industrial Research (ICBIR), pp. 172–176 (2018). https://doi.org/10.1109/ICBIR.2018.8391187

26. Tabia, H., Laga, H., Picard, D., Gosselin, P.H.: Covariance descriptors for 3D shape matching and retrieval. In: 2014 IEEE Conference on Computer Vision and Pattern Recognition, pp. 4185–4192 (2014). https://doi.org/10.1109/CVPR.2014.533
27. Tuzel, O., Porikli, F., Meer, P.: Region covariance: a fast descriptor for detection and classification. In: Leonardis, A., Bischof, H., Pinz, A. (eds.) ECCV 2006. LNCS, vol. 3952, pp. 589–600. Springer, Heidelberg (2006). https://doi.org/10.1007/11744047_45
28. Valdez, D.A.S., et al.: CUDA implementation of a point cloud shape descriptor method for archaeological studies. In: Blanc-Talon, J., Delmas, P., Philips, W., Popescu, D., Scheunders, P. (eds.) ACIVS 2020. LNCS, vol. 12002, pp. 457–466. Springer, Cham (2020). https://doi.org/10.1007/978-3-030-40605-9_39
29. Wang, G., Tie, Y., Qi, L.: Action recognition using multi-scale histograms of oriented gradients based depth motion trail Images. In: Falco, C.M., Jiang, X. (eds.) Ninth International Conference on Digital Image Processing (ICDIP 2017), vol. 10420, p. 104200I. SPIE (2017). https://doi.org/10.1117/12.2281553
30. Zhang, Y., et al.: Polarnet: an improved grid representation for online lidar point clouds semantic segmentation. In: Proceedings of the IEEE/CVF Conference on Computer Vision and Pattern Recognition, pp. 9601–9610 (2020)
31. Zhou, W., Gao, S., Zhang, L., Lou, X.: Histogram of oriented gradients feature extraction from raw Bayer pattern images. IEEE Trans. Circ. Syst. II Express Briefs **67**(5), 946–950 (2020). https://doi.org/10.1109/TCSII.2020.2980557

A 2D Cortical Flat Map Space for Computationally Efficient Mammalian Brain Simulation

Alexander Woodward[1(✉)], Rui Gong[1], Ken Nakae[2,3], and Patrice Delmas[4]

[1] Connectome Analysis Unit, RIKEN Center for Brain Science, Saitama, Japan
alexander.woodward@riken.jp
[2] Exploratory Research Center on Life and Living Systems, National Institutes of Natural Sciences, Aichi, Japan
[3] Graduate School of Informatics, Kyoto University, Kyoto, Japan
[4] Intelligent Vision Systems Lab, NAOInstitute, Auckland, New Zealand

Abstract. We present a method for computationally efficient cortical brain simulation by constructing a 2D cortical flat map space on a regular grid. Neuroscience data can be mapped into this space to provide experimental information and constraints for the simulation. Neuron locations can be determined probabilistically by treating neuron densities as empirical probability distributions that can be sampled from. Therefore, this approach can be used for specifying parameters for small-scale to large-scale brain simulations (that could simulate the true number of neurons in the brain). The spatial warping of the cortical surface, when going between the flattened 2D space back into 3D, is accounted for by an estimated scale factor. This can be used to scale properties such as diffusion rates of neural activity across the flat map. We demonstrate the approach using neuroimaging data of the common marmoset, a New World primate.

Keywords: Simulation · Connectomics · Brain · Image Processing

1 Introduction

Brain simulation is an active and promising area of research in computational neuroscience. The aim is to model the structure and function of the brain using biologically realistic or simplified neurons and synapses. Brain simulation has several potential applications, such as understanding the neural basis of cognition and perception, testing hypotheses about brain disorders and diseases, developing novel brain-inspired algorithms and technologies, and advancing the field of artificial intelligence. However, simulations face many challenges that can limit their feasibility and validity.

The first challenge is the scale of cortical brain simulation. The human cortex contains a huge number of neurons and even more synapses. Simulating such a large and complex system requires enormous computational resources and power.

J. Blanc-Talon et al. (Eds.): ACIVS 2023, LNCS 14124, pp. 322–331, 2023.
https://doi.org/10.1007/978-3-031-45382-3_27

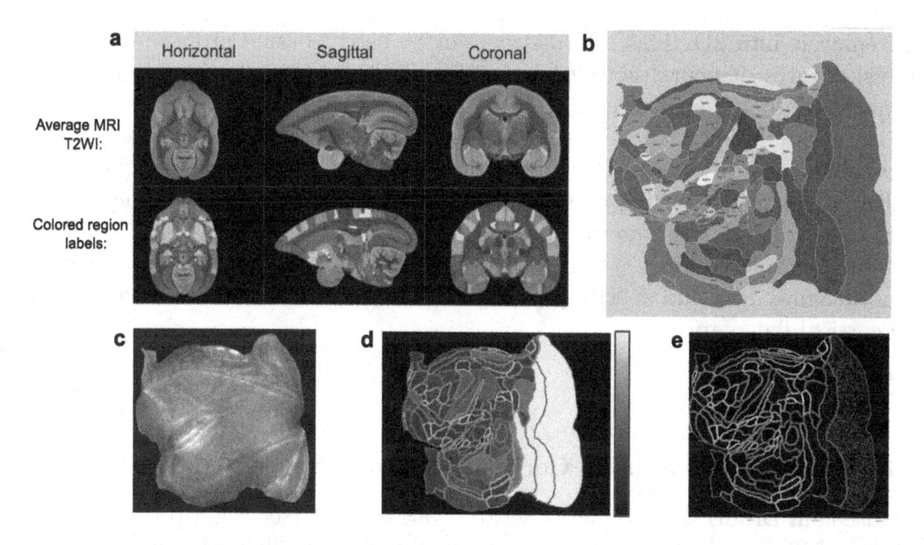

Fig. 1. (**a**) The Brain/MINDS marmoset brain atlas shown in horizontal, sagittal and coronal slices. (**b**) The cortical flat map generated from the atlas. (**c**) The map of the scaling at each point in space when going from 2D back to 3D. (**d**) The map of (normalized) neuron densities. (**e**) An example of neuron positions generated probabilistically based on the densities.

Another challenge is dealing with the complexity of the brain. The brain is not a homogeneous or static system, but rather a heterogeneous and dynamic one that exhibits multiple levels of organization and adaptation. The brain consists of different types of neurons and glia cells, which have diverse morphologies, electrophysiological properties, molecular profiles, synaptic connections, and plasticity mechanisms. Moreover, the brain changes over time due to development, learning, aging, injury, and disease. To capture all these aspects of brain complexity would require a vast amount of data and parameters that are not easily accessible or measurable, especially in humans. Therefore, most brain simulations rely on simplifications or assumptions that may compromise their biological realism or relevance.

One of the main goals and contributions of this paper is to propose a method for computationally efficient mammalian cortical brain simulation by constructing a 2D cortical flat map space on a regular grid. The rationale behind this is that in the mammalian brain the cortex forms a sheet with (up to) six layers. In principle it can therefore be flattened out onto a (2D) surface. A 2D space has a number of attractive properties, and treating computation on a regular grid maps well with computer hardware.

This method allows us to map neuroscience data into the flat map space and use it as experimental constraints for the simulation. Moreover, it enables us to determine neuron locations probabilistically by sampling from neuron density distributions, which can be used for parameterizing brain simulations at different scales. The method accounts for the spatial warping of the cortical surface when flattening it into 2D. This information can be used to scale properties such as diffusion rates of neural activity across the flat map. Finally, in this work we combine the flat map model with region-to-region connectivity information to create a simulation that accounts for the spread of local activity across the cortical sheet and long-range connectivity. We demonstrate the feasibility and utility of our method using a brain atlas and diffusion weighted imaging (DWI) data of the common marmoset brain.

Our method provides a computationally efficient and flexible way to map neuroimaging data and specify parameters for cortical brain simulation using a 2D cortical flat map space.

2 Methods

2.1 Construction of a 2D Cortical Flat Map Space

We used our Brain/MINDS marmoset brain atlas [2,10] (see Fig. 1a) to construct a cortical flat map using the Caret software [9]. The result is shown in Fig. 1b. This is in the form of a triangular mesh, so it must be processed in order to use it for computational purposes. We do this by mapping the mesh into a 2D image space. The domain of an image is the set of coordinates (x, y) that define the location of each pixel in the image. The domain can be represented by a regular grid of points that span the width and height of the image. We define an $M \times N$ arithmetic grid: $\mathbf{R}_{M,N} = (x, y) : 1 \leq x \leq M \wedge 1 \leq y \leq N$. An image model treats an image as a function: $I : \mathbf{R} \rightarrow \mathbf{V}$, where $\mathbf{V}$ is a set of signal values. For example, grey level intensities can be described as $I(\mathbf{p}) \equiv I(x, y) \in \mathbf{V} = \{0, 1, \cdots, G_{max}\}$. In our case, we can take any neuroscience data as our signal and map it into this image space. We map the flat map so that it fits within an arbitrarily chosen $M \times N$ grid (image). Each pixel of this 2D image is then assigned to a particular brain region based on the mapped information. We can then take advantage of the regular grid structure to accelerate computation through parallelization. This can speed up the processing time significantly when compared to applying the operations to the whole image sequentially by a single processor or core.

One of the challenges of constructing a 2D cortical flat map space is to account for the spatial warping of the cortical surface that occurs when it is flattened from 3D. This warping may affect the properties of neural activity that depend on the distance or area of the cortical regions, such as diffusion rates. To address this issue, we estimated a scale factor that quantifies the relative change in surface area of triangles in the brain surface mesh when going from the 2D cortical flat map space, back to 3D. The factor can be used to modify the diffusion rates of neural activity across the flat map, so that they are consistent with the 3D geometry. The result is shown in Fig. 1c.

As an example, we also mapped information about per-region neuron densities into the flat map space (obtained from [3]). These densities were then normalized to give an empirical probability distribution of relative neural densities (see Fig. 1d). If we sample this repeatedly we can generate N neuron locations that respect the true relative neuron densities of the cortex (see Fig. 1e). In this manner we can specify neuron counts up to the real number in the brain. Further information, such as region connectivity profiles derived from tracer studies or DWI, neuron types, or cortical layer information, could be mapped into the flat map space to enhance the realism of the simulation.

2.2 Generation of a Connection Matrix Using DWI Data and a Brain Atlas

For this work we used connectivity information derived from common marmoset brain in vivo DWI data, from a cohort of 126 individuals. This data was obtained from the NA216 dataset [5] at the Brain/MINDS Data Portal [2]. Diffusion weighted imaging (DWI) is a form of magnetic resonance imaging (MRI) that measures the random motion of water molecules within brain tissue. By applying different diffusion gradients along different directions, DWI can generate contrast based on the diffusion anisotropy of the tissue, which reflects its microstructural organization.

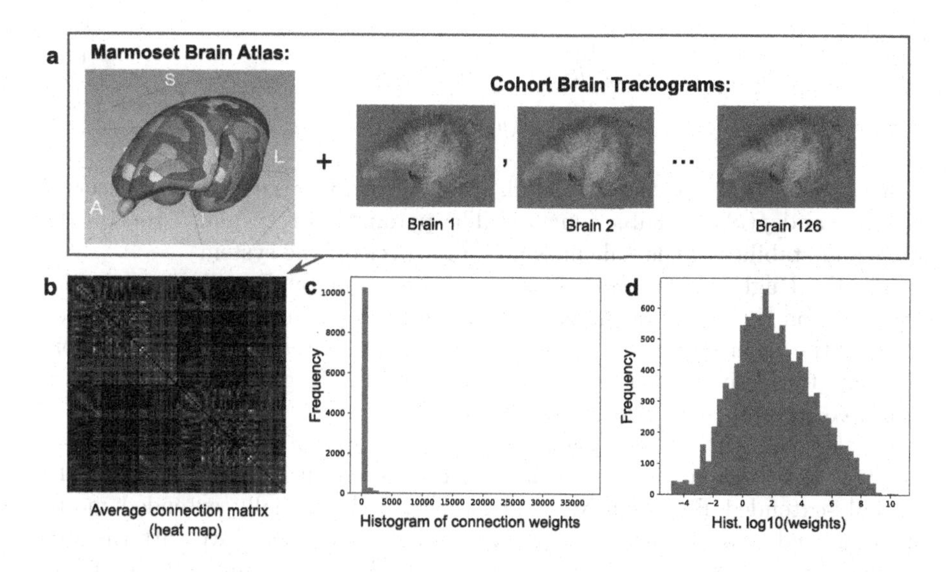

Fig. 2. (**a**) The brain atlas mapped with 126 tractograms used as input, to generate (**b**) the canonical (average) connection matrix. (**c**) Distribution of connection weights showing a long-tail. (**c**) Distribution of the logarithm to the base 10 of the connection weights.

The DWI data was processed to obtain the fiber orientation density (FOD) function at each voxel location. This was then used to generate a tractogram - a set of streamlines that estimates the axonal paths running through the brain. A simplified version of the Brain/MINDS marmoset brain atlas [2,10], with 52 regions per hemisphere, suitable for MRI studies, was registered with the DWI data using a nonlinear transformation (see Fig. 2a for a conceptual diagram). The number of streamlines connecting two regions were used as a measure of strength and stacked in a connection matrix. The matrices of the 126 individuals were averaged to generate a final average connection matrix (Fig. 2b). This symmetric 104×104 matrix describes the long range structural connectivity of the common marmoset brain.

The detailed methodology for data acquisition, preprocessing and connection matrix generation is described in [2,5,10,11]

We plotted the distribution of weights for the average connection matrix, shown in Figs. 2c–d. The distribution shows a long tail, a feature of real brains that is important for realistic brain simulation.

2.3 Simulation Equations and Parameters

As a starting point we examined the FitzHugh-Nagumo equations, a simplified model of the Hodgkin-Huxley equations that describe the electrical activity of neurons. They consist of two coupled ordinary differential equations:

$$\frac{du}{dt} = f(u) - v + I, \tag{1a}$$

$$\frac{dv}{dt} = \epsilon(u - \gamma v - \beta) \tag{1b}$$

where v is the membrane potential, w is a recovery variable, I is an external stimulus, and $f(v)$ is a cubic function. The parameters ϵ, γ, and β control the shape and stability of the nullclines and limit cycle of the system.

The FitzHugh-Nagumo equations exhibit bistability and oscillatory behavior depending on the values of the parameters and the stimulus. They can be used to model the generation and propagation of action potentials in neurons or, more generally, the activity of neural assemblies. The FitzHugh-Nagumo model has been used in brain simulation in works such as [8].

To describe our neural model we took the spatially extended forms of Eqs. (1a, 1b) to create a two-dimensional reaction-diffusion system (called the Spatially-extended FitzHugh-Nagumo model). This spatially extended model has been used in work such as [4] to study transient cortical wave patterns during migraines. For the 2D case, $u = u(x, y)$ and $v = v(x, y)$ and a diffusion term, $D_u \nabla^2 u$, is added (as described in [12])[1]. We additionally added a noise term, ω, to give:

[1] Some versions of the model add a diffusion term to w to give:

$$\frac{\partial w}{\partial t} = D_w \nabla^2 w + (v - \gamma w - \beta)$$

$$\frac{\partial u}{\partial t} = D_u \nabla^2 u + f(u) - v + I + \omega, \tag{2a}$$

$$\frac{\partial v}{\partial t} = \epsilon(u - \gamma v - \beta) \tag{2b}$$

Here, D_u is a positive constant that represents the diffusion coefficient of u. The diffusion term accounts for the spatial coupling of neighboring cells, and can lead to the formation of traveling waves or spiral waves in the system. We modify D_u on a per-pixel basis to include the scale factor describing the warping of space when going from 2D back to 3D. We used $f(v) = v(v - \alpha)(1 - v)$, as described in Eq. (5) of [5], with $\alpha = -0.3$, $\beta = 0$, $\gamma = 1e^{-8}$ and $\epsilon = 0.1$. Here I represents input into the neural assembly. In our case this comes from input based on the region-to-region long range (extrinsic) connectivity derived from the brain atlas and cohort DWI data. In terms of a reaction diffusion system this can be seen as a global feedback term. A scale factor G is applied to the input term (i.e. $I = G \times I_{extrinsic}$) in order to explore the affects of different levels of global coupling.

The Spatially-extended FitzHugh-Nagumo model is a special case of the 2D Generic Oscillator model used in The Virtual Brain (TVB) platform ([6,7]).

Simulation Loop. At each iteration of the simulation the steps are as follows: (1) Calculate the Laplacian of u to estimate the diffusion term, (2) update the values of u and v using Eqs. (2a, 2b) through integration, (3) based on the region-to-region connectivty matrix, sum the activity of u to calculate I for each region - this will be used for the next iteration.

3 Results and Discussion

For our experiments we used only the regions of the brain connection matrix contained in the cortex (as subcortical regions also existed). Using our approach it is very easy to carry out computations in the flatmap space, then map back onto the 3D surface using texture mapping (see Fig. 3a). For proof-of-concept we also restricted the simulation to a single hemisphere (one cortical flat map). The flat map space was sampled at 200 × 200 pixels to give a reasonable computation time for experimentation on a standard laptop computer (see Fig. 3b). A mask was also used to specify the region of interest and avoid unnecessary computations outside of the cortical region (Fig. 3c).

We used the Python Numba library to accelerate the computation of our method [1]. Numba is an open source, NumPy-aware optimizing compiler for Python that uses the LLVM compiler project to generate machine code from Python syntax. We used Numba's just-in-time (JIT) compilation feature and *parallel = True* option to automatically parallelize our loops over the rectangular grid. This allowed us to take advantage of the multiple CPU cores available in modern computers. Experiments were carried out on a MacBook Pro (13-inch, 2017) computer with a 3.5 GHz Intel Core i7 CPU and 16 GB of RAM. By

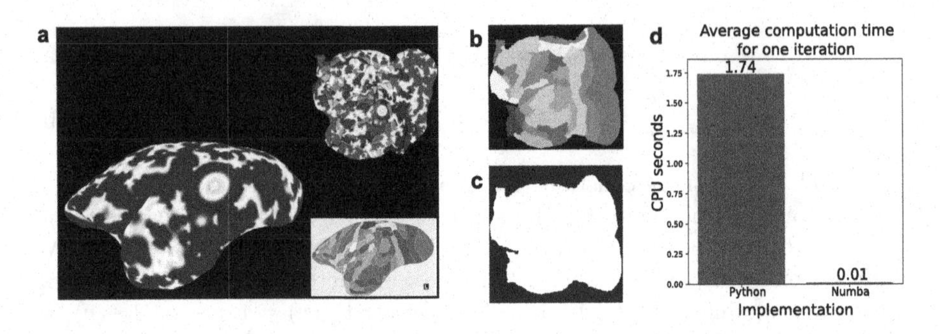

Fig. 3. (**a**) An example of mapping the 2D activity back onto the 3D brain surface. (**d**) The flat map space (sampled at 200×200 pixels) based on the 52 region per hemisphere atlas. (**c**) The mask used for specifying the region of interest for computations. (**d**) A comparison of the average computation time for one iteration of the simulation, for pure Python vs. Numba.

using Numba, we achieved a significant speedup - about 174 times - in our computations compared to pure Python, see Fig. 3d.

Figure 4 shows simulation results across a few iterations for different values of the global coupling parameter G. Figure 4a shows $G = 0.1$ where there is strong global coupling and the dynamics emphasize the region-to-region connectivity. This global feedback results in different brain regions showing activity and switching over time. Figure 4b shows reduced global coupling ($G = 0.005$) that generates perhaps the most interesting regime, with a mixture of local and global region-to-region (feedback) dynamics. Here, intricate wave formations are generated along with the larger scale whole region activity. Finally, in Fig. 4c, when $G = 0$ there is no global coupling only the spontaneous activity from the intrinsic local noise drives the dynamics. Overall, the system produces interesting dynamics but it remains future work to explore the possible parameter space, alternative neural equations, and comparisons with the dynamics in experimental data.

Our method has several advantages for cortical brain simulation. First, it is computationally efficient and scalable, as it can handle large-scale simulations with millions of neurons on a regular grid. Second, it is flexible and adaptable, as it can incorporate different types of neuroimaging data and account for spatial warping of the cortical surface. Third, using this approach we can simulate both the local and global dynamics of neural activity.

However, our method has a number of limitations and challenges. One limitation is that it does not model the subcortical structures or the white matter tracts, which are also important for brain function and communication. Another limitation is that we are projecting activity from one region into the entirety of another region, which is unrealistic. The incorporation of tracer study or DWI derived connectivity information at the sub-region level (e.g. voxel-to-voxel), could provide finer detailed information on projection patterns and make the

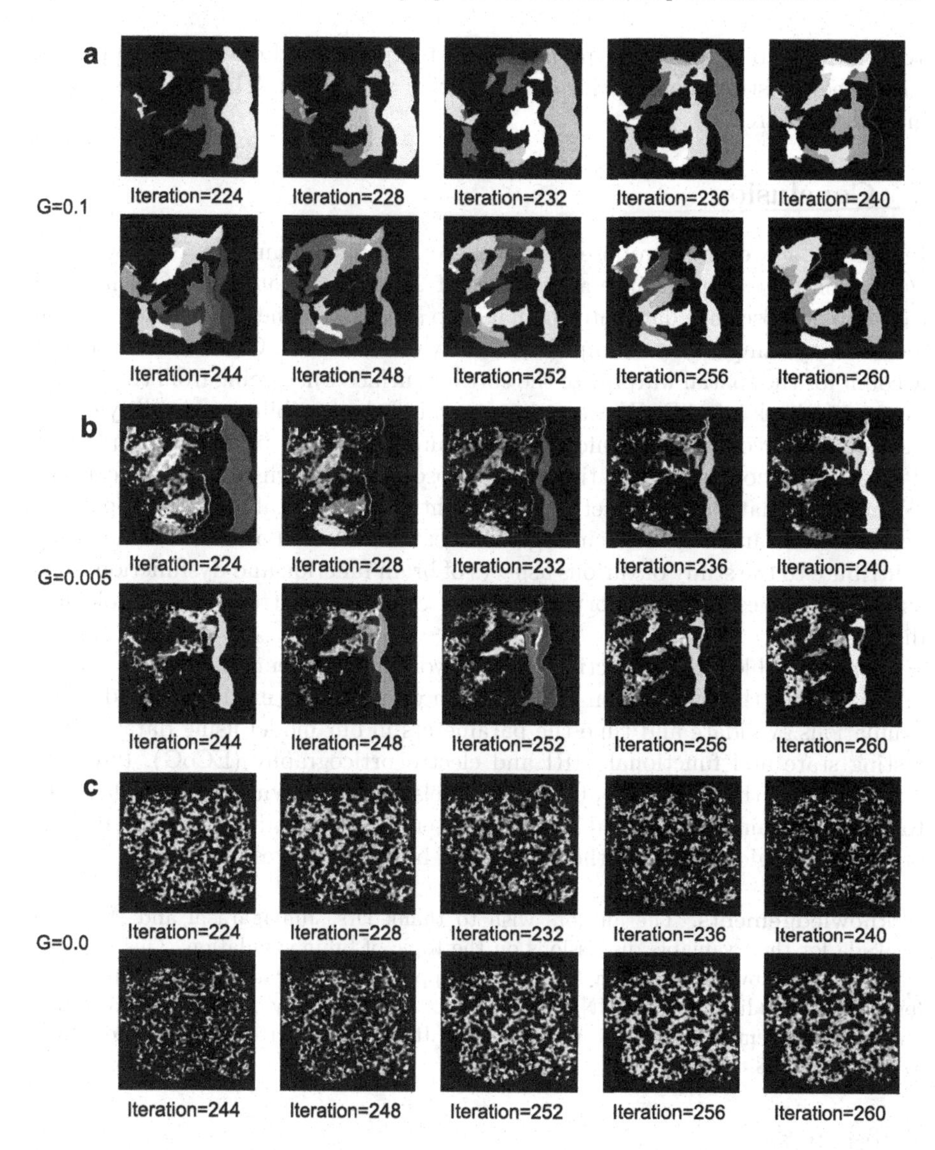

Fig. 4. Results for different global coupling parameter values. (**a**) Strong global coupling shows dynamics that emphasize global feedback, with different brain region activity switching over time. (**b**) Reduced global coupling shows a mixture of local and global feedback dynamics. (**c**) With no global coupling only the spontaneous activity from local noise drives the dynamics.

simulation more realistic. To explore this, we plan to take advantage of the available marmoset brain DWI data.

Finally, even though the model shows interesting dynamics, the overall system needs to be fine-tuned to better mimic the actual dynamics of the brain.

We will need to validate our method against experimental data and compare it with other existing methods for cortical brain simulation. This will be tackled in future research.

4 Conclusion

In this paper we have proposed a novel method for mammalian cortical brain simulation that leverages the advantages of 2D cortical flat maps. Our method can map neuroscience data into the flat map space and generate realistic neuron locations by sampling from empirical density distributions. Our method can also account for the spatial warping of the cortical surface when going from the 2D flat map space back into 3D. We have demonstrated the feasibility and utility of our method using connectivity information from a marmoset brain atlas and DWI data. Our method has implications for neuroscience research and its applications. We envision that such a model can shed light on how neural activity propagates across regions and how different regions interact with each other. This may then contribute to the study of various aspects of brain function and dysfunction, such as sensory processing, memory formation, cognitive control, and neurological disorders.

Some possible future directions for our work are: Implement a CUDA version of our method that can run on GPUs, which would enable even faster and larger simulations. Validate and tune the parameters in our model using data such as resting state and functional MRI and electrocorticography (ECoG). Integrate our method with other computational models of neural dynamics and plasticity to simulate brain function and learning. Extend our method into 3D and include other brain regions, such as the subcortical brain structures.

Acknowledgements. The authors wish to thank Drs. Jun Igarashi and Hiromich Tsukada for their valuable discussions on the topic of brain simulation. This research was supported by the program for Brain Mapping by Integrated Neurotechnologies for Disease Studies (Brain/MINDS) from the Japan Agency for Medical Research and Development, AMED. Grant number: JP15dm0207001 to A.W. and R.G., JP19dm0207088 to K.N.

References

1. Numba: A High Performance Python Compiler. https://numba.pydata.org/. Accessed 30 Apr 2023
2. The Brain/MINDS Data Portal. https://dataportal.brainminds.jp. Accessed 30 Apr 2023
3. Atapour, N., Majka, P., Wolkowicz, I.H., Malamanova, D., Worthy, K.H., Rosa, M.G.P.: Neuronal distribution across the cerebral cortex of the marmoset monkey (callithrix jacchus). Cereb. Cortex **29**(9), 3836–3863 (2019). https://doi.org/10.1093/cercor/bhy263
4. Dahlem, M.A., Isele, T.M.: Transient localized wave patterns and their application to migraine. J. Math. Neurosci. **3**(1), 7 (2013). https://doi.org/10.1186/2190-8567-3-7

5. Hata, J., et al.: Multi-modal brain magnetic resonance imaging database covering marmosets with a wide age range. Sci. Data **10**(1), 221 (2023). https://doi.org/10.1038/s41597-023-02121-2
6. Sanz Leon, P., et al.: The virtual brain: a simulator of primate brain network dynamics. Front. Neurosci. **7**, 10 (2013). https://doi.org/10.3389/fninf.2013.00010
7. Sanz-Leon, P., Knock, S.A., Spiegler, A., Jirsa, V.K.: Mathematical framework for large-scale brain network modeling in the virtual brain. NeuroImage **111**, 385–430 (2015). https://doi.org/10.1016/j.neuroimage.2015.01.002. https://www.sciencedirect.com/science/article/pii/S1053811915000051
8. Stefanescu, R.A., Jirsa, V.K.: A low dimensional description of globally coupled heterogeneous neural networks of excitatory and inhibitory neurons. PLoS Comput. Biol. **4**(11), 1–17 (2008). https://doi.org/10.1371/journal.pcbi.1000219
9. Van Essen, D.C.: Cortical cartography and caret software. NeuroImage **62**(2), 757–764 (2012). https://doi.org/10.1016/j.neuroimage.2011.10.077. https://www.sciencedirect.com/science/article/pii/S1053811911012419. 20 YEARS OF fMRI
10. Woodward, A.: The NanoZoomer artificial intelligence connectomics pipeline for tracer injection studies of the marmoset brain. Brain Struct. Funct. **225**(4), 1225–1243 (2020). https://doi.org/10.1007/s00429-020-02073-y
11. Woodward, A., et al.: The Brain/MINDS 3D digital marmoset brain atlas. Sci. Data **5**, 180009 (2018). https://doi.org/10.1038/sdata.2018.9
12. Xu, B., Binczak, S., Jacquir, S., Pont, O., Yahia, H.: Isolation and characterization of plasmid deoxyribonucleic acid from streptomyces fradiae. In: Annual International Conference of the IEEE Engineering in Medicine and Biology Society, pp. 4334–4337 (2014). https://doi.org/10.1109/EMBC.2014.6944583

Construction of a Novel Data Set for Pedestrian Tree Species Detection Using Google Street View Data

Martin Ooi[1(✉)], David Arturo Soriano Valdez[2], Mitchell Rogers[1], Rachel Ababou[3], Kaiqi Zhao[4], and Patrice Delmas[1]

[1] Intelligent Vision Systems Lab, University of Auckland, Private Bag 92019, Auckland 1142, New Zealand
{mooi002,mitchell.rogers,p.delmas}@auckland.ac.nz
[2] Strong AI Lab, University of Auckland, Private Bag 92019, Auckland 1142, New Zealand
[3] CREC, Academie Militaire de St-Cyr, Coetquidan, France
rachel.ababou@st-cyr.terre-net.defense.gouv.fr
[4] School of Computer Science, University of Auckland, Auckland, New Zealand
kaiqi.zhao@auckland.ac.nz

Abstract. Cities of the future will carefully manage their ecological environment, including parks and trees, as critical resources to balance the effects of climate change. As such, tree health has become an integral part of well-managed cities and urban areas, where tree censuses provide a critical source of ecological data. This data provides information to improve the ecological status of these areas; however, gathering this data is laborious and requires expert knowledge to accurately register each tree species included in the census. With recent advances in object-detection methods, automating this type of census is now possible. However, these approaches require training data to be gathered, labelled, and validated. This study merged data from a tree census in the Auckland region (New Zealand) with Google Street View image data and used a pre-trained model on specialised datasets released by prior authors to create a training dataset for pedestrian-view tree species detection. This approach can be used as the basis for wider data collection and labelling of New Zealand urban tree species, crucial for inventorying the state and health of its urban forests. Here, we demonstrated that training and deploying a fine-grained object detection model to an edge device for real-time inference on a video stream can achieve speeds of 25 frames per second (fps).

Keywords: Computer Vision · AI Detection · Tree · Pedestrian-View

1 Introduction

Computer vision applied to object detection has improved in performance with the development of larger deep-learning models such as the YOLO family of

J. Blanc-Talon et al. (Eds.): ACIVS 2023, LNCS 14124, pp. 332–344, 2023.
https://doi.org/10.1007/978-3-031-45382-3_28

methods [16]. The later development and addition of specialised processing units to these computing devices, such as tensor processing units (TPU) and neural processing units (NPU), have made it possible to add AI capabilities to small devices while maintaining a modest level of power consumption [12]. These developments have enabled the deployment of AI models for low-power edge devices [20]. One of the major limitations of deep-learning models is the need for a considerable amount of data. Although a vast amount of data is already available, every problem is different, and maximising detection rates under uncontrolled conditions requires domain-specific data. Therefore, constructing custom datasets is an essential step in training models for niche computer vision problems. However, labelling, reviewing, and validating the data requires considerable time and effort. It is preferable to avoid some of this work by using available datasets and customizing the data to meet specific requirements [3]. Once a custom dataset is constructed and validated using a state-of-the-art object detection method, it can be implemented using off-the-shelf hardware, according to the user's specific needs. One object detection application is tree censuses to assess the number of tree species in urban environments. Previous deep learning applications for tree censuses applied more cumbersome deep learning models [19] and often utilised Google Street View (GSV) data [9,15]. In this study, we train deep-learning models for arbitrary tree species detection from pedestrian view and apply one of these models for on-the-edge inference using an Nvidia Jetson device[1]. The main contributions of this study are as follows:

- We present a novel data set unique to urban trees within the New Zealand ecosystem. The data acquisition methodology described in this paper considers data collection at scale using easily accessible data sources, and we capture images and related metadata for localised urban trees linked to a systematic census.
- We demonstrate the effectiveness of this methodology combined with data sets from previous authors to pre-train a model to accelerate the annotation process.
- We deploy a trained model to an edge device that is capable of providing inference on live streams for detecting multiple urban tree species at near real-time speeds.

2 Methodology

2.1 Data Acquisition

Since there was (at the time of writing) no systematic, publicly available street-view image database of New Zealand urban trees, we acquired and annotated images of trees within the Auckland Central region (comprising the Auckland Central Business District (CBD) and central suburban areas using GSV images, with a specific focus on three tree species: (i) Metrosideros Excelsa (or

[1] https://www.nvidia.com/en-us/autonomous-machines/embedded-systems/.

Pohutukawa), (ii) Cordyline Australia (or Cabbage Tree), and (iii) Vitex Lucens (or Puriri). These species were selected because of their visual distinctness, proliferation in urban and suburban Auckland, and because they are evergreen species and hence are not affected by seasonal variation in appearance.

To build our street-view image database, we obtained an urban tree register from the University of Auckland GeoDataHub [14]. This register contains a list of identified tree assets within the Greater Auckland region. Each identified tree asset was recorded using a unique table key, a unique asset ID, the species and common name of the tree, and its geographical coordinates. From the 198,065 records in the urban tree register, we discarded all entries that did not record the species of a tree asset, leaving 178,215 records.

The geographic coordinates for each of the remaining tree assets were then submitted to the GSV Street View Static API [1] which searched for the nearest GSV panorama within a 50-metre radius and returned its metadata, including a unique panorama ID, the geographical coordinates of the panorama location, and the year and month of the panorama capture. A total of 12,353 tree assets within the register failed to return the nearest panorama ID, either because no panorama existed within 50 m of the location of the tree asset or because the GSV Street View Static API timed out, resulting in a total of 165,862 tree assets successfully matched to a panorama. Tree assets that contained invalid species labels (e.g., "Not Applicable" or "Unknown", totalling 5,720 records) were removed and 20,257 tree assets that were located within the Auckland CBD and central suburban areas (defined as areas that were within the 2014 Auckland Central General Electorate District [10], excluding islands or inlets) were selected. The purpose of this exclusion was to focus on tree specimens located in urban or suburban settings. Finally, the remaining tree assets were divided into specimens that belonged to species of interest to us (3,354) and other tree species (16,903). These tree assets were identified by their common names as a degree of variation in species spellings and granularity (e.g., identification of genus only, identification of species, and identification of specific cultivars). From the locations of the 3,354 tree records belonging to the three selected species within our region of interest, we selected 1,249 unique GSV panoramas associated with the locations of these tree records as our sample for further analysis.

Analysis of Selected Panoramas. The 1,249 selected panoramas were linked to 8,437 tree assets of various species in the urban tree register. We further discarded 188 user-submitted panoramas because these images would differ in quality and method of acquisition. The remaining panoramas in our sample included 6,461 tree assets from the urban tree register, the majority of which were either Pohutukawa or other tree species.

Of the 6,461 tree assets within our selected panoramas, 83.7% of tree assets were located within 30 m of the panorama. A significant proportion of the selected panoramas (51%) were captured during June, which is within the winter season in the Southern Hemisphere. As such, we expected the prevalence of tree assets captured within our panorama images to be at various stages of leaf abscission due to seasonal weather changes. The three selected tree species (Pohutukawas, Cabbage Trees and Puriris) are evergreen plants; therefore, we would not expect detection or classification tasks on these species to be significantly affected by seasonal variation. Finally, the majority of the selected panoramas (in excess of 86%) were acquired in 2022, indicating a good degree of recency in our resulting street-view image dataset.

Acquisition and Transformation of Panoramas. The 1,061 identified panoramas were acquired using Street View Download 360 [11], licenced software for downloading street view panoramas from Google Street View. The panoramas were obtained by providing the software with panorama IDs obtained through the Static Street View API. The software downloaded the panoramas for each panorama ID as 16384 by 8192 pixels images along with EXIF (Exchangeable Image File Format) metadata, which included the panorama zoom level, panorama ID, panorama latitude and longitude, year and month of panorama acquisition, panorama elevation and panorama rotation.

From the downloaded panoramas, a subset of 401 panoramas was sampled by visually inspecting the panoramas to validate that they contained at least one of the three selected species from the tree assets in the urban tree register tagged to the panorama ID. Because the downloaded panoramas are in an equi-rectangular projection, we used the equirectangular-perspective tool [5] to transform the panoramas to rectilinear Field of Views (FOVs), obtaining seven 100° FOVs from each panorama, totalling 2,807 images. Each FOV was taken at 45° yaw increments starting from −135° to 135° with a pitch of 10° to better capture the crowns of trees that were in close proximity to the camera.

2.2 Image Labelling and Annotation

For image annotation, we first trained the YOLO-v7 model on the urban tree dataset released by Wang et al. [18]. We used Wong et al.'s implementation of the YOLO-v7 model [7], with the default YOLOv7 model and hyper-parameter settings, and initialised the model with pretrained weights on the COCO dataset. The purpose of this trained model is to produce an initial set of bounding boxes for use in the labelling process.

We used an open-source data labelling solution, Label Studio [6], to edit the generated bounding boxes, produce new bounding boxes for trees that were not detected by the initial model, and provide class labels corresponding to tree species. An additional label, "other trees" was assigned to all bounding boxes of trees that did not belong to one of the three species of interest. Only trees containing a visible section of a branch (or trunk) and crown within the image were annotated. Along with the classification label, we also included some additional metadata with each bounding box indicating, if the tree was in a state of leaf abscission, was occluded (and if so whether it was occluded by a tree or by another object) and was only partially visible (and if so, which parts of the tree were visible).

2.3 Analysis of Annotations

Of the 2,807 images obtained from the sampled panoramas, 1,078 images (equivalent to 154 panoramas) were manually inspected and labelled. 130 images contained no trees, and in the 948 remaining images, there was an average of 4.36 trees annotated per image, with a standard deviation of 2.75. The maximum number of annotated trees in a single image was 17.

Of the 4,129 total annotations that were made, the most common tree species annotated was the Pohutukawa (1,627 trees), followed by the Cabbage Trees (403 trees) and the Puriri (280 trees). 1,819 trees which did not belong to the three species were labelled as "other tree". 338 annotated trees exhibited signs of leaf abscission, 2,852 trees exhibited some degree of occlusion by other trees while 2,033 trees exhibited some degree of occlusion by other objects. Finally, 3,514 trees were partially captured in the images. This is largely attributed to the number of trees that were positioned on the edges of the images or the bases or crowns of the trees, exceeding the vertical limits of the images.

2.4 Data Splitting

To obtain training, validation and testing datasets, the data was split by panorama using a roughly 80:10:10 split. We split by panoramas instead of FOVs to avoid potential target leakage caused by adjacent FOVs being shared between data splits. Table 1 lists the resulting statistics for the training, validation, and testing datasets. The distribution of the total annotations was fairly consistent; however, the distribution of annotations by species varied, especially in the distribution of Pohutukawa and Puriri annotations in the validation and test sets.

Table 1. Descriptive statistics by data set splits.

	Train	Val	Test
Panoramas	123	16	15
Images (FOVs)	861	112	105
Pohutukawa	1,340	133	154
Cabbage Tree	294	53	56
Puriri	228	45	7
Other tree	1,443	188	188
Total annotations	3305	419	405

2.5 Data Augmentation

Data augmentation is the process of creating new data by applying transforms and modifications to existing data. In machine learning tasks, data augmentation helps to introduce invariance to the dataset and increases the amount and diversity of data [4]. This is also a commonly used approach for reducing overfitting [8], allowing trained models to generalise better to out-of-sample data during inference. Augmentations applied to the training and validation sets included horizontal flips, cropping, hue and saturation adjustments, Gaussian noise, and affine scaling and translations. No augmentation was applied to the testing data to ensure that the test dataset resembled real-world data encountered during detection or inference.

3 Experimental Process and Evaluation

This section describes the process of training various object detection models to the collected data set for two tasks: single-class detection and multi-class detection (i.e. detection of trees by labelled species). YOLO-v7, a member of the You Only Look Once (YOLO) family of models, was selected for this task. First published by Redmon et al. [13] YOLO is a single-stage object detection model that treats the object detection problem as a regression problem, simultaneously predicting bounding box coordinates and class probability scores from a single convolutional network.

3.1 YOLO-V7

The changes proposed by Wang et al. [16] in YOLO-v7 reduced the required parameters and computation, providing a better balance between inference speed and accuracy compared with prior versions. We briefly discuss some of the contributions of Wang et al. [16] in their paper.

E-ELAN. Wang et al. [17] studied the effects of stacking increasingly large numbers of computational blocks using gradient path analysis, which was used to understand the impact on accuracy and convergence and create a new network design paradigm, gradient path driven design, which emphasised the source of gradients and how they were updated during training. Based on this network design strategy, the authors proposed E-ELAN (Extended ELAN) which, in addition to preventing the shortest gradient path from increasing too rapidly, used "expand, merge and shuffle cardinality" to allow the network to learn without destroying the original gradient path.

Model Scaling. Model scaling is a method of scaling a model architecture up or down to fit various computing devices and purposes, whether it is to optimise for inference speed on an edge device or for accuracy on a cloud GPU instance [16]. The authors adopted a compound scaling method that scaled the depth and width coherently, as the scaling of depth in concatenation-based models will also result in a change in the width of the output of the concatenation block.

Dynamic Label Assignment. Label assignment is the process of assigning targets to sampled regions in an image (e.g., labelling an object prediction). Modern label assignment methods make use of'soft' labels which use the quality of a prediction output (e.g., prediction probability scores or an IoU (Intersection Over Union) score) instead of a binary 'hard' label.

YOLO-v7 uses deep supervision which requires auxiliary heads in shallower regions of a network. Traditional methods for label assignment in a lead-auxiliary head design involve the lead and auxiliary heads to independently generate predictions and resulting losses (see Fig. 1a).

The author proposed two modifications: (i) first, only the lead head produces a prediction result and generates a soft label, which is then used to calculate the loss for both the lead and auxiliary heads; (ii) second, the lead head generates an additional set of soft labels with relaxed constraints, referred to as 'coarse' labels, to compensate for the auxiliary heads' weaker learning ability (see Fig. 1b). This design took advantage of the lead heads' stronger ability to learn and allowed the lead head to focus on learning features that were not learned by the shallower auxiliary heads.

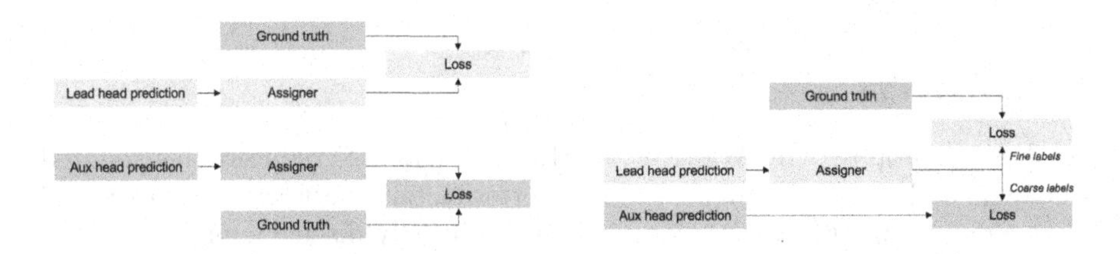

(a) Independent label assignment. (b) Coarse-to-fine label assignment

Fig. 1. Figure 1a illustrates traditional lead and auxiliary label assignment, with predictions and loss calculations being performed independently. Figure 1b illustrates the label assignment used by [16], showing the shared prediction between the lead and auxiliary heads and the use of fine and coarse labels.

3.2 Model Training

We trained normal-sized YOLO-v7 and compact YOLO-v7 tiny models on a system with an AMD Ryzen 9 CPU and 32 GB of RAM. All models were trained for 300 epochs using the default hyperparameter settings for both model sizes. YOLO-v7 also implements in-memory augmentation during the training. We retained all the default settings for these augmentations. Additionally, we trained a single and multi-class detection model using YOLO-v5s to serve as a benchmark against the YOLO-v7 models and for use in a proof-of-concept live detection setup on an edge-computing device (see Sect. 3.3).

We then performed hyperparameter evolution on the YOLO-v7 tiny models for both single- and multi-class tasks. The evolution was applied to all hyperparameter settings (including in-memory augmentations, except for the number of anchors and optimal transport allocation) for 300 generations to find optimal hyper-parameter settings using a base case of 10 epochs (owing to computation time limitations). The fitness function used for evaluating each generation is:

$$(0.1 \times \text{mAP}_{0.5}) + (0.9 \times \text{mAP}_{0.5:0.95}) \quad (1)$$

We then retrained both models for 300 epochs using the evolved hyperparameters.

3.3 Evaluation

Validation and Test Metrics. In this section, we present our evaluation of the trained models before and after hyper-parameter evolution. For each model iteration, we report the mAPs of the best epoch, defined as the epoch that produced the highest fitness value, based on the function defined in Eq. 1.

We can see from Table 2 that for single-class detection, the YOLO-v5s model exceeded the performance of the YOLO-v7 tiny model but performed poorly in the multi-class detection task. The YOLO-v7 model had the highest mAP in both tasks which was offset by slower inference speeds than either YOLO-v5s or YOLO-v7 tiny models. However, the incremental improvement in mAP for the YOLO-v7 model is relatively small compared to that of the YOLO-v7 tiny model. The relative success of the smaller and more parsimonious model suggests that the current volume and variety of training data are still insufficient to support a larger model, potentially leading to a more complex YOLO-v7 model overfitting.

Table 2. mAP of trained models using default hyper-parameters.

Model	Size	Detection	mAP@0.5	mAP@0.5:0.95
YOLO-v7	1280	Single-class	0.7807	0.5201
YOLO-v7 tiny	1280	Single-class	0.7597	0.4757
YOLO-v5s	640	Single-class	0.7668	0.5203
YOLO-v7	1280	Multi-class	0.7208	0.4868
YOLO-v7 tiny	1280	Multi-class	0.6957	0.4406
YOLO-v5s	640	Multi-class	0.6184	0.4086

In Table 3, we present the validation and test mAPs of the YOLO-v7 tiny models retrained using the evolved hyperparameters. Our results showed an uplift in the mAP@0.5:0.95 of both the single- and multi-class YOLO-v7 tiny models post-hyperparameter evolution (see Table 3), although the effect appeared to be smaller in the YOLO-v7 tiny model. Our results also showed that the test metrics for the single-class model were fairly similar to the validation metrics; however, atypically, the test metrics for the multi-class model showed better performance on the test set than on the validation set. This was likely due to the difference in the distribution of species between the validation and test sets (Table 1). This is especially pronounced in the different proportions of the annotated Pohutukawa and Puriri trees. If the multiclass detector exhibits greater precision in classifying one of these species, it would result in a test mAP that is greater than the validation mAP.

Table 3. mAP of YOLO-v7 tiny models with evolved hyper-parameters.

Model	Detection	mAP@0.5$_{val}$	mAP@0.5:0.95$_{val}$	mAP@0.5$_{test}$	mAP@0.5:0.95$_{test}$
YOLO-v7 tiny	Single-class	0.7614	0.4920	0.747	0.494
YOLO-v7 tiny	Multi-class	0.6851	0.4566	0.743	0.476

Out-of-Sample Detection. In this section, we use an in-house developed portable computer vision system to test our models under real-world conditions. This system is capable of acquiring pedestrian view footage that is processed in real time using the multiclass YOLO-v5s model with default detection settings. The core of this system is an NVIDIA Jetson NANO 8 GB and two Basler Cameras (acA2440-75uc). The cameras were synchronised using a propagation trigger shutter.

The results of our real-world testing were satisfactory in terms of real-time performance and tree detection and we achieved up to 25 fps. However, it is important to improve data acquisition because our model encountered difficulties in dealing with real-world light conditions. Figure 2 shows the results of one of the real-world experiments performed using the trained model along with the data acquisition with pedestrian view footage.

Additionally, we tested the evolved YOLO-v7 models with default detection settings on other out-of-sample sources comprising images obtained from Google Image Search results of the selected tree species.

Fig. 2. Portable computer vision system (Left) with example frames of tree detection results obtained using the multi-class YOLO-v5s model.

Figure 3 shows the selection of successful detections (a) and poor detections (b) on images retrieved from Google Image Search using multi- and single-class models, respectively. We see in the successful detections that there were no false positives and, in all but one case, the multi-class model successfully identified the correct species of trees. The false negatives in these detections are likely attributed to either the presence of occlusion by foreign objects (as in the traffic poles in image 3 of the single-class model) or occlusions by other trees or foliage, resulting in either major features of the tree being obscured by foliage (in image 5) or lighting being obscured (in image 3).

However, the poor detections exhibited several detection failures. In addition to false negatives, some of the detections failed to properly separate instances, produced false positives from green shrubbery, misclassified species, and produced overlapping predictions for the same object. In image 2, the single-class model failed to detect any objects, despite the clearly defined tree objects. Because these detections were performed using default sensitivity and IoU threshold settings (for non-max suppression), refinement of these settings would likely produce better results, particularly for false negatives or overlapping predictions. Overlapping predictions of different classes may also be improved using class-agnostic non-max suppression.

4 Discussion

Volume of Training Data. The data set curated and used in this study was acquired and labelled over a relatively short span of time by a limited number of people. A larger volume of data would reduce the amount of class imbalance across the dataset, preventing over-weighting of the larger classes and reducing the errors observed in Fig. 3. Additionally, a larger and more diverse dataset will also improve the model performance during inference by allowing it to learn a greater range of representations and patterns from the underlying data and reduce the generalisation error.

Data Validation. Whilst the specific species being labelled for fine-grained detection are visually distinct and every care was taken to provide accurate species labels, the resulting labels should be inspected and validated by a domain expert. This will help reduce inaccuracies due to label noise in the model training process.

Tree Appearance Variation Over Time. The tree species specified in this study have no seasonal variation in appearance. However, if the study is expanded to incorporate other prominent species, consideration would need to be given to the effect of season, age and health on tree appearance and how this can be efficiently tracked over time.

Fig. 3. Out-of-sample detections showing (a) high quality detections and classifications and (b) low quality detections and classifications by the multi and single-class models.

Performance Benchmarking. In order to properly evaluate our methodology, it is essential to benchmark the performance of the model and detection system used. Since existing methods in the current literature are evaluated on bespoke data sets and deployed on differing infrastructure, comparing accuracy and inference speed is not a trivial task and would require replication on the data set and devices used in this study.

Alternative Pre-trained Weights. Present day CNN backbones and detection models provide pre-trained weights as an alternative for random weight initialisation, typically obtained by training the model on an existing benchmark data set (e.g., COCO). With the release of the Auto Arborist dataset [2] by Google (containing 2.5 million labelled tree instances from street-level and aerial imagery across North America), it is likely that, upon wider distribution, this specialised dataset would better suit the niche domain of this study to obtain pre-trained weights.

Hierarchical Multi-label Classification. The task of fine-grained species detection can be formulated as a hierarchical multi-label classification (HMC) problem, where the classes or labels to be assigned contain a natural hierarchy (e.g., each instance can be labelled using a super-class "Tree", followed by its genus and species). Future work in this domain could explore adaptive approaches designed specifically for solving HMC problems in neural networks.

Preprocessing. During the real world test using the NVIDIA Jetson Nano device, we observed that the species labels assigned to detections were sensitive to ambient exposure. The introduction of a pre-processing step with adaptive exposure correction (or alternatively, adjustment of the camera aperture settings) would likely assist in obtaining greater stability in classification.

5 Conclusion

This study presented our methodology and approach for constructing a novel dataset for the task of fine-grained detection of select urban tree species in New Zealand. By employing a combination of pre-existing tree inventory data, we created a novel detection dataset of geographically unique urban tree species by employing a combination of pre-existing tree inventory data, an easily accessed dataset that allows the acquisition of image data and metadata at scale. We also demonstrated the effectiveness of the current generation of detection models for performing species detection and the viability of performing inference on edge devices at near real-time speeds.

References

1. Alphabet Inc: Google street view static API,. https://maps.googleapis.com/maps/api/streetview/. Accessed 25 Mar 2023
2. Beery, S., et al.: The auto arborist dataset: a large-scale benchmark for multiview urban forest monitoring under domain shift. In: Proceedings of the IEEE/CVF Conference on Computer Vision and Pattern Recognition (CVPR), pp. 21294–21307 (2022)

3. Braithwaite, J.M.: Chapter 17: challenges and payoffs of building a dataset from scratch, pp. 300–316. Edward Elgar Publishing, Cheltenham, UK (2022). https://doi.org/10.4337/9781839101014.00028
4. Cubuk, E.D., Zoph, B., Mané, D., Vasudevan, V., Le, Q.V.: Autoaugment: learning augmentation policies from data. CoRR (2018)
5. Fu-En, W.: Equirec2perspec. https://github.com/fuenwang/Equirec2Perspec. Accessed 9 Apr 2023
6. Heartex Inc: Label studio. https://labelstud.io/. Accessed 9 Apr 2023
7. Kin Yiu, W.: Yolov 7: implementation of paper (2022). https://github.com/WongKinYiu/yolov7. Accessed 9 Apr 2023
8. Krizhevsky, A., Sutskever, I., Hinton, G.E.: Imagenet classification with deep convolutional neural networks. Commun. ACM **60**(6), 84–90 (2017). https://doi.org/10.1145/3065386
9. Lumnitz, S.: Mapping urban trees with deep learning and street-level imagery. Ph.D. thesis, University of British Columbia (2019). http://dx.doi.org/10.14288/1.0387513
10. NZ, S.: General electoral district 2014. https://datafinder.stats.govt.nz/layer/104062-general-electoral-district-2014/. Accessed 3 Apr 2023
11. Orlita, T.: Steet view download 360 (2016). https://svd360.istreetview.com/. Accessed 9 Apr 2023
12. Pias, M., Botelho, S., Drews, P.: Perfect storm: DSAs embrace deep learning for GPU-based computer vision. In: 2019 32nd SIBGRAPI Conference on Graphics, Patterns and Images Tutorials (SIBGRAPI-T), pp. 8–21 (2019). https://doi.org/10.1109/SIBGRAPI-T.2019.00007
13. Redmon, J., Divvala, S., Girshick, R., Farhadi, A.: You only look once: unified, real-time object detection (2015). https://doi.org/10.48550/ARXIV.1506.02640
14. University of Auckland: Geodatahub. https://geodatahub.library.auckland.ac.nz/. Accessed 25 Mar 2023
15. Velasquez, L., Echeverria, L., Etxegarai, M., Anzaldi Varas, G., Miguel, S.D.: Mapping street trees using google street view and artificial intelligence (2022)
16. Wang, C.Y., Bochkovskiy, A., Liao, H.Y.M.: Yolov7: trainable bag-of-freebies sets new state-of-the-art for real-time object detectors (2022). https://doi.org/10.48550/ARXIV.2207.02696
17. Wang, C.Y., Liao, H.Y.M., Yeh, I.H.: Designing network design strategies through gradient path analysis (2022). https://doi.org/10.48550/ARXIV.2211.04800
18. Wang, Y., et al.: Utd dataset (2022). https://github.com/yz-wang/OD-UTDNet. Accessed 23 Oct 2022
19. Wegner, J.D., Branson, S., Hall, D., Schindler, K., Perona, P.: Cataloging public objects using aerial and street-level images - urban trees. In: 2016 IEEE Conference on Computer Vision and Pattern Recognition (CVPR), pp. 6014–6023 (2016)
20. Zou, Z., Chen, K., Shi, Z., Guo, Y., Ye, J.: Object detection in 20 years: a survey. Proc. IEEE **111**(3), 257–276 (2023). https://doi.org/10.1109/JPROC.2023.3238524

Texture-Based Data Augmentation for Small Datasets

Amanda Dash(✉) and Alexandra Branzan Albu

University of Victoria, Victoria, BC 08544, Canada
{adash42,aalbu}@uvic.ca

Abstract. This paper proposes a texture-based domain-specific data augmentation technique applicable when training on small datasets for deep learning classification tasks. Our method focuses on label-preservation to improve generalization and optimization robustness over data-dependent augmentation methods using textures. We generate a small perturbation in an image based on a randomly sampled texture image. The textures we use are naturally occurring and domain-independent of the training dataset: regular, near regular, irregular, near stochastic and stochastic classes. Our method uses the textures to apply sparse, patterned occlusion to images and a penalty regularization term during training to help ensure label preservation. We evaluate our method against the competitive soft-label Mixup and RICAP data augmentation methods with the ResNet-50 architecture using the unambiguous "Bird or Bicyle" and Oxford-IIT-Pet datasets, as well as a random sampling of the Open Images dataset. We experimentally validate the importance of label-preservation and improved generalization by using out-of-distribution examples and show that our method improves over competitive methods.

Keywords: data-independent augmentation · texture

1 Introduction

This paper proposes a domain-specific out-of-domain texture-based data augmentation technique for small dataset training. Data augmentation is a technique for supplementing small datasets by artificially generating "new" training images. The increased availability of large public general datasets has significantly contributed to the successful application of deep convolutional neural networks (CNNs) to difficult computer vision tasks. However, it is difficult to create large custom datasets for specific domains. For example, medical datasets are typically very small due to legal and privacy regulations in the medical world. When training models from scratch, undesirable behaviour occurs, such as overfitting. Recent studies indicate that overfitting is not an extensive issue [15,25], however, these studies are restricted to large datasets. When training on small datasets, issues such as slow convergence, vanishing gradients, sensitive parameter tuning, etc. are amplified during optimization. In fact, [21] notes that the

J. Blanc-Talon et al. (Eds.): ACIVS 2023, LNCS 14124, pp. 345–356, 2023.
https://doi.org/10.1007/978-3-031-45382-3_29

instabilities in optimization can be so extreme, that changing a single weight can induce variabilities equal to all other combined sources.

Regularization techniques such as Dropout and data augmentation can be used to mitigate overfitting. Data augmentation prevents overfitting by addressing the root of the problem, the lack of data, which can be defined as low sample size, unbalanced samples or insufficient data coverage. Hernandez-Garia and Zonig [7] also showed that data augmentation can replace explicit regularization, like dropout; data augmentation can be considered congruent with dropout performed on the image level.

Transfer learning (or pre-training) can also be used; this is a technique where a model is first trained on a large generic dataset, then fine-tuned on a smaller, task-specific dataset. Transfer learning provides the advantage of requiring less task-specific data and faster convergence. However, transfer learning may fail or perform poorly if the large generic and the smaller task-specific datasets are very different. Data augmentation can be used to solve this issue. Zoph *et al.* [28] studied transfer learning in conjunction to data augmentation. They note that strong data augmentation diminishes the value of pre-training via transfer learning.

Hernandez-Garcia and Konig [7] state there are two important categories when describing data augmentations: domain-specific augmentation and data-dependent regularization. Domain-specific augmentation (explicit regularization) are augmentations that generate new samples that are *perceptually plausible*, i.e. which humans can still recognize and categorize using the same label as for the original, non-augmented image. This type of augmentation is described as *label-preserving*. Augmentation methods which generate images that do not plausibly belong to the dataset, such as Mixup [26], but improve the optimization and generalizing of CNNs are called *data-dependent regularization* (implicit regularization). These methods often require additional logic to construct synthetic labels. Our method uses patterned occlusion, which occurs naturally and frequently in real-world images. As the textures that we work with are sparse, the occlusion is label preserving, and will allow for data independent generalization.

Our contribution is a domain-specific data augmentation using out-of-domain texture that improves generalization and optimization robustness over data-dependent regularization when applied to small datasets. Our method randomly samples domain-independent texture images to constrain the application of a perturbation hyperparameter, similar to adversarial examples [18].

Our method is particularly relevant for small datasets; we use textures to perform image-mixing augmentation on the light channel of a sampled L*a*b* image to create a label-preserving training image. The mixing is constrained by a perturbation hyperparameter, Δ, and a randomly sampled sparse texture map. Experiments show that our method converges quickly and stabilizes at high learning rates. By using the "Bird-or-Bicycle" dataset, which contains no label noise, we show that our method produces almost no label noise and outperforms the baseline when using near stochastic textures. We further examine our method by comparing the classification results when training on a small unbalanced dataset, Oxford IIIT-Pet, and evaluating against a subset of the Open Images Database to simulate large data variation in a small dataset.

2 Related Work

Image mixing is a type of data augmentation where entire images or patches are interpolated or replaced, thus obscuring or discarding some image areas; image mixing can be thought as performing dropout in the image space. Data augmentation does not affect inference, unlike other regularization methods like dropout. This section discusses related works on data augmentation based on image mixing.

Data disruption is a domain-specific image-mixing that uses out-of-domain samples to represent *object obstruction or occlusion* in the dataset. Cutout [4] performs a dropout-like process where a square region of the training image is zero-masked. The authors reported a 2.56% test error rate on the CIFAR-10 dataset [9]. Zhong *et al.* [27] use random pixel values or the ImageNet mean pixel value instead of a zero-mask (with random pixel values yielding better results). The authors reported an error rate reduction from 5.17 to 4.31% on the CIFAR-10 dataset.

In-domain image mixing techniques use multiple images from the training dataset to construct a unique training sample. Unlike conventional data warping techniques, such as random cropping or horizontal flipping, mixing images together seems counter-intuitive as the resulting images can be difficult for humans to interpret. However, these images successfully address misclassifications due to class competition.

Mixup [26] encourages a linear relationship between the soft labels of the training samples, allowing the prediction confidence to have a linear transition between classes, and thus improving optimization. The authors use alpha-blending to interpolate and superimpose two random training samples; the newly created training images increase data coverage, as well as emulate adversarial examples. They use soft labels by mixing class labels. Soft labels are intermediate probabilities, unlike binary or one-hot label encoding [8]. Mixup provides the soft labels of a probability equal to the mixing/blending hyperparameter α. The authors reported a decrease in the error rate of 2.7% on the CIFAR-10 dataset over previous state-of-the-art. Summers and Dinnen [20] explored non-linear mixed-example localized image mixing. The authors explored multiple methods, such as "VH-Mixup" and "VH-BC+" which concatenated two training samples in four grids, two of which use *Mixup* alpha-mixing. Summers and Dinnen reported a decrease in the error rate of 3.8% on the CIFAR-10 dataset for VH-Mixup/BC+ over previous state-of-the-art. Takashi *et al.* [22], with RICAP, depart from alpha-blending [20,26]; they concatenate four training samples, cropped together to form a new training sample. They also use soft labels, proportional to the relative random crop sizes. The authors reported a test error rate of 2.85% on CIFAR-10 over previous state-of-the-art.

Data disruption methods cause opaque obstruction of objects, which does not take into account partially translucent obstructions, such as shadows or overlaid classes (i.e. a dog behind a gate). Label noise can also be introduced if a key feature is obstructed for a given class. Label noise is also an issue for data-dependent regularization augmentation. Mixup employs an alpha-mixing strategy which may generate local features not present in the dataset. [5] noted that Mixup will cause underfitting resulting from the incongruities between the synthetic soft-labels and training labels. RICAP [22], which generates non-linear four-grid training images and corresponding soft-labels, also exhibits label conflicts due to randomized crop selection on the training samples. RICAP preserves local features by using random crops (spatial blending), but there are no guarantees that the foreground object is predominantly displayed in the crop, which may cause low confidence logits due to class competition. For datasets that contain classes with features very similar to common background scenes (i.e. shower curtains), if the random crop contains too much background then not only will not enough object features be learned by the CNNs, but the image would essentially be mis-annotated (*i.e.* the wrong label is applied to that image).

Our proposed approach is domain-specific, but applies an out-of-domain localized sparse disruption to the entire image globally, unlike [4,27]. The localized sparse disruptions are generated using a set of *textures*. The textures provide varying local and global patterns to the obstructions. The chosen textures occur naturally in the real-world (see Fig. 1) mimicking complex backgrounds and natural occlusion.

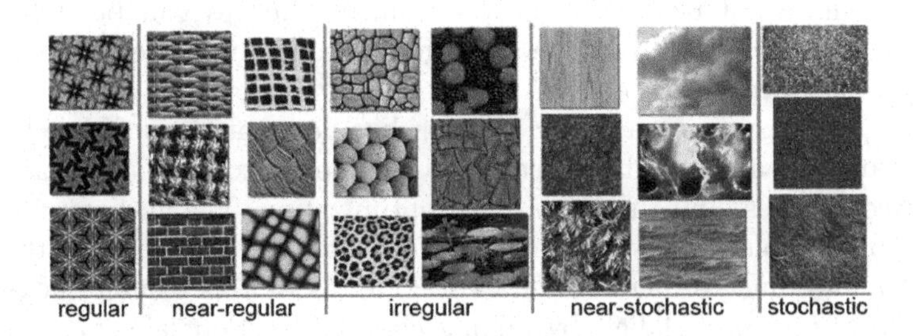

Fig. 1. Spectrum of texture categories from highly structured (left) to random (right). Regular textures have regular patterns. The regularity of these pattern vary from repeating duplicate structures, to repeating similar structures. Stochastic textures look like noise, with very small repeating primitives.

3 Proposed Method

We propose a texture-based method for Deep Convolutional Networks (CNNs) which performs label-preserving augmentation for small datasets. Label noise is usually present in large datasets, which have limited expert curation due to their size. With small datasets label noise and label preservation become critical to training generalizable and robust models; small datasets allow for careful curation but this advantage is lost if the data augmentation contributes significantly to label noise.

Our approach consists of three modules: (1) generation of the texture-based perturbation maps, (2) augmentation via image mixing and (3) modification of the loss function using a perceptual difference regularization.

3.1 Texture-Based Perturbation Maps

The first step is the conversion of natural texture images into templates used for augmentation. We consider five classes of natural textures: regular, near regular, irregular, near stochastic, and stochastic. The texture spectrum is shown in Fig. 1. At one end of the spectrum, regular textures are arrangements of texels (elementary texture elements) exhibiting a high spatial order, while at the other end of the spectrum, stochastic textures are random arrangements of texels. Standard image processing steps are taken to generate the *texture-based perturbation maps* from the original textured images. The gray-scale texture image is processed using histogram equalization to enhance subtle textural patterns. A bilateral filter [23] removes noise before an edge detection step using the Laplacian operator is applied. The texture is re-scaled to unit variance $[0, 1]$ so that pixels belonging to uniform regions are zero-valued and thus have no effect on the training image. The edges are important for our algorithm, as they represent local regions of rapid change, and represent potential features to "fool" the model.

3.2 Augmentation via Image Mixing

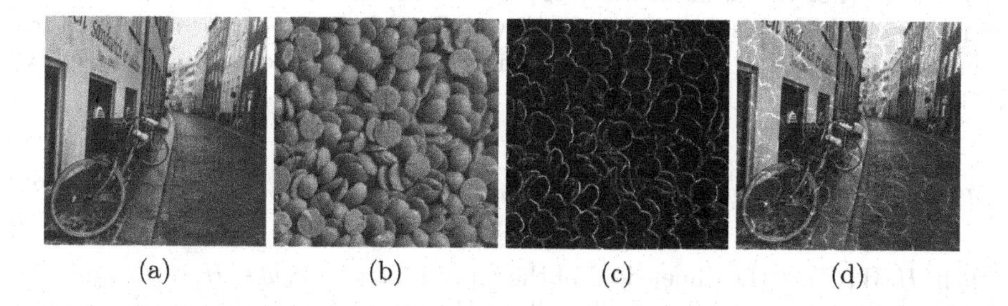

Fig. 2. Example of texture-based augmentation. (a) Randomly sampled *bicycle* training image from the "Bird-or-bicycle" dataset; (b) randomly sampled *near regular* texture; (c) processed texture constraint image with $\Delta = 16$ (normalized for visual presentation); (d) final textured augmented image.

A random pair consisting of a training image from the original dataset and a texture-based perturbation map is selected and resized to matching spatial dimensions. The training sample is converted to the L*a*b* colorspace. The L*-channel represents the *perceptual lightness* of the image. We maintain the *plausibility of the data augmentation* by only generating "textured" images, as shown in Fig. 2. This method is inspired by *adversarial training*, which modifies image inputs by applying small, intentional perturbations designed to "trick" the neural network into outputting high confidence mis-classification [18]. Attack images can be constructed by minimizing L_0, L_2 or L_∞ distances between the adversarial image and benign original image to preserve perceptual similarity between the two images [19]. In our method, the spatial perturbation regions and weights are texture-based, where the texture is a *constraint* on the augmentation represented by the weighted perturbation mask, $T(x)$. We use L_∞-norm bound where the Δ is the maximum perturbation applied to any given pixel. This relationship is shown in Eq. 1.

$$\hat{I} = \text{round}(I(x) + \Delta * T(x)) \tag{1}$$

where (I, T) is the randomly sampled training image and texture template, x is the current pixel location, $\Delta > 0$ is the perturbation hyperparameter, round$(\cdot)$ is a rounding and clipping function [0,255] and $\hat{I}$ is the augmented training image.

3.3 Perceptual Difference Regularization

Deep learning is robust to label noise *if* the increase in dataset size is proportional to the ratio by which correct labels have been diluted by label noise [16]. Therefore, label-preservation is critical when performing strong data augmentation. Our proposed method adds a penalty regularization term, independent of the labels, that evaluates the usefulness based on its *plausibility*. We assume that an image is useful if it plausible which we define as having minimal label error caused by the augmentation.

A characteristic of natural images is that the variance between neighbouring pixels is low (i.e. similar intensities). A plausible image can therefore be defined as an image in which this property is preserved. This property can be quantified by using Total Variation [17] (TV, Eq. 2), which is the measure of spatial complexity with respect to pixel intensities:

$$TV(\hat{I}) = \frac{1}{HWC} \sum_{uvw} [(\hat{I}(v, u+1, w) - \hat{I}(v, u, w)^2) + (\hat{I}(v+1, u, w) - \hat{I}(v, u, w)^2)] \tag{2}$$

where H, W, C are the dimensions of the input image, $u, v, w \in H, W, C$, and $\hat{I}$ is the augmented image. This penalty allows for perturbations that are L_∞ norm-bound by Δ, or greater than *just noticeable difference* to be used during training. Our proposed method adds a total variation loss as a penalty regularization term to the cost function as follows:

$$Loss(I(x), \hat{I}(x)) = \frac{1}{4e^\lambda} Loss(I(x)) + \delta TV(\hat{I}(x)) \tag{3}$$

where $Loss$ is the task loss function (i.e. cross-entropy loss), λ is the learning rate, $\delta \in (0, 1]$ is a tuning hyperparameter, and $TV(x)$ is the penalty regularization term (Eq. 2). The $\frac{1}{4e^\lambda}$ acts as an adaptive weighting term for the task loss function and was determined experimentally.

4 Experimental Results

In this section we describe the experimental methodology for evaluating our approach against RICAP and Mixup for two-class classification tasks using the "Bird-or-Bicycle" [2] and Oxford-IIIT-Pet [14] datasets. We also evaluated on a large random selection of "dog" or "cat" images from the Open Images Dataset [10] ("Dog-or-Cat").

4.1 Training Parameters

All experiments were performed using PyTorch[1] and NumPy[2] with a common randomization seed and no CUDA benchmarking [21] to reduce variation between tests. To reduce external factors when evaluating our approach, we do not perform any conventional augmentation techniques, other than normalization and image resizing. The ResNet-50 [6] model with cross-entropy loss as $Loss(.)$ is used as it is a sufficiently complex model to solve the classification tasks but unlike larger models, like WideResNet [24], is less prone to overfitting.

Optimization is done using Stochastic Gradient Descent (SGD) with Nesterov momentum [13]. Hyperparameters were determined using hyperparameter optimization [1] for each dataset.

4.2 Classification: Bird-or-Bicycle

The "Bird-or-Bicycle" dataset was proposed as an adversarial robustness challenge. We selected this dataset because (a) the images (train and validation set) are carefully curated, using multiple annotators per image, and are humanly unambiguous, (b) it is completely solvable on the train and validation splits using machine learning, and (c) a large test split is available. The standard data augmentation benchmark dataset, CIFAR-10/100 [9] was not suitable for our method, which requires images larger than 32×32.

The dataset consists of two classes: "bicycle" and "bird". Each class is balanced, consisting of 250 229×229 RGB images for each class, split evenly between the train and validation splits. The test split (extras) consists of 27,500 images, evenly split between the two classes, and is uncurated.

Hyperparameters: These same hyperparameters were used for all experiments: momentum = 0.5, weight decay = 1e−4, image size = 224, batch size = 64 and training epochs = 200.

4.3 Classification: Oxford-IIIT-Pet and Open Images

The Oxford-III-Pet dataset was proposed for breed classification, but we have simplified it to a two-class set of "cat" and "dog". This dataset (OxfordPet) was chosen as its a simple classification task, but the task is more complex than

[1] pytorch.org.
[2] numpy.org.

the "Bird-or-Bicycle" dataset as it contains more images and is unbalanced. To evaluate model generalization in real-world conditions where the data distribution and sources are unknown and may differ from the training data, we simulate these conditions by having balanced and unbalanced data split sets. The Oxford-Pet unbalanced train split consists of 2,937 images: 941 "cat" and 1996 "dog" (1:2 ratio). From this, we created a balanced subset of 1,500 images (OxfordPet-B). The validation sets were created with the same distribution properties, consisting of 736 and 400 images, respectively.

To further analyze possible real-world uncurated conditions, we create a large uncurated test dataset, "Dog-or-Cat". We randomly sampled 20,000 images (10,000 of each class, "dog" and "cat"), from the Open Images database.

Hyperparameters: These same hyperparameters were used for all experiments: momentum = 0.5, weight decay = 1e−4, image size = 224, batch size = 8, and training epochs = 600.

4.4 Texture Category Analysis

For the texture-based augmentation, we use five different classes of natural textures: regular (R), near regular (NR), irregular (IR), near stochastic (NS), and stochastic (S). The texture spectrum is shown in Fig. 1; we use textures that vary from having high local structures (high local patch similarity) to textures that are very noisy. We constructed a small textures dataset from the following datasets: Kylberg [11], University of Illinois Urbana Champaign (UIUC) texture dataset [12], and Describe Textures Dataset (DTD) [3]. Two DTD classes were chosen per texture classification, with each texture category consisting of 20 images, for a total of 100 images. To evaluate the behaviour of the network when using our method with different texture categories, we trained a model using the training hyperparameters discussed in the previous section.

When using the same hyperparameters, the textures have a mean test error rate of 0.2035 ± 0.0132. We observed that there were cases in which the test error rate for the textured validation/test augmented image outperformed the test error rate for the unaugmented sample image (see "regular (R)" in Table 1). This seems to indicate that when using low local variance, high global structure, neurons are co-adapting too much and overfitting. This will cause non-data domain features (i.e. texture features) to be learned.

The behaviour is greatly reduced at high λ, Δ and lower δ parameter settings, resulting in a decrease in the test error rate. The best hyperparameter set found was $\lambda = 1.0, \Delta = 32, \delta = 0.6$ for the *near stochastic (NS)* class. When using all (A) textures the λ, Δ needed to be reduced due to the higher variation in training samples to 0.1, and 16 respectively, with δ increased to 0.7. As the δ value decreases, the results converge to the baseline, as the data augmentation becomes negligible. If the Δ becomes too large, then smooth regions have more influence on the augmentation and the data distortion causes label noise. Since the stochastic textures have smaller smooth regions, the better performance at higher Δ seems to confirm this theory.

Table 1. Test Error Rates (%) of the best set of hyperparameters (λ,Δ,δ) for each texture classifications trained on the ResNet-50 model with the "Bird-or-Bicyle" dataset. The test error rate is calculated for unaugmented (original) images and our texture-based augmented images.

Texture	Unaugmented	Texturized	λ	Δ	δ
regular (R)	18.73	18.43	0.1	8	0.8
near regular (NR)	18.67	18.97	0.1	16	0.8
irregular (IR)	18.95	18.73	0.1	16	0.8
near stochastic (NS)	**17.95**	22.44	1.0	32	0.6
stochastic (S)	18.02	23.67	0.1	32	0.8
all textures (A)	18.70	18.28	0.1	16	0.7

Table 2. Test Error Rates (%) of ResNet-50 on the "Bird-or-Bicycle" dataset.

Method	$\lambda = 0.01$	$\lambda = 0.1$	$\lambda = 1.0$
Baseline	**18.47**	20.24	39.02
+ mixup ($\alpha = 0.2$)	19.63	19.50	44.19
+ RICAP ($\beta = 0.2$)	23.69	20.94	41.91
+ Ours+all ($\Delta = 16, \delta = 0$)	25.25	22.60	21.38
+ Ours+NS ($\Delta = 32, \delta = 0$)	20.15	20.68	19.79
+ Ours+all ($\Delta = 16, \delta = 0.7$)	21.04	19.78	**18.94**
+ Ours+NS ($\Delta = 32, \delta = 0.6$)	20.32	21.69	**17.95**

4.5 Comparative Evaluation and Discussion

In this section, we evaluated our method using the results in Sect. 4.4 against two competitive strong data augmentation methods, Mixup [26] and RICAP [22].

Label Noise. Hyperparameter optimization was used to determine the best α, β values for Mixup and RICAP, respectively. Both hyperparameters are lower than the reported results, which was to be expected as the higher the α,β values, the greater the potential label noise added during training. As an ablation experiment, we set the δ value to zero, effectively turning off the regularization penalty term and training only using the weighted cross-entropy function. The results are summarized in Table 2.

The best result we achieved was 17.95% test error rate over the next best result, which was the baseline implementation at 18.47%. Since the "Bird-or-Bicycle" contains no label noise, this shows that our method does not introduce label noise while training, as opposed to Mixup and RICAP. Our method converged to near zero faster than the baseline. At lower learning rates, this is indicative of overfitting but when increasing λ to 1.0 the regularization penalty stabilizes the oscillating loss due to the adaptive weighting term for the task loss function. This resulted in an average 2–3 percentage points (pp) decrease in the test error rate when using the regularization penalty.

Table 3. Evaluating the generalization to out-of-distribution examples by varying *class and data distributions*. The top column header row indicates the **training set**. The second row indicates the **evaluation dataset** and the Test Error Rates (%) for ResNet-50 are reported for each training and evaluation pair. The *class distribution independent* (CI) pairs are OxfordPet/OxfordPet-B, OxfordPet/"Dog-or-Cat". The *data independent pairs* (DI) are OxfordPet(-B)/"Dog-or-Cat". The CI pairs represent the ability of the data augmentation to generalize to out-of-distribution examples with unknown class distributions. The DI pairs represent the ability of the data augmentation to generalize to out-of-distribution examples with unknown data distribution. Our method achieved the best result for CI & DI out-of-distribution examples with a test error rate of 27.59%

	OxfordPet-B			OxforPet		
Method	OxfordPet-B	OxfordPet	Dog-or-Cat	OxfordPet	OxfordPet-B	Dog-or-Cat
Baseline	**5.25**	5.22	28.60	5.22	7.00	31.07
+ mixup ($\alpha = 0.2$)	6.75	5.57	31.06	4.32	9.11	29.60
+ RICAP ($\beta = 0.2$)	5.55	**4.64**	**28.37**	5.16	7.05	31.83
+ Ours+all ($\Delta = 16, \delta = 0.6$)	8.75	7.92	31.15	**3.99**	**4.70**	**27.59**
+ Ours+R ($\Delta = 16, \delta = 0.6$)	7.20	7.18	29.46	4.78	5.30	28.91

Generalization to Out-of-Distribution Examples. Deep learning models are often trained with the closed-world assumption, i.e. the train and test data distributions are equal. However, often this assumption is false and the real-world distribution of data does not match the training data, particularly with small datasets. To evaluate the effectiveness of our texture-based data-independent augmentation, we constructed a toy example of out-of-distribution examples using the OxfordPet, OxfordPet-B, and "Dog-or-cat" datasets. The OxfordPet and OxfordPet-B datasets have different *class distributions independent* (CI) (2:1 vs 1:1) but are *data dependent.* The "Dog-or-Cat" dataset is *data independent* (DI) but has the same class distribution as OxfordPet-B (1:1). Each augmentation was trained and validated on OxfordPet or OxfordPet-B and evaluated against "Dog-or-Cat" and the remaining dataset (OxfordPet or OxfordPet-B). The results are summarized in Table 3.

The best results were achieved when the test and train data were in-distribution and data dependent; our method improved over Mixup by 0.33 pp RICAP performed the best when the datasets are data dependent, achieving a test error rate of 4.64% on the OxfordPet(-B) datasets. Our method generalizes to out-of-distribution examples when the training and test data sets are independent (OxfordPet/"Dog-or-Cat" train/test); we improved our results by 2.01 pp over Mixup and achieved the best test error rate of 27.59% for the "Dog-or-Cat" dataset. We observed similar results for all textures, achieving an averaged test error rate of 29.31 ± 0.94%. Our method was able to leverage the larger, unbalanced, OxfordPet data on the "Dog-or-Cat" dataset by improving on average 1.77 ± 1.17% when training on the smaller, balanced, OxfordPet-B subset. For this experiment, we observed that the *regular* texture category was the next highest performer, as opposed to the "Bird-or-Bicycle" experiments. This indicates that as the dataset increases in size, the cost-benefit of using a less sparse but more structure texture increases. In Fig. 3, we provide images from the "Dog-or-Cat" dataset with complex background textural patterns that were misclassified when using the other data augmentation methods but correctly classified when using our best method, trained on OxfordPet.

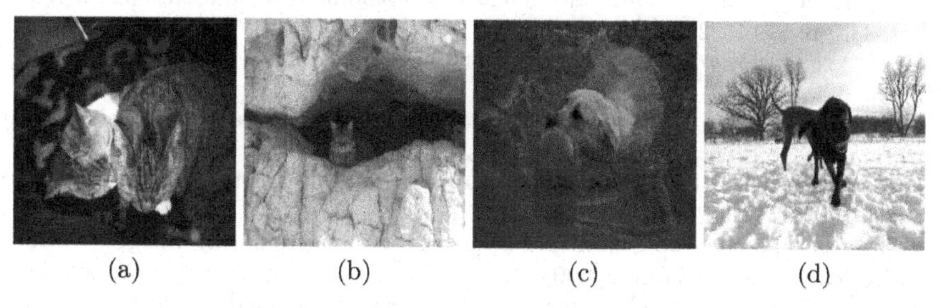

(a) (b) (c) (d)

Fig. 3. Examples of images from the "Dog-or-Cat" dataset that were misclassified by the ResNet-50 model trained using no augmentation (baseline), Mixup, and RICAP but correctly classified using our method (using all textures) on the OxfordPet dataset (see Table 3). Image class left to right: (a) Cat; (b) Cat; (c) Dog; (d) Dog

5 Conclusion

This paper proposed a texture-based domain-specific data augmentation technique applicable when training on small datasets for deep learning classification tasks. Our method focuses on label-preservation to improve generalization and optimization robustness over data-dependent augmentation methods using textures. Naturally occurring textures are used to apply patterned occlusion to training images. Image-mixing augmentation on the light channel of a sampled L*a*b* image creates a label-preserving training image. The mixing is constrained by a perturbation hyperparameter, Δ, and a randomly sampled sparse texture map. We explored different texture taxonomies: regular, near regular, irregular, near stochastic and stochastic. For tiny datasets like "Bird-or-Bicycle", we achieved a test error rate of 17.95% using the *near stochastic* texture, improving over the baseline by 0.52 pp. Experimentally, we evaluated the generalization to out-of-distribution examples using Oxford-IIIT-Pet and Open Images database ("Dog-or-Cat"). Using all available textures, we improved our test error rate by 0.33 pp on Oxford-IIIT-Pet and 2.01 pp on "Dog-or-Cat" over RICAP. In future work, we aim to overcome the weakness of using texture images by using generated synthetic textures based on gradient information.

References

1. Bergstra, J., et al.: Making a science of model search: Hyperparameter optimization in hundreds of dimensions for vision architectures. In: ICML, pp. 115–23. PMLR (2013)

2. Brown, T.B., et al.: Unrestricted adversarial examples. arXiv preprint arXiv:1809.08352 (2018)
3. Cimpoi, M., et al.: Describing textures in the wild. In: CVPR, pp. 3606–13 (2014)
4. DeVries, T., Taylor, G.W.: Improved regularization of convolutional neural networks with cutout. arXiv preprint arXiv:1708.04552 (2017)
5. Guo, H., et al.: MixUp as locally linear out-of-manifold regularization. In: AAAI, vol. 33, pp. 3714–22 (2019)
6. He, K., et al.: Deep residual learning for image recognition. In: CVPR, pp. 770–8 (2016)
7. Hernández-García, A., König, P.: Data augmentation instead of explicit regularization. arXiv preprint arXiv:1806.03852 (2018)
8. Hinton, G., et al.: Distilling the knowledge in a neural network. arXiv preprint arXiv:1503.02531 (2015)
9. Krizhevsky, A.: Learning multiple layers of features from tiny images. Master's thesis, University of Tront (2009)
10. Kuznetsova, A., et al.: The open images dataset V4: unified image classification, object detection, and visual relationship detection at scale. Int. J. Comput. Vis. **128**(7), 1956–1981 (2020)
11. Kylberg, G.: Kylberg texture dataset v. 1.0. Centre for Image Analysis, Swedish University of Agricultural Sciences and Uppsala University (2011)
12. Lazebnik, S., et al.: A sparse texture representation using local affine regions. IEEE Trans. Pattern Anal. Mach. Intell. **27**(8), 1265–78 (2005)
13. Nesterov, Y.: A method of solving a convex programming problem with convergence rate O (1/k 2) O (1/k2). In: Soviet Mathematics. Doklady, vol. 27, no. 2
14. Parkhi, O.M., et al.: Cats and dogs. In: CVPR, pp. 3498–3505. IEEE (2012)
15. Rahaman, N., et al.: On the spectral bias of neural networks. In: ICML, pp. 5301–10. PMLR (2019)
16. Rolnick, D., et al.: Deep learning is robust to massive label noise. arXiv preprint arXiv:1705.10694 (2017)
17. Rudin, L.I., et al.: Nonlinear total variation based noise removal algorithms. Phys. D: Nonlinear Phenom. **60**(1–4), 259–268 (1992)
18. Shafahi, A., et al.: Are adversarial examples inevitable? In: ICLR (2019)
19. Sharif, M., et al.: On the suitability of Lp-norms for creating and preventing adversarial examples. In: CVPRW, pp. 1605–13 (2018)
20. Summers, C., Dinneen, M.J.: Improved mixed-example data augmentation. In: WACV, pp. 1262–70. IEEE (2019)
21. Summers, C., Dinneen, M.J.: Nondeterminism and instability in neural network optimization. In: ICML, pp. 9913–22. PMLR (2021)
22. Takahashi, R., et al.: Data augmentation using random image cropping and patching for deep CNNs. IEEE Trans. Circuits Syst. Video Technol. **30**(9), 2917–31 (2019)
23. Tomasi, C., Manduchi, R.: Bilateral filtering for gray and color images. In: ICCV, pp. 839–846. IEEE (1998)
24. Zagoruyko, S., Komodakis, N.: Wide residual networks. In: BMVC. BMVA (2016)
25. Zhang, C., et al.: Understanding deep learning (still) requires rethinking generalization. Commun. ACM **64**(3), 107–15 (2021)
26. Zhang, H., et al.: MixUp: beyond empirical risk minimization. arXiv preprint arXiv:1710.09412 (2017)
27. Zhong, Z., et al.: Random erasing data augmentation. In: AAAI, vol. 34, pp. 13001–13008 (2020)
28. Zoph, B., et al.: Rethinking pre-training and self-training. In: NeurIPS, pp. 3833–3845. ACM (2020)

Multimodal Representations for Teacher-Guided Compositional Visual Reasoning

Wafa Aissa(✉), Marin Ferecatu, and Michel Crucianu

Cedric laboratory, Conservatoire National des Arts et Metiers, Paris, France
{wafa.aissa,marin.ferecatu,michel.crucianu}@lecnam.net

Abstract. Neural Module Networks (NMN) are a compelling method for visual question answering, enabling the translation of a question into a program consisting of a series of reasoning sub-tasks that are sequentially executed on the image to produce an answer. NMNs provide enhanced explainability compared to integrated models, allowing for a better understanding of the underlying reasoning process. To improve the effectiveness of NMNs we propose to exploit features obtained by a large-scale cross-modal encoder. Also, the current training approach of NMNs relies on the propagation of module outputs to subsequent modules, leading to the accumulation of prediction errors and the generation of false answers. To mitigate this, we introduce an NMN learning strategy involving scheduled teacher guidance. Initially, the model is fully guided by the ground-truth intermediate outputs, but gradually transitions to an autonomous behavior as training progresses. This reduces error accumulation, thus improving training efficiency and final performance. We demonstrate that by incorporating cross-modal features and employing more effective training techniques for NMN, we achieve a favorable balance between performance and transparency in the reasoning process.

Keywords: Visual reasoning · Neural module networks · Multi-modality

1 Introduction

Visual reasoning, the ability to reason about the visual world, encompasses various canonical sub-tasks such as object and attribute categorization, object and relationship detection, comparison, and spatial reasoning. Solving this complex task requires robust computational models that can effectively capture visual cues and perform intricate reasoning operations. In recent years, deep learning approaches have gained prominence in tackling visual reasoning challenges, with the emergence of foundation models playing a key role in advancing the field. Among the deep learning techniques employed for visual reasoning tasks, integrated attention networks, such as transformers [16], have demonstrated remarkable success in natural language processing and computer vision applications, including image classification and object detection. These models leverage attention mechanisms to capture long-range dependencies and contextual

J. Blanc-Talon et al. (Eds.): ACIVS 2023, LNCS 14124, pp. 357–369, 2023.
https://doi.org/10.1007/978-3-031-45382-3_30

relationships for language and vision inputs. However, despite their impressive performance, reasoning solely based on integrated attention networks may be susceptible to taking "shortcuts" and relying heavily on dataset bias. There is an increasing need to address the issue of interpretability and explainability, enabling reaserchers to understand the reasoning behind the model's predictions. By leveraging transformer models as a backbone for VQA systems to encode language and visual information, we can adopt modular approaches that enable an improved understanding and transparency in the visual reasoning process. Modular approaches, such as neural module networks (NMNs), break down the problem (question) into smaller sub-tasks which can be independently solved and then combined to produce the final answer. The modular design confers the advantage of greater transparency and interpretability, as the model explicitly represents the various sub-tasks and their interrelationships.

In this paper we aim to bridge the gap between accuracy and explainability in visual question answering systems by providing insights into the reasoning process. To accomplish this, we propose an enhanced training approach for our modular network using a teacher forcing training technique [17], where the ground truth output of an intermediate module is used to guide the learning process of subsequent modules during training. By employing this approach, our module gains the ability to learn its reasoning sub-task both in a stand-alone and end-to-end manners leading to improved training efficiency.

To evaluate the effectiveness of our approach, we conduct experiments using training programs sourced from the GQA dataset. This dataset provides a diverse range of scenarios, enabling us to thoroughly assess the capability of our approach to reason about the visual world. Through experimentation and analysis, our results demonstrate the effectiveness of our proposed method in achieving explainability in VQA systems while maintaining a high degree of effectiveness.

In summary, this work makes two key contributions: first, the utilization of decaying teacher forcing during training, which enhances generalization capabilities, and second, the incorporation of cross-modal language and vision features to capture intricate relationships between text and images, resulting in more accurate and interpretable results.

The remaining sections of the paper are structured as follows: Sect. 2 provides a discussion on related work, Sect. 3 presents our cross-modal neural module network framework, and Sect. 4 introduces our teacher guidance procedure. In Sect. 5, we outline the validation protocol, followed by the presentation of experimental results in Sect. 6. Finally, in Sect. 7, we conclude the paper by synthesizing our findings and discuss potential future developments.

2 Related Work

In this section, we begin by examining integrated and modular approaches employed in visual reasoning tasks. We then introduce the teacher forcing training method and its application to modular neural networks.

Transformer Networks. Transformers [16] have been widely applied as foundation models for various language and vision tasks due to their remarkable performance. They have also been adapted for reasoning problems like Visual Question Answering (VQA). Notably, models such as ViLBERT [13], Visual-BERT [12] and LXMERT [15] have demonstrated interesting performance on popular VQA datasets like VQA2.0 [6] and GQA [8]. These frameworks follow a two-step approach: first, they extract textual and image features. Word embeddings are obtained using a pre-trained BERT [5] model, while Faster RCNN generates image region bounding boxes along with their corresponding visual features. Subsequently, a cross-attention mechanism is employed to align the word embeddings with the image features, leveraging training on a diverse range of multi-modal tasks.

Despite the benefits of the integrated approaches, these models also have notable drawbacks. One prominent limitation is their lack of interpretability, making it challenging to understand—and debug, when necessary—the underlying reasoning process. Moreover, these models often rely on "shortcuts" in the reasoning, which means learning biases present in the training data. Consequently, their performance tends to suffer when confronted with out-of-distribution data, as shown on GQA-OOD [9]. This research also emphasizes the importance of employing high-quality input representations for the transformer model.

To address interpretability concerns, we use features produced by an off-the-shelf cross-modal transformer encoder in a step-by-step explainable reasoning architecture. This approach balances the power of the transformer model in capturing relationships between modalities with the ability to understand the reasoning process.

Neural Module Networks. To enhance the transparency and emulate a human-like reasoning, compositional Neural Module Networks (NMNs) such as those introduced by [7] and [11] break down complex reasoning tasks into more manageable subtasks through a multi-hop reasoning approach. A typical NMN comprises a generator and an executor. The generator maps a given question to a sequence of reasoning instructions, known as a program. Subsequently, the executor assigns each sub-task from the program to a neural module and propagates the results to subsequent modules.

In a recent study by [4], a meta-learning approach is adopted within the NMN framework to enhance the scalability and generalization capabilities of the resulting model. The generator decodes the question to generate a program, which is utilized to instantiate a meta-module. Visual features are extracted through a transformer-based visual encoder, while a cross-attention layer combines word embeddings and image features. Although the combination of a generator and an executor in NMNs may appear more intricate compared to an integrated model, the inherent transparency of the "hardwired" reasoning process in NMNs has the potential to mitigate certain reasoning "shortcuts" resulting from data bias.

A more recent study [1] has investigated the effects of curriculum learning techniques in the context of neural module networks. The research demonstrated that reorganizing the dataset to begin training with simpler programs and progressively increasing the difficulty by incorporating longer programs (based on the number of concepts involved in the program) facilitates faster convergence and promotes a more human-like reasoning process. This highlights the importance of curriculum learning in improving the training dynamics and enhancing the model's ability to reason and generalize effectively.

Interestingly, [10] demonstrated that leveraging the programs generated from questions as additional supervision for the LXMERT integrated model led to a reduction in sample complexity and improved performance on the GQA-OOD (Out Of Distribution) dataset [9].

Building upon this, our work aims to capitalize on both the transparency offered by NMN architectures and the high-quality transformer-encoded representations by implementing a composable NMN that integrates multimodal vision and language features.

Teacher Forcing. Teacher forcing (TF) [17] is a widely used technique in sequence prediction or generation tasks, especially in RNNs with an encoder-decoder architecture. It involves training the model using the true output as novel input, which helps improve prediction accuracy. However, during inference, the model relies on its own predictions without access to ground-truth information, leading to a discrepancy known as exposure bias.

Scheduled sampling (SS) is a notable approach to mitigating the train-test discrepancy in sequence generation tasks [2]. It introduces randomness during training by choosing between using ground truth tokens or the model's predictions at each time step. This technique, initially developed for RNN architectures, has also been adapted for transformer networks [14], aiding to align the model's performance during training and inference.

NMNs, on the other hand, are trained using only the output of a module as input for the next module, which has drawbacks. Errors made by an intermediate module can propagate to subsequent modules, leading to cumulative bad predictions. This effect is particularly prominent during the early stages of training when the model's predictions are close to random.

NMNs can leverage the TF strategy to enhance their training process. Initially, training begins with a fully guided schema, where the true previous outputs are used as input. As training progresses, the model gradually transitions to a less guided scheme, relying more on the generated outputs from previous steps as input. This gradual reduction in guidance and increased reliance on the model's own predictions, named decaying TF, helps NMNs better learn and adapt to the complexity of the task. With decaying TF, modules can conform to their expected behavior for their respective sub-tasks.

3 Cross-Modal Neural Module Network

Our model takes an image, question, and program triplet as input and predicts an answer. We extract aligned language and vision features for the image and question using a cross-modal transformer. The program, represented as a sequence of modules, is used to build an NMN, which is then executed on the image to answer the question (refer to Fig. 1). In the next subsections we detail the feature extraction process and describe the program executor.

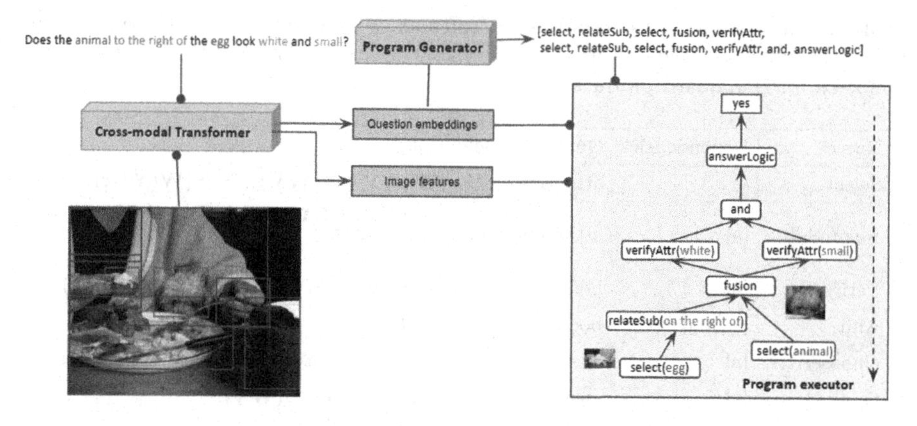

Fig. 1. The proposed modular VQA framework. Plain arrows represent the output flow, while dotted arrows represent the Multi-Task loss backward flow.

Cross-Modal Features. Compositional visual reasoning involves the ability to make logical and geometric inferences on complex scenes by leveraging both visual and textual information. This requires accurate representations of objects and questions. To address this, we employ LXMERT [15], a pretrained transformer model specifically designed for multi-modal tasks. LXMERT has demonstrated impressive performance across various tasks and serves as our feature extractor. In our approach, we discard the answer classification component and freeze the model's weights. To extract cross-modal representations, we process the image I through the object-relationship encoder and the question Q through the language encoder. Then, the Cross-Modality Encoder aligns these representations and produces object bounding box features v_j for each object b_j in I, as well as word embeddings txt_i for each word w_i of Q.

Neural Modules. Our NMN approach tackles complex reasoning tasks by decomposing them into simpler sub-tasks, inspired by human reasoning skills like object detection, attribute identification, object relation recognition, and object comparison. We developed a library of modules tailored to address specific sub-tasks. These modules are designed to be intuitive and interpretable, using simple building blocks like dot products and MLPs. They are categorized into three groups: attention, boolean, and answer modules. For instance, the `Select` attention module focuses on detecting object bounding boxes by applying an attention vector to the available bounding boxes within an image. On the other hand, boolean modules like `And` or `Or` make logical inferences, while answer modules such as `QueryName` provide probability distributions over the vocabulary of possible answers. To get a glimpse of the variety within our module library you can refer to Table 1, which showcases an example from each module category.

Modular Network Instantiation. A program in our framework consists of a sequence of neural modules (Table 1). These modules are instantiated within a larger Neural Module Network (NMN) following the program sequence. Each

Table 1. Sample module definitions. S: softmax, σ: sigmoid, r: RELU, $\mathbf{W}_i$: weight matrix, $\mathbf{a}$: attention vector (36×1), $\mathbf{V}$: visual features (768×36), $\mathbf{t}$: text features (768×1), $\odot$: Hadamard product.

Name	Dependencies	Output	Definition
Select	–	attention	$\mathbf{x} = r(\mathbf{W}\,\mathbf{t}), \mathbf{Y} = r(\mathbf{W}\mathbf{V}), \mathbf{o} = S(\mathbf{W}(\mathbf{Y}^T\mathbf{x}))$
RelateSub	[a]	attention	$\mathbf{x} = r(\mathbf{W}\,\mathbf{t}), \mathbf{Y} = r(\mathbf{W}\mathbf{V}), \mathbf{z} = S(\mathbf{W}(\mathbf{Y}^T\mathbf{x}))$ $\mathbf{o} = S(\mathbf{W}(\mathbf{x} \odot \mathbf{y} \odot \mathbf{z}))$
VerifyAttr	[a]	boolean	$\mathbf{x} = r(\mathbf{W}\,\mathbf{t}), \mathbf{y} = r(\mathbf{W}(\mathbf{V}\,\mathbf{a}), \mathbf{o} = \sigma(\mathbf{W}(\mathbf{x} \odot \mathbf{y}))$
And	[b$_1$,b$_2$]	boolean	$\mathbf{o} = \mathbf{b}_1 \times \mathbf{b}_2$
ChooseAttr	[a]	answer	$\mathbf{x} = r(\mathbf{W}\,\mathbf{t}), \mathbf{y} = r(\mathbf{W}(\mathbf{V}\,\mathbf{a}), \mathbf{o} = S(\mathbf{W}(\mathbf{x} \odot \mathbf{y}))$
QueryName	[a]	answer	$\mathbf{y} = r(\mathbf{W}(\mathbf{V}\,\mathbf{a})), \mathbf{o} = S(\mathbf{W}\,\mathbf{y})$

module has dependencies, denoted as d_m, which allow it to access information from the previous modules, and arguments, denoted as txt_m, which condition its behavior. For example, the `FilterAttribute` module, which relies on the output of the `Select` module, aims to shift attention to the selected objects by considering the attribute that corresponds to the provided text argument. To handle module dependencies, the program executor employs a memory buffer to store the outputs, further used as inputs for subsequent modules. This approach also enables the computation of multi-task losses (see Sect. 4) by comparing the outputs produced by the modules with the expected ground-truth outputs.

4 Teacher Guidance for Neural Module Networks

To achieve explainable reasoning, we use teacher forcing (TF) to guide the modules by providing them with ground-truth inputs. We also employ a multi-task (MT) loss to provide feedback and correct their behaviors towards the expected intermediate outputs. This process is illustrated in Fig. 2.

Given a program p, a question q and image I, the modular network executes p on I with the textual arguments txt_m encoded in q, producing an answer a. This can be represented as $a = p(I, q)$, where $p = m_1 \circ m_2 \circ ... \circ m_n$ denotes the sequential execution of n modules within the program. Each module m_t inputs the output of the previous module m_{t-1} and performs a specific computation or reasoning step to contribute to the final answer. The NMN is trained by minimizing the cross entropy loss L_{CE} over the set of (p, q, I, a) examples. In fact, when a module is provided with its golden input and expected output, it is independently optimized to perform its specific sub-task. However, when modules are jointly trained in a sequential manner, they learn to adapt their behaviors to work together and engage in explicit reasoning without taking shortcuts. This collaborative approach enables the modules to develop a deeper understanding of the task and enables them to perform complex reasoning operations.

From a back-propagation perspective, during the early stages of training, the gradients are computed based on the losses of individual modules when processing correct inputs. As a result, the backward gradient flow of the MT loss is

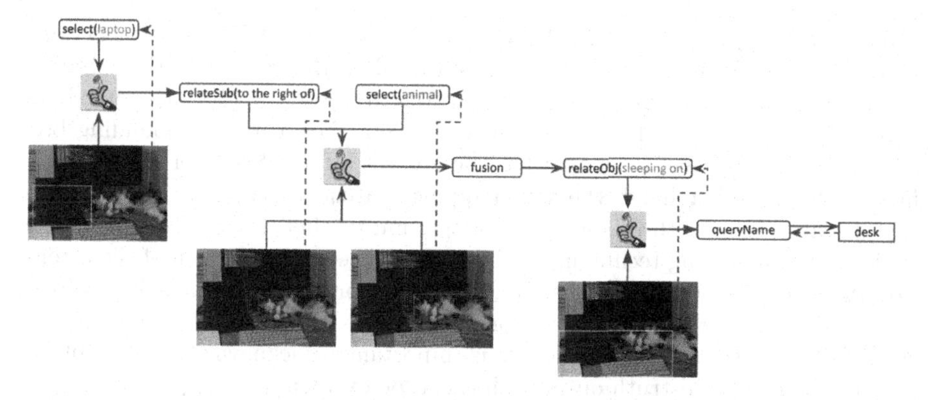

Fig. 2. The teacher guidance for the program execution process related to the question 'On what is the animal to the right of the laptop sleeping?'. Plain arrows represent input guidance and dotted arrows represent the output feedback.

interrupted at the first ground truth input. However, in the case of collaborative module interactions without TF, the full back-propagation can be computed. The intermediate outputs are preserved in continuous form throughout the program execution, enabling the flow of backward gradients between modules. Errors and updates can be propagated through the entire network, facilitating effective learning and enhancing the overall performance of the NMN. We give details about our guidance mechanism in the following subsections.

Input Guidance. The modules receive input guidance through decaying teacher forcing (TF). As shown in Fig. 2, at each reasoning step t the executor randomly decides whether to use the predicted output $\hat{o}_{t-1}$ or the ground-truth output o^*_{t-1} from the previous module m_{t-1} as its input. This decision is made by flipping a coin, where o^*_{t-1} is chosen with a probability of ϵ_e and $\hat{o}_{t-1}$ is chosen with a probability of $1 - \epsilon_e$. The coin-flipping process for input selection occurs at each reasoning step, allowing the model to train on various sub-programs. The probability ϵ_e of selecting o^*_{t-1} depends on the epoch number e. As training progresses and the epoch number increases, ϵ_e decreases, giving more preference to the module's predictions over the ground-truth intermediate outputs.

Output Feedback. We employ a multi-task (MT) loss approach to provide feedback to the modules based on their outputs. The loss consists of a weighted sum $L = \alpha L_{att} + \beta L_{bool} + \gamma L_{answer}$ of individual losses for the attention modules, boolean modules and answer modules, with α, β and γ scaling factors. Each module is assigned its own average loss, considering its frequency of appearance, to prevent overemphasis on frequent modules at the expense of infrequent ones.

For Boolean modules, we rely on the provided answer to infer the module's output and generate the intermediate Boolean outputs. However, for attention modules, it is necessary to establish correspondences between the bounding boxes in the image graph and those obtained from Faster-RCNN, which brings us to the issue of the ground-truth intermediate outputs, detailed in the following.

Soft Matching and Hard Matching. Two mapping techniques, namely hard matching and soft matching, are employed to align the ground-truth bounding boxes with those obtained from the feature extractor. In the hard matching approach, a ground-truth bounding box b_g is matched with the bounding box o_i^* from the feature extractor that has the highest Intersection over Union (IoU) factor. On the other hand, the soft mapping matches b_g with all o_i^* that have an IoU value above a threshold, resembling a multi-label classification task. The choice of the matching technique directly affects the representation of the attention intermediate output vectors. Hard mapping produces one-hot-like vectors, while soft mapping multi-label vectors, with one(s) for positive matching and zeros for negative matching boxes. It is important to acknowledge that not all modules have ground-truth outputs that can be extracted.

5 Protocol Design

Dataset and Metrics. The GQA balanced dataset [8] consists of over 1 million compositional questions and 113,000 real-world images. The questions are represented by functional programs that capture the reasoning steps involved in answering them. To ensure consistent evaluation, the dataset authors suggest using the `testdev` split instead of the `val` split when utilizing object-based features due to potential overlap in training images. In line with the latter and following LXMERT, our model is trained on the combined `train+val` set. For testing, we evaluate the model's performance on the `testdev-all` split from the unbalanced set. This allows us to gather additional examples and gain a comprehensive understanding of the NMN's behavior. To simplify the module structure in the GQA dataset, we consolidate specific modules into more general ones based on similar operations. For example, modules like `ChooseHealthier` and `ChooseOlder` are combined into `ChooseAttribute` module, with an argument txt_m specifying the attribute to select. This reduces the number of modules from 124 to 32. Our experiments directly utilize the pre-processed GQA dataset programs, with a specific focus on evaluating the teacher forcing training on the Program Executor module. While our system employs a transformer model as a generator to convert the question into its corresponding program, this task is relatively straightforward compared to the training of the executor. As in previous studies [4,11], we achieve nearly perfect translation results on `testdev-all`.

We assess the performance of our approach by measuring answer accuracy. Additionally, we conduct a qualitative evaluation of the intermediate outputs, visualized through plotted images in Sect. 6.

Evaluated Methods. As presented in the previous sections, we propose two contributions to improve neural module networks for VQA. First, we use teacher guidance during training, leading to better generalization. Second, we leverage cross-modal language and vision features to capture complex relationships between text and images, resulting in more accurate and interpretable results.

We use the following notations to describe the various experimental setups:

- **LXV**: Employ the cross-modal representations from the LXMERT model [15].
- **TF**: Apply decaying teacher forcing to guide the inputs of the modules.
- **MT**: Apply multi-task losses to guide the expected outputs of the modules.
- **Soft**: Use the soft matching technique described in Sect. 4.
- **Hard**: Employ the hard matching technique described in Sect. 4.
- **BertV**: Use unimodal contextual language and vision representions, where contextual text embeddings are extracted by the BERT model [5] and Faster-RCNN bounding boxes features are provided by the GQA dataset [8].
- **FasttextV**: Employ unimodal non-contextual fastText embeddings [3] along with Faster-RCNN bounding boxes features.

6 Results Analysis

In our evaluation, we begin by comparing the different teacher-guided training strategies. We also compare the impact of the addition of the multi-task losses and its correlation with the decaying TF along with the soft and hard matching techniques. Later, we compare the usage of multi-modal representations against uni-modal representations.

Analysis of the Teacher Guided Training. We aim to enable modular reasoning for visual question answering on the GQA dataset. We evaluate the effectiveness of our approach by measuring the answer accuracy of several models (described in Sect. 5), and report the results in Table 2. Overall, our findings demonstrate that using a combination of input guidance (denoted as **TF**) and output guidance (**MT**) achieves the highest accuracy, with a score of 63.2%.

When comparing **LXV-TF** (decaying teacher forcing) with **LXV-MT** (multi-task loss), we observe that the multi-task loss alone achieves higher accuracy than using decaying teacher forcing alone. This can be attributed to the fact that when using TF alone, the final loss L is solely determined by the answer modules loss L_{answer} and during early training stages, the application of TF limits the backpropagation process, preventing it from reaching the first modules of the programs. As a result, the impact of L_{answer} on initial modules is limited.

Table 2. Performance of various training methods on the `testdev-all` set.

Model	accuracy
LXV-TF-hard	0.548
LXV-MT-hard	0.598
LXV-TF-MT-hard	0.630
LXV-TF-soft	0.536
LXV-MT-soft	0.563
LXV-TF-MT-soft	**0.632**

Interestingly, the combination of multi-task loss and decaying teacher forcing exhibits complementary effects, leveraging the strengths of both techniques to enhance training dynamics and overall performance.

To assess the effectiveness of the decaying teacher forcing guidance, we compare **LXV-MT** against **LXV-TF-MT**. The **TF** guidance has led to accuracy improvements in both **soft** and **hard** matching settings for NMN. This technique can be viewed as a form of curriculum learning, where the model trains on programs of increasing length and complexity. During training, we observed a faster increase in accuracy for the models using **TF** compared to those without TF, as the answer modules receive ground truth inputs in the early stages. As training progresses, the training performance continues to improve until it reaches a peak, after which it slightly degrades due to the reduced use of TF and the modules adjusting to collaborative functioning. Nonetheless, as training continues, the testing performance surpasses that of the models without TF.

When combining the **MT** loss with **LXV-TF**, modules are optimized based on their intermediate outputs losses and they can benefit from the additional guidance provided by the back-propagation of L_{att} and L_{bool}. We reach the best performances outlined by **LXV-TF-MT-soft** and **LXV-TF-MT-hard**. The increase in accuracy ranges from +8.2% in the **hard** matching setting to +9.6% in the **soft** matching setting.

Unimodal *vs* Cross-Modal Representations. We measure the impact of different input representations on the performance (see Table 3). For unimodal embeddings we encode the question with fastText word embeddings or BERT language model, and the image with Faster-RCNN features. For cross-modal representations, we encode the question and the image with LXMERT. The experiments are conducted using our best training strategies from the previous section, i.e. we employ the TF guidance and the MT loss for all the experiments.

When comparing **fastText** and **BERT**, empirical observations indicate that **BERT** tends to achieve better performance when utilizing hard matching, which involves a focused and selective attention mechanism. Conversely, **fastText** demonstrates improved performance with the soft matching mechanism, enabling a multi-label approach. The choice between these matching mechanisms relies on the inputs of the models and the training strategy, as each model may demonstrate superior performance in different scenarios.

Table 3. Language and vision representations results on `testdev-all`.

Model	accuracy
FasttextV-TF-MT-hard	0.495
BertV-TF-MT-hard	0.506
LXV-TF-MT-hard	0.630
BertV-TF-MT-soft	0.485
FasttextV-TF-MT-soft	0.511
LXV-TF-MT-soft	**0.632**

Cross-modal aligned features provided by LXMERT (denoted as **LXV**) have shown a significant increase in accuracy, with a +12.1% improvement when using soft matching and a +12.4% improvement when using hard matching. This validates our intuition that leveraging cross-modal features pretrained on diverse tasks and large datasets can greatly benefit NMNs. By incorporating these features, the modular reasoning process is performed with a better understanding of word embeddings and bounding box features, leading to enhanced performance and more accurate predictions.

Qualitative Analysis of the Modular Approach. In Fig. 3, we illustrate the reasoning process for three different questions. We highlight the bounding boxes with the highest attention values from the attention output vector. For boolean modules, we display the output probability and finally the predicted answer. Taking "Question 2" as an example, the first step successfully selects the skateboard as the object of focus. In the second step, the attention shifts to white objects. Since the skateboard is not white, the attention is then redirected to the white building. The "exist" module assesses if there is an object with a high attention value and produces a probability score based on which the answer is predicted. These examples demonstrate the explainability of our approach and the ability to trace the model's decision-making process.

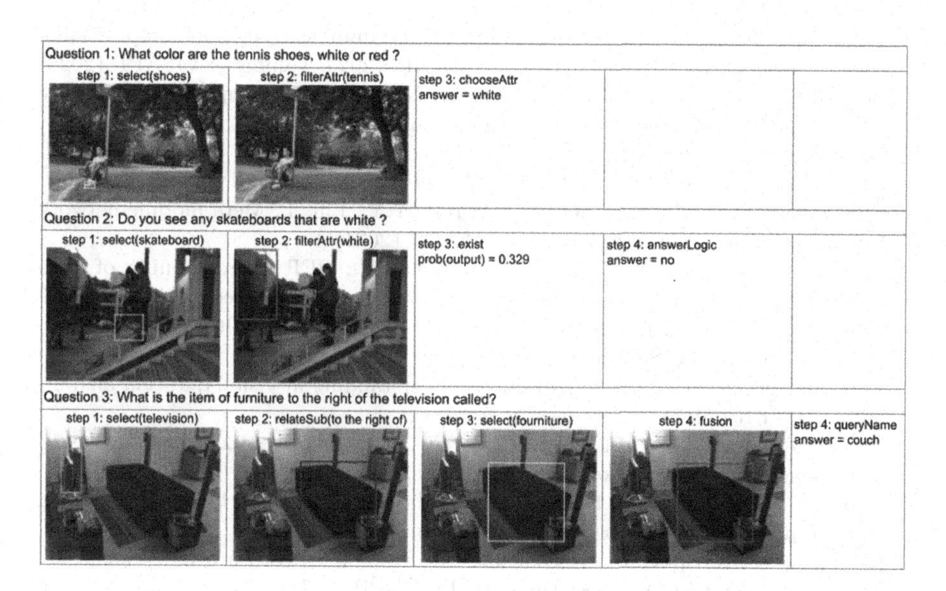

Fig. 3. Visualization of the reasoning process.

7 Conclusion

We have presented a neural module framework trained using a teacher guidance strategy, which has demonstrated several key contributions. First, our approach enhances generalization and promotes a transparent reasoning process, as evidenced by the experimental results on the GQA dataset. Additionally, the utilization of cross-modal language and vision features allows to capture intricate relationships between text and images, leading to improved accuracy. By harnessing our proposed approach, the neural modules acquire the capability to learn their reasoning sub-tasks both independently and in an end-to-end manner. This not only enhances training efficiency but also increases the interpretability of the system, allowing for a better understanding of the underlying reasoning processes. In addition to the aforementioned contributions, our work paves the way to a better understanding of NMNs for the task of visual reasoning. Promising directions include extending our approach to other visual reasoning datasets for a broader evaluation, and exploring alternative training strategies to enhance performance and efficiency.

Acknowledgments. We thank Souheil Hanoune for his insightful comments. This work was partly supported by the French Cifre fellowship 2018/1601 granted by ANRT, and by XXII Group.

References

1. Aissa, W., Ferecatu, M., Crucianu, M.: Curriculum learning for compositional visual reasoning. In: Proceedings of VISIGRAPP 2023, Volume 5: VISAPP (2023)
2. Bengio, S., Vinyals, O., Jaitly, N., Shazeer, N.: Scheduled sampling for sequence prediction with recurrent neural networks. CoRR (2015)
3. Bojanowski, P., Grave, E., Joulin, A., Mikolov, T.: Enriching word vectors with subword information. Trans. ACL **5**, 135–146 (2016)
4. Chen, W., Gan, Z., Li, L., Cheng, Y., Wang, W.Y., Liu, J.: Meta module network for compositional visual reasoning. In: WACV (2021)
5. Devlin, J., Chang, M.W., Lee, K., Toutanova, K.: BERT: pre-training of deep bidirectional transformers for language understanding. In: Proceedings of the 2019 Conference of the NAACL: Human Language Technologies, Volume 1 (2019)
6. Goyal, Y., Khot, T., Summers-Stay, D., Batra, D., Parikh, D.: Making the V in VQA matter: elevating the role of image understanding in visual question answering. In: CVPR (2017)
7. Hu, R., Andreas, J., Rohrbach, M., Darrell, T., Saenko, K.: Learning to reason: end-to-end module networks for visual question answering. In: ICCV (2017)
8. Hudson, D.A., Manning, C.D.: GQA: a new dataset for real-world visual reasoning and compositional question answering (2019)
9. Kervadec, C., Antipov, G., Baccouche, M., Wolf, C.: Roses are red, violets are blue... but should VQA expect them to? In: CVPR (2021)
10. Kervadec, C., Wolf, C., Antipov, G., Baccouche, M., Nadri, M.: Supervising the transfer of reasoning patterns in VQA, vol. 34. Curran Associates, Inc. (2021)
11. Li, G., Wang, X., Zhu, W.: Perceptual visual reasoning with knowledge propagation. In: ACM MM, p. 530–538. MM 2019, ACM, New York, NY, USA (2019)
12. Li, L.H., Yatskar, M., Yin, D., Hsieh, C.J., Chang, K.W.: VisualBERT: a simple and performant baseline for vision and language. In: Arxiv (2019)
13. Lu, J., Batra, D., Parikh, D., Lee, S.: VilBERT: pretraining task-agnostic visiolinguistic representations for vision-and-language tasks. In: NeurIPS (2019)

14. Mihaylova, T., Martins, A.F.T.: Scheduled sampling for transformers. In: Proceedings of ACL: Student Research Workshop. Florence, Italy (2019)
15. Tan, H., Bansal, M.: LXMERT: learning cross-modality encoder representations from transformers. In: Proceedings of EMNLP-IJCNLP (2019)
16. Vaswani, A., et al.: Attention is all you need. In: NeurIPS, vol. 30 (2017)
17. Williams, R.J., Zipser, D.: A Learning algorithm for continually running fully recurrent neural networks. Neural Comput. **1**(2), 270–280 (1989)

Enhanced Color QR Codes with Resilient Error Correction for Dirt-Prone Surfaces

Minh Nguyen(✉)

School of Engineering, Computer and Mathematical Sciences Auckland University of Technology, Auckland 1010, New Zealand
minh.nguyen@aut.ac.nz

Abstract. This study focuses on overcoming the limitations of traditional QR codes, specifically their vulnerability to damage and insufficient error correction capabilities. We introduce an innovative approach, the Enhanced Color QR (CQR) code, which strengthens the error correction ability of QR codes by employing red, green, and blue color channels. This pioneering technology removes critical zones, enabling damage tolerance anywhere on the code and allowing for up to 50% damage to the code area, considerably surpassing the performance of existing QR code systems. Importantly, our CQR code maintains backward compatibility, ensuring readability by current QR code scanners. This state-of-the-art improvement is especially useful in scenarios where conventional 2D barcodes face challenges, such as on non-flat or reflective surfaces frequently encountered on fruits, cans, bottles, and medical equipment like blood test sample tubes and syringes. Furthermore, our CQR code's four corners and boundary can be estimated without requiring corner visibility, offering potential advantages for augmented reality applications. By addressing the key issues associated with traditional QR codes, our research presents a significant advancement in the field of computer vision and provides a more resilient and versatile solution for a wide range of real-world applications.

Keywords: QR codes · error correction · self-correcting ability · 2D barcodes · computer vision · real-world applications

1 Introduction

QR (Quick Response) codes have become the most widely recognized and internationally standardized 2D barcodes in today's world. They are commonly found on websites, magazines, and various marketing materials. QR codes offer numerous benefits across industries, as they can digitally store information about a product or service, allowing users to efficiently scan and transfer this information to smartphones, tablets, and other electronic devices. The primary advantage of QR codes lies in their ability to eliminate tedious typing and searching for information. During the COVID-19 pandemic, QR codes have become instrumental in contact tracing efforts in many countries [9], enabling contactless QR code scanning for check-ins at workplaces and other locations [6].

J. Blanc-Talon et al. (Eds.): ACIVS 2023, LNCS 14124, pp. 370–381, 2023.
https://doi.org/10.1007/978-3-031-45382-3_31

This research proposes a color version of the QR code that significantly improves damage resistance. Figure 1 displays the traditional binary QR code (left) alongside our proposed Colour QR Code (right). Unlike traditional QR codes, where certain critical patterns must remain undamaged for successful decoding, this innovative color QR code remains readable even if large portions are torn off. The new code can still be read by conventional barcode scanners, ensuring usability. This development holds potential to greatly enhance the reliability of information retrieval via QR codes across different applications and environments, due to its increased robustness and damage-resistance.

Fig. 1. A traditional black and white QR code (left) and our proposed CQR Code (right); both codes store the same data.

1.1 Self Correction Needed for 2D Barcodes

Two-dimensional (2D) barcodes, such as QR codes, have the advantage of storing significantly more data compared to traditional one-dimensional (1D) barcodes [4]. In the case of 1D barcodes, it is common practice to include the encoded data as text or numbers beneath the barcode, allowing for manual entry in the event of damage. However, this approach is rarely seen with 2D barcodes. The primary reason is that the sheer amount of data contained within a 2D barcode may be too extensive to print in plain text form. Furthermore, even if the information were imprinted, manually entering it would be both time-consuming and challenging for most users. This characteristic of 2D barcodes presents a unique challenge: if a scanner fails to read the code, manual data retrieval becomes nearly impossible.

In certain scenarios, damages to 2D barcodes are unavoidable, necessitating a reliable error correction mechanism to handle such instances. The current QR code standard incorporates Reed-Solomon error correction, which enables data restoration in cases of dirt or damage. QR codes offer four distinct error correction levels [5]:

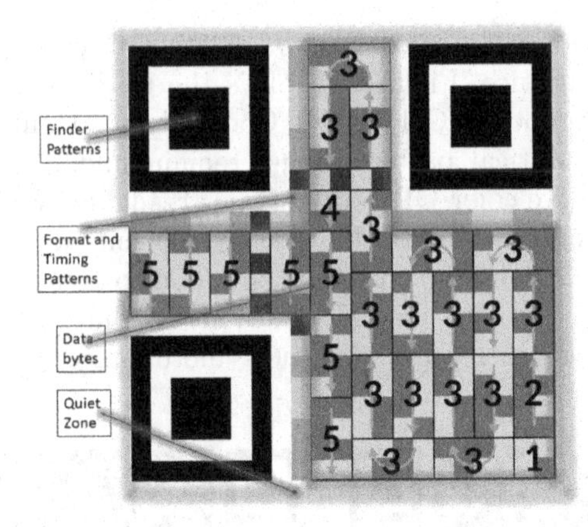

Fig. 2. The patterns in a standard binary QR Code.

- L (Low): Allows restoration of 7% of data bytes.
- M (Medium): Allows restoration of 15% of data bytes.
- Q (Quartile): Allows restoration of 25% of data bytes.
- H (High): Allows restoration of 30% of data bytes.

Increasing the error correction level enhances the code's resilience, albeit at the cost of a larger code size. Typically, Level M (15%) is the most commonly used. However, these percentages can be misleading. The data bytes are stored within a fish-shaped area, as illustrated in Fig. 2. Damages occurring within this area can be restored, as demonstrated by the damaged but decodable QR codes in Fig. 3. It is crucial to note that damages can occur anywhere on the code, and there is no guarantee that they will be confined to this fish-shaped area.

A more detailed examination of the QR code structure is provided in Fig. 2, which outlines the crucial patterns that must be preserved for accurate QR code detection and decoding. Among these patterns, the three "Finder Patterns" situated in the top-left, top-right, and bottom-left corners are arguably the most critical. Damage to any of these patterns would render the QR code undetectable, resulting in incorrect decoding. Maintaining the "Quiet Zone" clear is also essential for proper QR code functionality .

Conventional QR codes may be susceptible to minor damages affecting the "Finder Patterns." For example, in Fig. 4-left, the QR code becomes unreadable when a single yellow dot is placed inside one of its "Finder Patterns." Fig. 4-middle illustrates a real-world scenario where the top-right corner of the QR code is torn off, rendering it unscannable.

1.2 Challenges with Glossy, and Irregular Surfaces

QR codes work optimally when printed on flat, matte surfaces. However, many real-world situations require QR codes to be displayed on reflective LCD screens, adhered behind glass doors, or printed on plastic bags, tubes, bottles, syringes,

Fig. 3. One demonstration of a damaged but decodable QR code shows how the codes can still be read despite having a quarter of the codes missing, demonstrated on *wikipedia.org*

beer mugs, coffee cups, and fruits. Figure 4-right demonstrates an example where a QR code is printed on a bottle, and the curved surface prevents it from being properly decoded using a standard scanner. Several attempts have been made to address this issue, such as [7,10]. However, they can only handle uniformly curved or nearly curved shapes, and the entire QR code must be visible. Currently, for applications involving printing QR codes on such surfaces, most companies opt for smaller versions of the codes. However, smaller QR codes also mean reduced data capacity, and the customer's camera might not capture enough detail to decode it correctly. In summary, there are still no adequate solutions for printing and scanning 2D barcodes on products with shiny, glossy, and uneven surfaces.

Fig. 4. Three situations where minor damages or uneven surfaces could create non-decidable codes: (left) a yellow point added, (middle) a torn corner, (right) curved surface. (Color figure online)

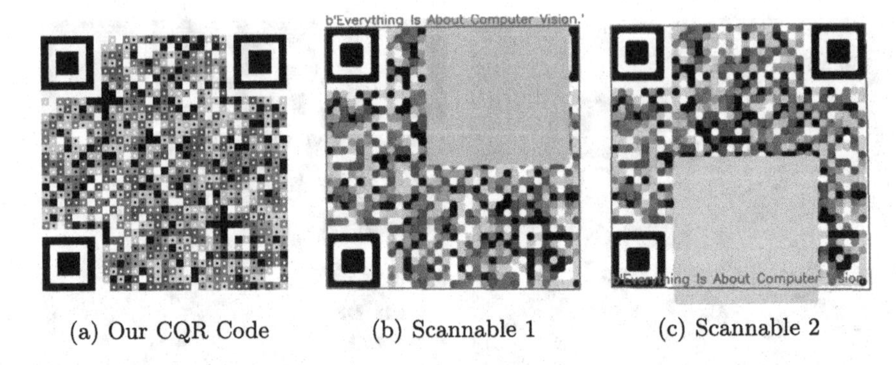

(a) Our CQR Code (b) Scannable 1 (c) Scannable 2

Fig. 5. Two demonstrations of our proposed Colour QR Code, which remain readable even when large parts of them are covered.

1.3 Similar/Related Works on Adding Colors to QR Codes

Our objective is to develop a more reliable version of the QR code that is more resistant to incidental damage. There have been several attempts to use color channels to improve traditional binary QR codes. Notable examples include the High Capacity Colored 2-Dimensional (HCC2D) Code [11], the Colour QR Code [8], the JAB code [1], and others [2,3,8,12,13]. All of these efforts aimed to increase the data density and storage capacity of the 2D codes; however, to the best of our knowledge, none have successfully addressed the aforementioned issue. Most of these codes still inherit the properties of traditional QR codes, meaning that the "Finder Patterns" must remain visible to be valid. In contrast, Fig. 5 showcases our proposed Colour QR Code in two situations where large portions of the codes are covered. The codes are successfully decoded in both cases, and the boundaries of the Colour QR Code are also accurately detected. This feature is particularly important in AR applications, where such coordinates are necessary to render the 3D graphic models precisely on the real-world scene. Section 2 will detail the design and implementation of our proposed code and explain how damage resistance is facilitated.

2 Design and Implementation

This section describes, in detail, the design and implementation of two key components: the encoding of CQR in Sect. 2.1 and the decoding of CQR in Sect. 2.2. The significance of this research lies in its potential to significantly increase the damage resistance of QR codes, as demonstrated through various test cases.

2.1 Encoding of CQR

It is well established that any color image can be decomposed into three corresponding channels: Red, Green, and Blue. Each channel can be utilized to store

Fig. 6. The steps to create our proposed Colour QR Code from an original QR Code: (1) extract the red channel, (2) extract the green channel and flip diagonally around the top-left bottom-right axis, (3) extract the blue channel and flip diagonally around the top-right bottom-left axis, (4) reconstruct a dot-version of the code, (5) merge all four to construct the CQR code. (Color figure online)

an independent monochrome QR code. If the three channels store the same QR code with different orientations or rotations, we can replicate each bit of the original QR code three times in three distinct locations. In other words, for every single point of the original QR code P_i, we have two other backup locations available, P_i' and P_i''. Suppose P_i in the red image is damaged; we can find its backup, P_i', from the green image. If that location is also damaged, the last backup, P_i'', can be found in the blue image. This idea forms the primary mechanism of automatic correction for our proposed CQR code. The aim of this mechanism is to significantly increase the damage resistance of the newly proposed CQR code compared to the original one. This paper will describe several testing cases to determine the magnitude of the improvement in error correction. Further details will be covered in Sect. 3. The creation of our proposed CQR code is relatively straightforward, as demonstrated in Fig. 6; it includes five steps. It starts with a traditional binary QR code $I_{QR} = (I_R, I_G, I_B)$, which encodes some binary data. I_{QR} is encoded with the highest error-correction H for the best damage resistance. I_{QR} is a square image; we first separate the barcode image's three Red, Blue, and Green channels: I_R, I_G, I_B. The red channel I_R is kept as it is (step 1):

$$QRC_{Red} = QR_{Code} \tag{1}$$

The green channel I_G is flipped diagonally around the top-left bottom-right axis (step 2). This can be achieved with the following:

$$QRC_{Green} = flip\_at\_top\_left\_corner(QR_{Code}) \tag{2}$$

The blue channel I_B is flipped diagonally around the top-right bottom-left axis (step 3), achieved by:

$$QRC_{Blue} = flip\_at\_top\_right\_corner(QR_{Code}) \tag{3}$$

At step 4, a filled-dot version of the original QR code is constructed. The purpose of having this is to maintain the backward compatibility of the newly created QRC, e.g. an ordinary barcode scanner could still read our new QRC code:

$$QRC_{Dot} = make\_dots(QR_{Code}) \tag{4}$$

The last step is to merge all four images together to make the proposed CQR code I_{QRC} (as shown Fig. 6 - right):

$$QRC = QRC_{Red} + QRC_{Green} + QRC_{Blue} + QRC_{Dot} \tag{5}$$

The merged QRC contains four layers. Each of the three R, G, and B layers holds a copy of the original QR Code QR_{Code} in different orientations, while the fourth layer is there to maintain backward compatibility. This means most available QR code scanners can read this CQR Code. We tested several available scanner apps on Apple and Android markets; all could read this CQR code correctly. The encoding process for the proposed CQR code offers increased robustness against damage while maintaining compatibility with existing QR code scanners. By leveraging the redundancy provided by the three layers of differently oriented QR codes, the CQR code can effectively recover information even when some portions are damaged.

2.2 Decoding of CQR

The decoding process is designed to extract the information stored in the CQR code by analyzing each channel and determining the correct orientation based on the QRC_{Dot} reference. The decoding process involves the following steps: As the image captured by a camera may exhibit reduced saturation of colors, the first step is to maximize the saturation of the input image to restore the CQR code to its original form. To achieve this, each channel of the RGB image is set to either 0 or 255, as represented by the following equation:

$$f(x) = \begin{cases} 255, & \text{if } x \geq 128 \\ 0, & \text{otherwise} \end{cases} \tag{6}$$

In this equation, x represents the input color channel value (either red, green, or blue) for a given pixel. If the value is greater than or equal to 128, it is set to 255; otherwise, it is set to 0. This function is applied independently to each

Fig. 7. A demonstration of converting a general image (left) into a maximized saturation image (right).

color channel of every pixel in the input image to produce an output image with maximized saturation (Fig. 7).

Subsequently, we eliminate the dots by applying a median filter. Following this, we separate the CQR code into its Red, Green, and Blue channels:

$$(QR_{Red}, QR_{Green}, QR_{Blue}) = split(CQR) \tag{7}$$

We then determine the correct orientation of each QR code by inverting the steps shown in Fig. 6. However, after correct orientation, the three QR codes are not overlapping due to the original rotation. We can make them overlap by identifying an affine transformation by matching two images. Thus, we take any two input images from $((QR_{Red}, QR_{Green}, QR_{Blue}))$ and find the homography (a transformation matrix) between them using ORB (Oriented FAST and Rotated BRIEF) feature detection and matching. It filters out matches with a distance greater than a specified maximum distance and checks if there are enough good matches. If the number of good matches is above a threshold, the function computes the homography using RANSAC, applies a perspective transformation, and returns the transformed image. Otherwise, it prints a message indicating insufficient matches and returns the second input image unaltered. Once transformed, the three QR codes are combined using the *addWeighted* function with equal weights. Finally, it overlays them to create the final output QR Code thus obtaining the original QR code. Decoding the reconstructed QR code using a standard QR code decoder to obtain the original information:

```
decoded_data = decode_QR(QR_Code)
```

The proposed decoding process is robust to various types of damage and can effectively recover the information stored in the CQR code.

3 Results and Evaluations

In the previous section, we demonstrated the successful decoding of our proposed CQR code even when a significant portion was obscured. In this section, we will carry out various experiments to determine the damage resistance of the CQR code and compare it to a standard QR code. We begin with a QR code and a CQR code, both encoding the phrase "Everything Is About Computer Vision." The codes are generated with the highest level of error correction, which is 'H'. We then introduce damage to each barcode by overlaying a colored rectangle at a random position on the code. The extent of damage varies from 0% to 60%. For each level of damage, we create 200 samples with slight rotation, distortions, and noise. As a result, our testing dataset consists of 12,000 damaged QR codes and 12,000 damaged CQR codes, totaling 24,000 damaged codes. Examples of damaged QR codes can be seen in Fig. 8 (top row), and damaged CQR codes are showcased in Fig. 8 (bottom row).

Fig. 8. Top: QR codes with 10, 20, 30, 40, 50, and 60% damages. Bottom: CQR codes with 10, 20, 30, 40, 50, and 60% damages.

After scanning all 24,000 barcode samples, we gathered data on the accuracy of the decoded information. The first 44 rows of the results, which represent the average percentage of correct scanning outcomes for all barcodes with varying levels of damage, are presented in Table 1. These findings are also illustrated graphically in Fig. 10. Both the graph and the table's data clearly demonstrate that our proposed CQR code outperforms the standard QR code in terms of error correction. The red curve (representing the CQR) is noticeably higher than the green curve (representing the QR code). When damage to the QR code ranges from 0 to 1%, its readability immediately falls from 100% to 94%. In contrast, the CQR code maintains nearly 100% readability with damage levels of up to 20%. Furthermore, the QR code's readability approaches zero with approximately 35% damage, whereas the CQR code's readability persists above 55%. The CQR code appears to fail completely when damage exceeds 50Ȯverall, these results emphasize the superior error correction capabilities of our proposed CQR code compared to traditional QR codes (Fig. 9).

Table 1. Detection rates of CQR Codes versus original QR codes.

Damage (%)	QR Readable (%)	CQR Readable (%)
0	100	100
1	94.0	99.5
2	91.0	99.0
3	89.0	99.5
4	84.5	100
5	85.0	100
6	85.5	99.5
7	84.5	99.0
8	87.5	99.0
9	79.5	100
10	82.0	99.5
11	81.5	99.5
12	77.0	99.0
13	76.5	99.0
14	80.5	99.5
15	77.0	98.5
16	77.5	99.0
17	68.5	100
18	66.5	98.5
19	61.0	97.0
20	44.5	97.0
21	29.5	94.5
22	29.0	92.5
23	23.0	93.0
24	11.0	90.5
25	16.0	85.5
26	11.5	88.0
27	7.5	82.0
28	6.0	77.5
29	5.5	80.0
30	3.0	69.5
31	1.5	67.0
32	1.5	67.0
33	3.0	66.5
34	1.0	56.0
35	2.5	55.5
36	1.0	47.0
37	0	45.5
38	0.5	40.5
39	0	33.0
40	0	24.5
41	0	26.5
42	0	21.0
43	0	13.5

Fig. 9. Detected CQR codes with damages in real photo capture.

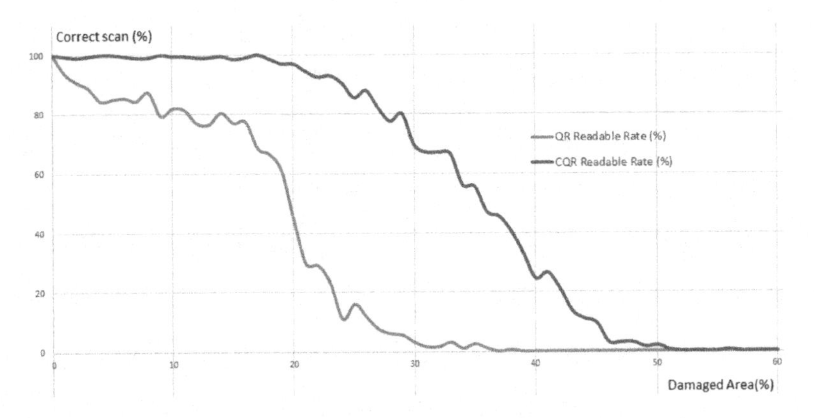

Fig. 10. Successfully decoded rates of CQR Codes versus original QR codes.

4 Conclusion and Future Works

In this paper, we presented the Colour QR code (CQR), an enhanced version of the conventional QR code that leverages the red, green, and blue color channels to improve its error correction capabilities. This novel approach addresses several challenges, such as the dependency on the "Finder Pattern" and the requirement for "Quiet Zones" to be visible for successful detection and decoding. Furthermore, the CQR code demonstrates a significant improvement in error correction, as it can still be read even when up to 50% of the code area is damaged. Additionally, the four corners and boundaries of the CQR code can be approximated without the need for visible corner markers. This feature may prove beneficial for augmented reality systems that rely on these corners for rendering 3D models on real-world surfaces. The enhanced error correction capabilities of the CQR code make it a promising solution for scenarios where conventional QR codes fail, such as on non-flat and shiny surfaces, including fruits, cans, bottles, medical blood test sample tubes, and syringes. The significance of this research lies in its potential to expand the use of QR codes in various applications, providing a more robust and reliable means of encoding and decoding information in challenging environments. Future work will involve further testing of the CQR code on the mentioned surfaces and exploring additional applications where this technology could provide a distinct advantage.

References

1. Berchtold, W., Liu, H., Steinebach, M., Klein, D., Senger, T., Thenee, N.: Jab code-a versatile polychrome 2D barcode. Electron. Imaging **2020**(3), 207-1 (2020)
2. Bhardwaj, N., Kumar, R., Verma, R., Jindal, A., Bhondekar, A.P.: Decoding algorithm for color QR code: a mobile scanner application. In: 2016 International Conference on Recent Trends in Information Technology (ICRTIT), pp. 1–6 (2016). https://doi.org/10.1109/ICRTIT.2016.7569561
3. Bulan, O., Blasinski, H., Sharma, G.: Color QR codes: increased capacity via per-channel data encoding and interference cancellation. In: Color and Imaging Conference, vol. 2011, pp. 156–159. Society for Imaging Science and Technology (2011)
4. Kato, H., Tan, K.: 2D barcodes for mobile phones. In: 2005 2nd Asia Pacific Conference on Mobile Technology, Applications and Systems, pp. 8-pp. IEEE (2005)
5. Kieseberg, P., et al.: QR code security. In: Proceedings of the 8th International Conference on Advances in Mobile Computing and Multimedia, pp. 430–435 (2010)
6. Lee, C.Y., Mohd-Mokhtar, R.: Contactless tool for COVID-19 surveillance system. In: 2021 IEEE 19th Student Conference on Research and Development (SCOReD), pp. 52–57. IEEE (2021)
7. Liu, P., Duan, M., Liu, W., Wang, Y., Li, Q., Dai, Y.: Research on the graphic correction technology based on morphological dilation and form function QR codes. In: 2016 International Conference on Network and Information Systems for Computers (ICNISC), pp. 323–327. IEEE (2016)
8. Melgar, M.E.V., Zaghetto, A., Macchiavello, B., Nascimento, A.C.: CQR codes: colored quick-response codes. In: 2012 IEEE Second International Conference on Consumer Electronics-Berlin (ICCE-Berlin), pp. 321–325. IEEE (2012)
9. Nakamoto, I., Wang, S., Guo, Y., Zhuang, W.: A QR code-based contact tracing framework for sustainable containment of COVID-19: evaluation of an approach to assist the return to normal activity. JMIR Mhealth Uhealth **8**(9), e22321 (2020)
10. Qian, J., Xing, B., Zhang, B., Yang, H.: Optimizing QR code readability for curved agro-food packages using response surface methodology to improve mobile phone-based traceability. Food Packag. Shelf Life **28**, 100638 (2021)
11. Querini, M., Grillo, A., Lentini, A., Italiano, G.F.: 2D color barcodes for mobile phones. Int. J. Comput. Sci. Appl. **8**(1), 136–155 (2011)
12. Querini, M., Italiano, G.F.: Reliability and data density in high capacity color barcodes. Comput. Sci. Inf. Syst. **11**(4), 1595–1615 (2014)
13. Ramya, M., Jayasheela, M.: Improved color QR codes for real time applications with high embedding capacity. Int. J. Comput. Appl. **91**(8) (2014)

Author Index

J. Blanc-Talon et al. (Eds.): ACIVS 2023, LNCS 14124, pp. 383–384, 2023.
https://doi.org/10.1007/978-3-031-45382-3

Printed in the United States
by Baker & Taylor Publisher Services

Printed in the United States
by Baker & Taylor Publisher Services